Proceedings of the Third Conference

Localization & Energy Transfer in Nonlinear Systems

Proceedings of the Third Conference

Localization & Energy Transfer in Nonlinear Systems

June 17-21 2002, San Lorenzo de El Escorial Madrid

editors

Luis Vázquez,
Universidad Complutense and Centro de Atrobiología CSIC/INTA, Spain

Robert S. MacKay,
Warwick University, UK

María Paz Zorzano,
Centro de Astrobiología CSIC/INTA, Spain

LOCALIZATION AND ENERGY TRANSFER
IN NONLINEAR SYSTEMS
JUNE 17–21 2002, SAN LORENZO de EL ESCORIAL MAD

World Scientific
New Jersey • London • Singapore • Hong Kong

Published by

World Scientific Publishing Co. Pte. Ltd.

5 Toh Tuck Link, Singapore 596224

USA office: Suite 202, 1060 Main Street, River Edge, NJ 07661

UK office: 57 Shelton Street, Covent Garden, London WC2H 9HE

British Library Cataloguing-in-Publication Data
A catalogue record for this book is available from the British Library.

LOCALIZATION AND ENERGY TRANSFER IN NONLINEAR SYSTEMS

ISBN 981-238-296-8

Printed in Singapore by World Scientific Printers (S) Pte Ltd

PREFACE

The contributions of this volume are based on invited reviews and original work presented at the Conference *Localization and Energy Transfer in Nonlinear Systems*, held at El Escorial, Spain, in June 2002. The Meeting was organized to foster communication among scientist involved in the European LOCNET Project (Localization by Nonlinearity and Spatial Discreteness and Energy Transfer in Crystals, Biomolecules and Josephson Arrays. Contract HPRN-CT-1999-00163).

The Proceedings are a selection of the main contributions presented in the Conference together with a set of tutorial reviews. This volume can serve to initiate the non-expert into the field of nonlinearity, discreteness, and energy localization. The topic of the Conference was the localization of vibrational energy, electric charge and magnetic flux in nonlinear, spatially discrete systems (Hamiltonian, dissipative and quantum systems) from theoretical, numerical and experimental perspectives. In the context explained above, the key issues represented are the classical breathers, quantum breathers in molecular crystals, vibrational energy distribution and flow, polarons and bipolarons, Josephson unction arrays, friction and nonlinear conductivity, and the breathers in biomolecules.

We would like to thank the substantial financial support from the LOCNET Project. We also thank the financial support received from the Centro de Astrobiología (CSIC-INTA) and the Ministerio de Ciencia y Tecnología of Spain. We are very grateful to Professors S. Jiménez, P.J. Pascual, C. Aguirre and Dr. D. Usero of the Organizing Committee for their help and assistance. We are indebted to Mr. C. Peña for the photo and press covering of the Conference. Finally, we would like to extend our thanks to all speakers, participants, and referees of the manuscripts for their contribution to a successful Conference and Proceedings.

Luis Vázquez
Robert S. MacKay
María Paz Zorzano

CONTENTS

A NON ADIABATIC THEORY FOR ULTRAFAST AND CATALYTIC TRANSFER OF ELECTRONS AT LOW TEMPERATURE

S. AUBRY

Laboratoire Léon Brillouin (CEA-CNRS), CEA Saclay
91191-Gif-sur-Yvette Cedex, France
E-mail: aubry@llb.saclay.cea.fr

G. KOPIDAKIS

Department of Physics, University of Crete, P.O. Box 2208, 71003 Heraklion, Crete, Greece

Electron transfer (ET) between weakly interacting molecules is a ubiquitous elementary process of chemical reactions generally well described by the Marcus theory as a thermally activated process. In the vicinity of the inversion point where the activation energy becomes small, ET is faster and should occur by quantum tunnelling at low temperature. Then, the standard adiabatic approximation used in the Marcus theory looses its validity and improvements are needed.

We construct a non-adiabatic theory of ET using the complex amplitudes on each molecule of the electronic wave-functions as Kramers variables. The effective dynamics of ET is then described by a nonlinear equation with dissipative terms and colored Langevin forces globally modeling the interaction with the thermalized environment.

Far from the inversion point, our model reproduces essentially the standard Marcus results but close to it, a correct description of the quantum tunnelling of the electron in its deformable environment requires to take into account nonlinearities and damping.

We analyze quantum ET on the base of the recently proposed nonlinear concept of Targeted Transfer which extends but also qualitatively modifies the well-known concept of linear resonance and tunnelling. In addition, we predict and numerically confirm spectacular catalytic effects if ET occurs in the presence of a third molecule chosen very special.

Then even in situations with a large energy barrier, where a direct ET between donor and acceptor cannot occur at low temperature, a weak coupling with an extra appropriately tuned (catalytic) site can trigger selectively an ultrafast ET at low temperature. New perspectives on ET and more generally on selective quantum transitions in complex systems are opened and seems to be highly relevant for biosystems (e.g. the photosynthetic reaction center).

1. Introduction-Standard Marcus theory of ET

According to transition state theory, chemical reactions decompose into elementary reactions among which electron transfer is ubiquitous [1]. Most electron transfer (ET) are well-described by the Marcus theory [2]. This theory considers the transfer of an electron from a donor site to an acceptor site at lower energy connected by a transfer integral supposed to be small. It is essentially an adiabatic theory where the atomic fluctuations are supposed to be slow at the scale of the characteristic time of the electron dynamics. Then, the wavefunction of the electron should remain practically an eigenstate of the time dependent potential created by the nuclei (polaron effect). There is an energy barrier for ET at zero degree K for transferring the electron from the donor to the acceptor due to the self trapping of the electron initially on the donor site. Consequently, ET may only occur at nonvanishing temperature and is driven by the potential fluctuations due to the thermalized environment. In the Marcus theory, ET is essentially a thermally activated process where the characteristic energy is the energy barrier between the donor and the acceptor. The characteristic time required for ET between the donor and the acceptor depends on temperature according to a standard Arhenius law. The Marcus theory well describes ET between two reacting molecules in many real systems.

Since the transfer integral between the donor and the acceptor is small, the eigen electronic state are localized either on the donor or the acceptor (except during the time where the electron tunnel from the donor to the acceptor but this time is short compared to the characteristic time of the electron transfer). It is then sufficient to consider only two free energy curves of the system as a function of the reaction coordinates (describing the configuration of the nuclei and the environment) when the electron is either on the donor molecule or the acceptor molecule. One obtain the well-known Marcus scheme represented fig.1 (see ref.[2] for details). $\Delta G^{\star}$ appearing fig.1 at the intersection of these two free energy curves, is the energy barrier which has to be overcome by the electron and is the characteristic energy of the thermally activated process of ET.

Two regimes appears within this description. In the normal regime, the direct electronic transition from the donor to the acceptor at fixed Reaction Coordinates (i.e. at fixed nuclei and environment) requires a positive energy $\hbar\omega_{el} = -\Delta_{el}$. On the opposite, in the inverted regime, this energy $-\Delta_{el}$ is negative. Although in that case, ET could be achieved at low temperature by a photon emission at frequency (energy) $\hbar\omega_{el} = \Delta_{el}$ (photoluminescent chemical reaction [2]), an activation process above the

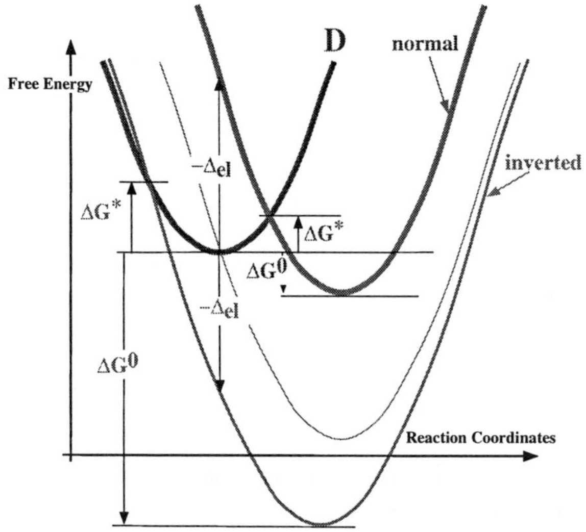

Figure 1. Free energy versus Reaction Coordinates of the system donor-acceptor when the electron is on the donor (top left curve D) or on the acceptor for several redox potentials in the normal regime (top right curve), at the inversion point (middle right curve) and in the inverted regime (bottom right curve). The chemical reaction energy is the distance between the energy minima ΔG^0. The energy barrier is $\Delta G^\star$. The electronic excitation energy on the donor at fixed reaction Coordinates is Δ_{el}.

energy barrier $\Delta G^\star$ become by far more efficient and prevalent at higher temperature [a])

$\Delta G^\star$ turns out to be just zero at the inversion point when $\Delta_{el} = 0$ (see fig.1). This is the regime where ET is expected to become effective at low temperature. Because of the absence of energy barrier, no thermal fluctuations are required and at low temperature ET is essentially reduced to a direct quantum tunnelling (followed by a relaxation of the environment once the electron is on the acceptor).

Since the transfer integral between donor and acceptor is supposed to be small (weak reactants), the characteristic time of this quantum tunnelling is nevertheless relatively long and could be comparable to the phonon characteristic times. In that case, the deformability of the self consistent local potential followed by the electron during the intermediate stage of the tunnelling should be taken into account. It is equivalent to note that in the

[a]At the top of the energy barrier, the electron has to tunnel from the donor to the acceptor. This time for this process and its probability participate as a coefficient for the reaction rate. It could treated either adiabatically (strong reactants) or diabatic corrections could be made (weak reactants) [1].

vicinity of this inversion point, the validity of the adiabatic hypothesis necessarily breaks down since the characteristic energy $|\Delta_{el}|$ of the direct electronic excitation (at fixed environment) becomes small and necessarily comparable to the phonon characteristic energies (note that Δ_{el} is calculated at zero transfer integral but is not significantly changed if this transfer integral is small).

It is well-known that in general the adiabatic assumption requires to be valid that the largest phonon energy $\hbar\omega_{ph}$ which is involved, be much smaller than the smallest excitation energy of the electron. In our case, this electronic excitation energy is essentially the distance Δ_{el} between the two electronic levels at fixed environment. This condition is well fulfilled in many real ET which are far enough from the inversion point but fails close to it.

Although it is clear that the characteristic time for ET will be minimum in the vicinity of the Marcus inversion point, and that the conditions for ultrafast ET will require to be in this vicinity, ultrafast ET is not well described within the original adiabatic Marcus theory because of its adiabatic assumption. We propose a non adiabatic theory for the dimer model which recovers the Marcus theory far from the inversion point (when adiabaticity is recovered) and which improves it close to it. Moreover, our approach can be extended when more electronic sites are involved and opens new perspectives for understanding enzymatic catalysis. In particular, we show that a weakly interacting but well tuned third site (catalyst) could trigger selectively ultrafast ET a low temperature which otherwise would require a huge activation energy and would not occur at low temperature. We shall also briefly discuss extensions of this new approach to quantum transitions in general and the new perspectives opened.

2. A Non-adiabatic Model for ET

We consider the simpler problem where a single electron tunnels between donor (D) and acceptor (A) systems representing large molecules with many vibrational degrees of freedom (see fig.2). Each of these molecules α ($\alpha = D$ or A but more molecules may be involved) is supposed to involve for simplicity a single electronic state with a wavefunction $|\Psi_\alpha >= \Psi_\alpha(\mathbf{r}; \{u_i^\alpha\})$, where $\mathbf{r}$ is the space coordinate, u_i^α the phonon coordinates, and we assume the Born-Oppenheimer (adiabatic) approximation is valid for the isolated molecule. This approximation holds when the other electronic states on this molecule are far apart in energy from the considered state $|\Psi_\alpha >$ (at the scale of the maximum phonon energy). Then this electronic wavefunction can be considered as a function of the molecule phonon coordinates

$\{u_i^\alpha\}$, including the whole environment and in particular the solvent.

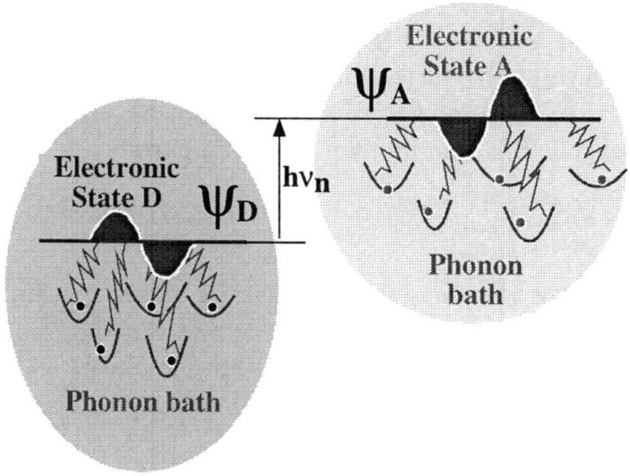

Figure 2. Schematic representation of a donor-acceptor pair with electronic levels interacting with phonon baths in the normal Marcus regime

Within a standard tight-binding representation, the state of the electron in the whole system, has the form $\sum_\alpha \psi_\alpha|\mathbf{\Psi}_\alpha>$. We use as Kramers reaction coordinates [3] the complex amplitudes ψ_α of the electronic wave-function with the normalization condition $\sum_\alpha |\psi_\alpha|^2 = 1$.

It is convenient to define first the minimum energy of the system of the two coupled electronic states $H_T(\{\psi_\alpha\})$ at fixed collective variables $\{\psi_\alpha\}$. Then, the Hamiltonian of the interacting electron-phonon system can be written as

$$\mathbf{H} = H_T(\{\psi_\alpha\}) + \sum_\alpha H_\alpha^{ph}(\{|\psi_\alpha|^2, u_i^\alpha, p_i^\alpha\}) \tag{1}$$

The electronic density $|\psi_\alpha|^2$ on each molecule α couples to the coordinates of the same molecule assumed to be harmonic and thus consisting of a collection of independent harmonic oscillators i with position-momentum coordinates $u_{\alpha,i}, p_{\alpha,i}$, mass $m_{\alpha,i}$, and frequencies $\omega_{\alpha,i}$

$$H_\alpha^{ph}(\{|\psi_\alpha|^2, u_{\alpha,i}, p_{\alpha,i}\}) = \sum_i \frac{1}{2} m_{\alpha,i}\omega_{\alpha,i}^2 \left(u_{\alpha,i} - k_{\alpha,i}|\psi_\alpha|^2\right)^2 + \frac{1}{2m_{\alpha,i}} p_{\alpha,i}^2 \tag{2}$$

In principle this coupling energy involves all possible interactions with the atomic coordinates and in particular, the chemical energies and the electrostatic energies. The latter ones could be especially important in biomolecules which are polyelectrolytes surrounded by ions and highly polarizable water.

In general $H_T(\{\psi_\alpha\})$ has not the quadratic Hermitian form $<\{\psi_\alpha\}|H_T|\{\psi_\alpha\}>$ where H_T is a linear operator but is highly **nonlinear** as a consequence of the surrounding electric field and the molecule and environment reorganization generated by the density variations of ψ_α. It is convenient to split this Hamiltonian in several parts

$$H_T(\{\psi_\alpha\}) = \sum_\alpha H_\alpha(|\psi_\alpha|^2) + H_f(\{|\psi_\alpha|^2\}) + H_t(\{\psi_\alpha\}) \tag{3}$$

where $H_\alpha(|\psi_\alpha|^2)$ is the energy of the isolated molecule α (in its environment) which depends only on its electron density $I_\alpha = |\psi_\alpha|^2$. It is sufficient to expand this energy at second order in electronic density which is the lowest order where nonlinearity shows up, for obtaining the essential physical features

$$H_\alpha(|\psi_\alpha|^2) = \mu_\alpha|\psi_\alpha|^2 + \frac{1}{2}\chi_\alpha|\psi_\alpha|^4 \tag{4}$$

μ_α is the linear electronic level at zero occupation. $\chi_\alpha = \chi_\alpha^C + \chi_\alpha^R$ is the sum of two contributions. χ_α^C is the positive coefficient for the energy of the electric field generated by the charge I_α without lattice reorganization (capacitive energy). This coefficient takes into account the local electronic dielectric constant ϵ_∞ of the environment.

χ_α^R is the negative coefficient of the energy gain from the local reorganization of the surrounding nuclei due to the presence of the electron. This energy involves for a part chemical bond energies which could be broken or created and electrostatic terms due to ions displacements (which are involved in the local static dielectric constant ϵ_0). Actually, in our model (2) the reorganization energy can be explicitly calculated from the coupling with the phonon bath which describes in the harmonic approximation the energy generated by all of the displacements of the surrounding nuclei. Then, we get $\chi_\alpha^R = -\sum_i m_{\alpha,i}\omega_{\alpha,i}^2 k_{\alpha,i}^2$.

It is essential to remark that that the sum of the two contributions χ_α might be positive or negative depending whether the electrostatic energies or the chemical energies are prevalent. We expect for example that when the electronic state α belong to the inner shell of a transition metal ion (which could be embedded in the core of a large biomolecule), the electrostatic energy is prevalent so that χ_α is positive. When it belongs to a chemical bond (or a ring of bonds), it is more likely negative.

However, these rough intuitive arguments are not reliable for a precise calculation of this coefficient χ_α. More sophisticated *ab initio* calculations which takes into account all the interactions in the complete quantum system should estimate more precisely the values of these coefficients χ_α. Indeed, our coefficients χ_α can be related to the more general hardness matrix defined within the Density Functional Theory in [5] as coefficients

$$\eta_{i,j} = \frac{\partial^2 E}{\partial n_i \partial n_j} \tag{5}$$

where E is the total energy of the system and n_i and n_j are occupation numbers of eigenstates i, j in the Kohn-Sham effective potential. In our simple model where the transfer integral between the different molecules sustaining the electronic sites α is a perturbation supposed to be very small, the eigenstates $i, j, \ldots$ are localized on single molecule sites $\alpha, \beta, \ldots$ and the crossed terms $\eta_{\alpha,\beta}$ with $\alpha \neq \beta$ are small. The hardness matrix becomes practically diagonal with diagonal coefficients $\eta_{\alpha,\alpha} = \chi_\alpha$. In general these coefficients are not zero and could be positive or negative. Thus, the most recent *ab initio* calculations reveals the existence of intrinsic nonlinearities in the electron dynamics.

The existence of nonlinearity manifests physically by the fact the electronic level on each molecule α

$$E_\alpha = \frac{\partial H_\alpha}{\partial I_\alpha} = \mu_\alpha + \chi_\alpha |\psi_\alpha|^2 \tag{6}$$

is not fixed rigidly but depends on its electronic occupation density.

The next term $H_f(\{|\psi_\alpha|^2\})$ in (3) is the correction of energy due the direct molecule interactions which not related to the electron transfer and which could originate from Van de Waal, dipolar, Coulomb etc... interactions $H_f(\{|\psi_\alpha|^2\}) =$
$\sum_{\alpha,\beta} C_{\alpha,\beta} |\psi_\alpha|^2 |\psi_\beta|^2$ where $C_{\alpha,\beta}$ are mutual capacitance coefficients.

Finally, $H_t(\{\psi_\alpha\})$ in (3) is the small energy term due to overlaps of the quantum orbitals of the electron on neighboring molecules. This term is the one which allows the quantum tunnelling of the electron and permits ET. We may choose for simplicity the form $H_t(\{\psi_\alpha\}) =$
$\lambda_{\alpha,\beta} \psi_\alpha^\star \psi_\beta + c.c.$ These coefficients $\lambda_{\alpha,\beta}$ are obtained within the standard by tight binding representations as overlap integrals and thus are called transfer integrals. They are assumed to be small and to range within the order of phonon energies $\hbar\omega_{\alpha,i}$.

At the ideal anti-adiabatic regime which however is physically nonrealistic, these transfer integrals $\lambda_{\alpha,\beta}$ between different molecules are much smaller than the phonon energies $\hbar\omega_{\alpha,i}$ and then ET is much slower than the phonon dynamics. It becomes exact in that limit to consider the dynamics of the fast phonon variables $\{u_{\alpha,i}\}$ is described by Hamiltonian (2) where the slow electron variables $\{\psi_\alpha\}$ are supposed to be static variables. Then at zero temperature, $H_T(\{\psi_\alpha\})$ becomes the exact Hamiltonian describing the electron dynamics through the set of nonlinear Hamilton equations $i\hbar\dot{\psi}_\alpha = \partial H_T/\partial\psi^\star_\alpha$.

In some cases, nonlinearities in $H_T(\{\psi_\alpha\})$ may generate energy barriers. ET at zero temperature requires the absence of energy barriers but then since the dynamics of the electron is purely Hamiltonian with no energy dissipation, the electron cannot moves the state on the acceptor which has an energy lower than on the donor. Consequently, ET at zero temperature cannot be obtained at this antiadiabatic limit (neither at the adiabatic limit). As a result, ET at zero temperature requires an efficient energy dissipation which can be obtained only in the regime intermediate between adiabatic and anti adiabatic

Note the dynamical coupling of the electron with the electromagnetic field is not considered here but it also generates in principle energy dissipation. However, this coupling is very weak and is in the regime where the Fermi golden rule holds. Indeed, we may have quantum emission of photons but the life time of the electronic states is very long at the time scale we are interested in. Therefore, we can neglect this energy dissipation for describing ultrafast ET.

Unlike this situation, the lattice reorganization due to the presence of an electron on a molecule is large and involves large nuclei displacements much larger than their zero point quantum motion (that is it involves the coherent creation of many quantum phonons). It is thus a good approximation to consider that the phonon variables $\{u_{\alpha,i}\}$ in the Hamiltonian are classical variables while the electron variables ψ_α still represent quantum variables. The dynamical equations of the coupled system (1,2) are

$$i\hbar\dot{\psi}_\alpha = \frac{\partial H_T}{\partial\psi^\star_\alpha} - \sum_i m_{\alpha,i}k_{\alpha,i}\omega^2_{\alpha,i}(u_{\alpha,i} - k_{\alpha,i}|\psi_\alpha|^2)\psi_\alpha \tag{7}$$

$$\ddot{u}_{\alpha,i} + \omega^2_{\alpha,i}(u_{\alpha,i} - k_{\alpha,i}|\psi_\alpha|^2) = 0 \tag{8}$$

The harmonic motions $u_{\alpha,i}(t)$ can be explicitly obtained from the linear equations (8) [4] as the sum of functions of the time dependent driving force $|\psi_\alpha(t)|^2$ and a solution of the equation without driving force. Actually, this term physically corresponds to thermal fluctuations of $u_{\alpha,i}$ and thus is random. Then, substituting $u_{\alpha,i}(t)$ in eq.(7) yields the fundamental equation for non-adiabatic electron dynamics (which preserves the norm $\sum_\alpha |\psi_\alpha|^2$)

$$i\hbar\dot{\psi}_\alpha = \frac{\partial H_T}{\partial\psi^\star_\alpha} + \left(\int_{-\infty}^{t} \Gamma_\alpha(t-\tau)\frac{d|\psi_\alpha|^2}{d\tau}d\tau + \zeta_\alpha(t)\right)\psi_\alpha \tag{9}$$

where $\Gamma_\alpha(t) = \sum_i m_{\alpha,i}\omega_{\alpha,i}^2 k_{\alpha,i}^2 \cos(\omega_{\alpha,i}t)$. If there are many phonon modes with a rather uniform distribution, $\Gamma_\alpha(t)$ can be assumed to be a smooth decaying function of time. It generates energy dissipation as a kernel in eq.(9) (the absorption rate in energy of a charge fluctuation at site α at frequency ω is nothing but the product of the square of its amplitude with the Fourier transform of $\Gamma_\alpha(t)$). We also have $\Gamma_\alpha(0) = -\chi_\alpha^R$. The time dependent potential $\zeta_\alpha(t)$ is produced by the thermal fluctuations of the lattice. It is a colored random Langevin force with correlation function which fulfills $< \zeta_\alpha(t+\tau)\zeta_\alpha(\tau) >_\tau = \Gamma_\alpha(t)\ k_B T$ at temperature T [4].

Thus, the effect of non-adiabaticity is to transform the standard linear Schrödinger equation describing the dynamics of the electron into a **non-linear** Schrödinger equation (9) with norm preserving **energy dissipation** terms and with **random colored** time dependent potentials generated by atomic thermal fluctuations.

3. ET in the Dimer Model

We first show that far from the Marcus inversion point we essentially recover the basic result of the standard theory [2]. The initial Hamiltonian (1,2) restricted to a dimer model ($\alpha = D$ or A) readily yields the energy surfaces schematically shown in fig.1. Neglecting the small interaction energy terms between the molecules we obtain the essential parameters of this theory which are $\Delta G^0 = \mu_D + \frac{1}{2}\chi_D - \mu_A - \frac{1}{2}\chi_A$, $\Delta_{el} = \Delta G^0 + \frac{1}{2}(\chi_D^R + \chi_A^R) = \mu'_D + \chi_D^R - \mu'_A$ and $\Delta G^\star = -\Delta_{el}^2/(2(\chi_D^R + \chi_A^R))$ where $\mu'_\alpha = \mu_\alpha + \chi_\alpha^C/2$.

The energy variation $E_T(I_A) = H_D(1-I_A) + H_A(I_A) - H_D(1)$ of our dimer as a function of the electron density $I_A = |\psi_A|^2$ on the acceptor is

$$E_T(I_A) = (\mu_A - \mu_D - \chi_D)I_A + \frac{1}{2}(\chi_D + \chi_A)I_A^2 \qquad (10)$$

and $-E_T(1) = \Delta G^0$ is the chemical reaction energy. There is always an energy barrier between donor and acceptor when $\frac{dE_T}{dI_A}(0) > 0$ or equivalently $\mu_A < \mu_D + \chi_D$. Otherwise, the minimum of energy is not necessarily obtained for a total transfer at the acceptor when $\chi_D + \chi_A > 0$. The derivative $dE_T/dI_A = E_A - E_D$ (see eq.(6)) is the difference of the electronic levels on the acceptor and the donor at the transfer I_A. Thus, resonance between donor and acceptor implies a zero derivative.

For recovering the same results as the Marcus theory from our equation, it is essential to note that the phonon spectrum has a cut-off at relatively small frequencies ω_c. Beyond this frequency, the Fourier spectrum of $\Gamma(t)$ is zero. Thus, when the characteristic energy of the electron dynamics, which is the energy difference $E_D - E_A$ between the electronic levels, eq.(6), is

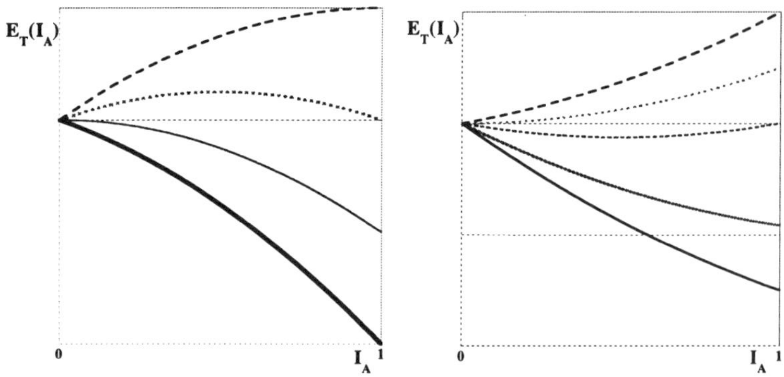

Figure 3. Several energy profiles (10) of the system donor-acceptor versus electron density on the acceptor in the soft case $\chi_D+\chi_A < 0$ (left) or in the hard case $\chi_D+\chi_A > 0$ (right)

larger than the phonon energy $\hbar\omega_c$, there is no more energy dissipation [4]. Then ET cannot be achieved at zero degree K but requires thermal fluctuations. Actually, this is the regime of validity of the adiabatic approximation.

The electron density which is initially on the donor $|\psi_D|^2 = 1$ and $|\psi_A|^2 = 0$ remains practically constant. Eq.(7) gives in this case the random potential as due to phonon variables, $\zeta_\alpha(t) = -\sum_i m_{\alpha,i} k_{\alpha,i} \omega^2_{\alpha,i}(u_{\alpha,i} - k_{\alpha,i}|\psi_\alpha|^2)$. The slowly varying potential $\zeta_D(t)$ makes the energy level of the electron $E_D(t) = \mu_D + \chi_D + \zeta_D(t)$ on the donor time dependent. The unoccupied energy level on the acceptor $E_A(t) = \mu_A + \zeta_A(t)$ fluctuates similarly. It occurs statistically that $E_D(t) \approx E_A(t)$ or

$$\mu_D + \chi_D - \mu_A + \zeta_D(t) - \zeta_A(t) \approx 0 \tag{11}$$

induces an almost resonance between donor and acceptor so that the electron could tunnel (see fig.4). If we discard the details concerning the probability of this tunneling process and neglect its intrinsic time which is generally short compared to the characteristic time for reaching the resonance, the characteristic time for ET is mostly related to the time required to reach the resonance [b].

Condition 11 may be compared with the condition for the intersection

[b]The electron tunneling problem has been considered differently in the literature which distinguishes between adiabatic processes (strong reactants) and diabatic processes (weak reactants) [1]. Our non-adiabatic theory (potentially) describes both cases as well as the intermediate cases.

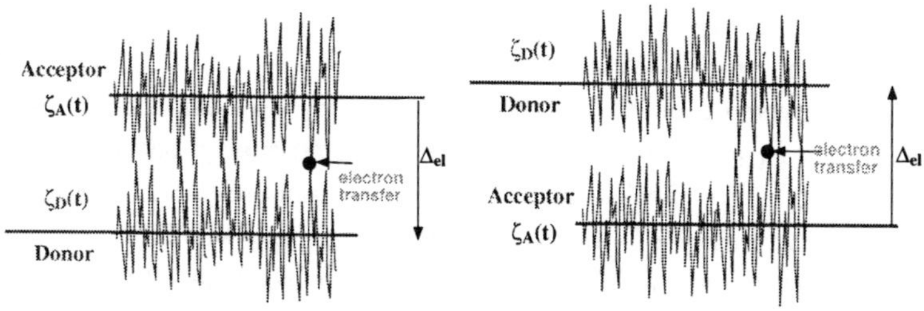

Figure 4. Sketch of the electronic level fluctuations in the normal regime (left) and in the Marcus inverted regime(right).

of the two free energy surfaces shown in fig.1, which can be written as

$$\mu'_D + \chi^R_D - \mu'_A + \zeta_D(t) - \zeta_A(t) = \Delta_{el} + \zeta_D(t) - \zeta_A(t) = 0 \qquad (12)$$

This condition is similar but different from our resonance condition (11). The reason is that the energy level variations due to the Coulomb energies are not taken into account in the Marcus theory unlike the reorganization energy (if $\chi^C_\alpha = 0$ conditions (11) and (12) become identical). Nevertheless, we can also interpret the probability of reaching the resonance as shown in fig.4 [4] in terms of an activation process with an activation energy $\Delta' G^\star = -\Delta'_{el}/(2(\chi^R_D+\chi^R_A)) \neq \Delta G^\star$ where $\Delta'_{el} = \Delta_{el}+(\chi^C_D+\chi^C_A)/2 = \mu_D+\chi_D-\mu_A$ is different from Δ_{el} because of the Coulomb terms.

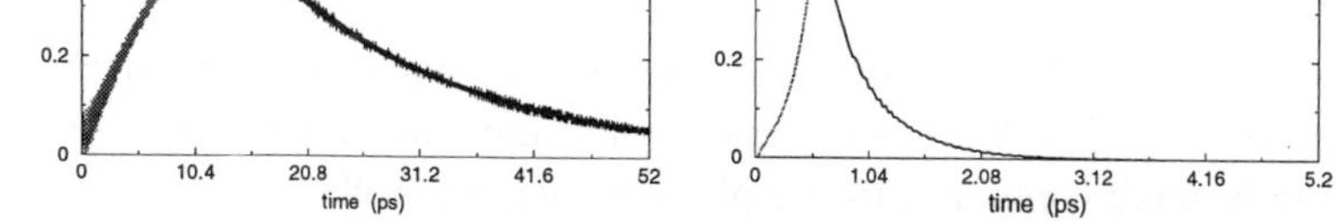

Figure 5. Electron density on the donor and the acceptor versus time for the dimer model at the inversion point and zero degree K where $\mu_D = 2$, $\chi_D = -1$, $\mu_A = 1$, $\chi_A = -0.75$, $\lambda_{AD} = 10^{-2}$, $\gamma_D = \gamma_A = 1.$ (left) or 20. (right) (the time unit is 1 ps$=10^{-12}$ s for energy in units of eV)

Nevertheless, our approach confirms the existence of an inversion point when resonance is obtained at zero degree K for $\zeta_D = \zeta_A = 0$. Then, (11) yields $\Delta'_{el} = 0$ or $\mu_D + \chi_D = \mu_A$ which again is different from the condition $\Delta_{el} = 0$ in the original Marcus theory.

At our inversion point, the energy profile $E_T(I_A)$ has a zero derivative at the origin $I_A = 0$. Fig.3 shows that there is no energy barrier only when $\chi_D + \chi_A \leq 0$ [c]. We check that the initial resonance $E_D = E_A$ triggers ET. If the derivative dE_T/dI_A remains smaller than $\hbar\omega_c$ which is equivalent to a small reaction energy $\Delta G^0 < \hbar\omega_c/2$, ET can be achieved at zero degree K without thermal fluctuations. Fig.5 shows two examples. The speed of ET strongly depends on the damping which is related to the coupling to the phonon bath. In these examples and in the following we chose for simplicity $H_f(\{|\psi_\alpha|^2\}) = 0$ in (3). If the phonon frequency cutoff ω_c is large compared to the characteristic electronic frequencies, a reasonable approximation is to assume that $\Gamma(t) = 2\gamma_\alpha\delta(t)$ is a Dirac function. Then, $\int_{-\infty}^{t} \Gamma(t-\tau)(d|\psi_\alpha|^2/d\tau)d\tau \approx \gamma_\alpha d|\psi_\alpha(t)|^2/dt$ in eq.(9). There is an optimal damping constant ($\gamma_\alpha \approx 40$) where the characteristic time required for ET is minimum.

ET is triggered at zero degree (but slows down) when escaping only on one side of the inversion point when $\mu_A < \mu_D + \chi_D$. On the other side, it is blocked at zero degree K because of the appearance of an energy barrier. However it is complete only when dE_T/dI_A remains always negative with a modulus which never exceeds the phonon cutoff energy $\hbar\omega_c$ (for having efficient energy dissipation). In summary, our approach yields results which are qualitatively similar to those of the Marcus theory far from the inversion point but with a redefinition of characteristic parameters. It yields more detailed features not predicted by the original Marcus theory, close to the inversion point.

4. Principle of Catalytic ET in a trimer model

ET could be fast for the simple donor-acceptor system only in special conditions close to the Marcus inversion point and when the chemical reaction energy is small compared to the phonon energy cutoff.

We now show that we can take advantage of a third catalytic site weakly coupled to the donor for triggering at zero temperature an ultrafast ET from the donor to the acceptor while in the absence of catalyst, a large energy barrier would prevent any transfer at zero degrees.

It is clear that ET could become fast only in case of resonance or almost resonance within the phonon energy range. Otherwise, the electronic level

[c]The only case implicitly considered in the Marcus theory is for negative $\chi_\alpha = \chi_\alpha^R < 0$.

E_α on a molecule depends on its occupation density $E_\alpha = \partial H_\alpha(I_\alpha)/\partial I_\alpha = \mu_\alpha + \chi_\alpha I_\alpha$

How to get complete ultrafast ET between a pair donor and acceptor at zero degree K which is not at the Marcus inversion point? We suggest to take advantage of the nonlinearities for inducing relatively slow oscillations of the electronic level on the donor (or on the acceptor) which could produce a resonance between donor and acceptor. To be efficient, the characteristic time associated with the transfer energy to the acceptor should be shorter or at least comparable with the period of the oscillation.

These energy oscillations may be obtained by thermal fluctuations of the nuclei environment as shown fig.4. Actually, this is implicitly the wellknown standard situation which is well described by the Marcus theory as we explained above.

Another way is to induce artificially these electronic level oscillations by exciting some specific phonon well coupled to the electronic level. Note that this is what happens systematically when creating an exciton by the absorption of a photon before the Franck-Condon relaxation. This situation very likely occurs in some real systems and then could induce ultrafast ET for example which could be photoinduced in some doped polymers [15] but we shall not discuss these cases here.

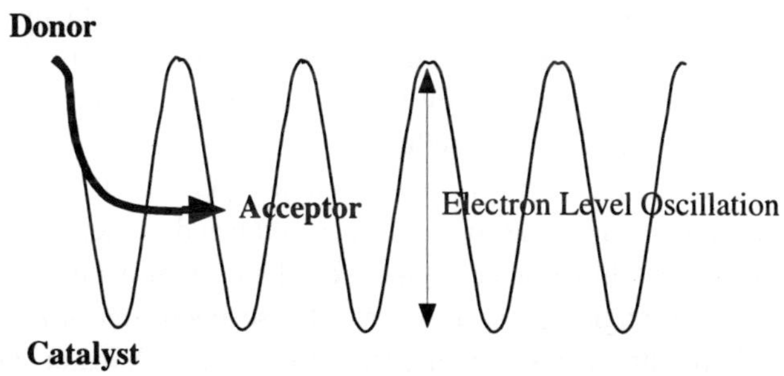

Figure 6. Sketch of electronic level oscillations on the donor system inducing resonance with the acceptor level.

Another efficient way to induce these level oscillations, is based on the effect of Targeted Transfer between the donor and a third site (we call "catalyst") which is appropriately tuned on the donor. This is the situation we describe which could relevant for understanding biochemistry which involves highly selective enzymatic catalysis [d] We need first to recall what is Targeted Transfer.

4.1. *Targeted Energy Transfer between nonlinear oscillators*

The concept of Targeted Energy Transfer [6,7] extends to the nonlinear case the well-known and ancient concept of linear resonance but with some essential differences.

When two linear oscillators are weakly coupled and in resonance (that is with the same frequency), any amount of energy which is injected to the first harmonic oscillator is totally transferred to the second harmonic oscillator initially at rest. All the energy oscillates back and forth between the two oscillator with a frequency proportional to the small coupling. Very soon when the frequencies of the two oscillators becomes unequal, the resonance is broken and the energy remains trapped on the oscillator which initially excited (see fig.7). This picture is also a model for quantum tunnelling which describes the oscillation of the electronic wave function between two weakly coupled degenerate states (see fig.8. Then the electronic energy levels corresponds to the oscillator frequencies. The frequencies of the two harmonic oscillators (or the electronic levels) remains equal and constant during the slow wave oscillation.

In the nonlinear case, the frequency of a anharmonic oscillator depends on its energy. Thus in general even if the frequency of the first oscillator at the injected energy is initially resonant with the frequency of the second oscillator at rest, this resonance generally does not persist since both frequencies varies. Then, energy transfer stops very soon and the energy remains trapped on the first oscillator.

The special situation where this resonance persists, is called called Targeted Transfer. This situation has been analyzed in details in [6] and the general conditions have been explicit for the oscillator hamiltonians oscillators in action-angle representation. These conditions require the first anharmonic oscillator is soft while the second one is hard or vice versa.

Despite the energy oscillates as well as in the linear resonant case, targeted energy transfer only occurs at a selected initial energy for the first oscillator when it is resonant with the second one at rest. In addition, the frequencies of both oscillators which remain equal during the transfer, oscillates with the same period as the energy transfer in the interval deter-

[d]This model concerns enzymatic catalysis where the catalytic effect is induced by one of few active sites unlike for example surface catalysis where collective effects are likely necessary.

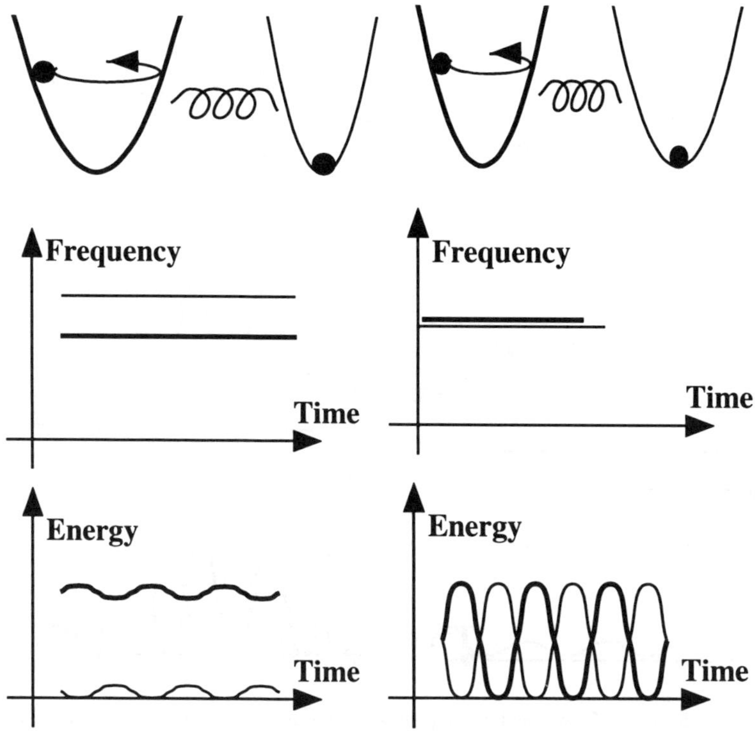

Figure 7. Scheme of two weakly coupled linear oscillators. When they are not resonant (left), there is almost no energy transfer and when they are (right), any amount of energy on the first oscillator is totally transferred and slowly oscillates between the two oscillators which keeps constant and equal frequencies.

mined by the frequencies of the oscillators at maximum amplitude and at rest which sharply contrast with the linear resonant case.

It has been shown that Targeted Energy Transfer persists for well-tuned pairs of discrete breathers. Then, energy could be observed to oscillate back and forth between two distant regions of a complex system. These selected sites could be detected by our techniques [7].

4.2. *Targeted Transfer of Electron between donor and catalyst*

The same theory holds for electronic resonances. Since the electronic transfer integral between the donor and the acceptor has been assumed to be small (weak reactants), the electron tunnelling is slow and occurs at a time scale in which the environment is changing. The electronic levels vary as

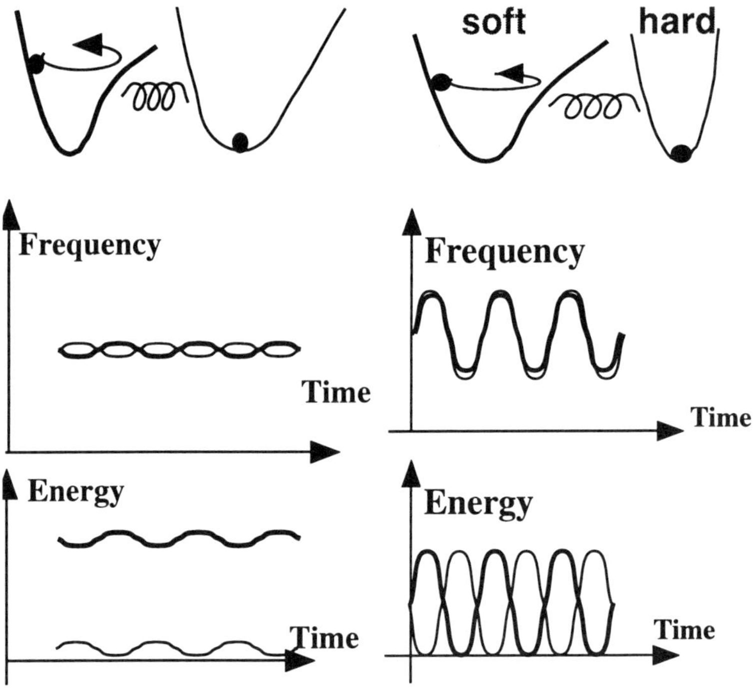

Figure 8. Scheme of two weakly coupled anharmonic oscillators. In general, no substantial energy transfer occurs at weak coupling even if they are initially resonant (left). When the nonlinearities of the two oscillators, their resonance persists all over the transfer. But then although the energy of the first oscillator is totally transferred and oscillates back and forth between the two nonlinear oscillators as in the linear resonant case, this effect only occurs if the injected energy on the donor is selectively chosen and then the frequencies of both oscillators although remaining equal, oscillates with the same period as the energy oscillation.

a function of their occupation density due to the reorganization of the environment. This is the origin of the nonlinearity (and damping) in our quantum model.

Although a donor and an acceptor at the Marcus inversion point are initially in resonance but in general they do not remain in resonance during the transfer. When the nonlinearities are expanded at the lowest significant order for the hamiltonians of the donor (D) and the donor we now call catalyst (C) are $H_D(I_D) = \mu_D I_D + \chi_D I_D^2/2$ and $H_C(I_C) = \mu_C I_C + \chi_C I_C^2/2$ where $I_D = |\psi_D|^2$ and $I_C = |\psi_C|^2$ are the electron densities. The electronic levels depends on the electronic density as $E_D(I_D) = \partial H_D/\partial I_D = \mu_D + \chi_D I_D$ and $E_C(I_C) = \mu_C + \chi_C I_C$.

We have targeted transfer for the electron initially on the donor $I_D(0) =$

1 and $I_C(0) = 0$, when $\mu_C = \mu_D + \chi_D$ and $\chi_D + \chi_C = 0$ [6]. This situation requires to associate a soft electronic level $\chi_D < 0$ with a hard electronic level $\chi_C = -\chi_D > 0$ on the catalyst (which could involve a metallic ion as suggested above). The role of the donor and the catalyst could be reversed. Then, it is clear that the electron resonance persists during the whole electron transfer because the total electronic density $I_D + I_C = 1$ remaining unity implies the electronic levels remains equal $E_D(I_D) = E_C(I_C)$. Actually, the dynamical trajectory of this model at TET can be explicitly calculated [6]).

However, note the electron transfer from the donor to the catalyst has a reaction energy for the electron transfer which is zero, since $H_D(I_D) + H_D(I_C)$. There is no irreversible chemical reaction but nevertheless (in the absence of damping), the electron initially on the donor will slowly oscillates back and forth between the donor and the catalyst (C), with a frequency corresponding to the transfer integral λ_{CD}. The electronic level oscillates $\mu_D + \chi_D I_D = \mu_C + \chi_C I_C$ as well with the same period.

If the tuning is not perfect, the fraction of the electron charge which is transferred to the catalyst sharply drop to zero and there is no electronic level oscillations. This is illustrated by fig.9 from [8] which shows the maximum density of the electron which is transferred in a model for ET versus the initial density (note this initial density could be normalized to 1 by an appropriate rescaling of the parameters of the model and then it appears as a detuning parameter)

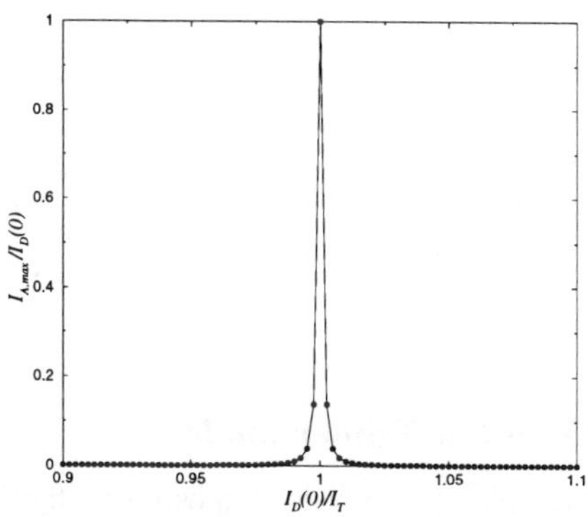

Figure 9. Ratio of the maximum electron density transferred onto the catalyst versus the initial electron density on the donor at $\chi_D = 0.5$, $\chi_C = -0.5$, $\mu_D = -1$, $\mu_C = 1$ and $\lambda_{CD} = 0.001$

Actually, because of the small transfer integral λ_{CD}, the degeneracy between donor and catalyst at $\lambda_{CD} = 0$ is raised. The electron ground-state is a covalent state with equal density of the electron on donor and catalyst with a small binding energy $|\lambda_{CD}|$ which can be considered negligible. However, when there is energy dissipation, this oscillation is damped and converges to (see fig.10). In any case, at small temperature $k_B T > \approx |\lambda_{CD}|$, the thermal noise in the dynamical equations should make this composite mode involving electron and environment oscillations should persistently excited.

These electron oscillations generate variation of in the electron energy level within the interval $[\mu_D + \chi_D, \mu_D]$ (it could be reduced to the interval $[\mu_D + \chi_D/2, \mu_D]$ at very low temperature in the overdamped case). The consequence is that small thermal fluctuations are sufficient to generate giant charge fluctuations. This donor and catalyst do not bind chemically but could trigger ultrafast ET to another molecule with an electronic level in the range of variation of these fluctuations. This is a catalytic effect which in the absence of the weakly coupled catalyst could occur.

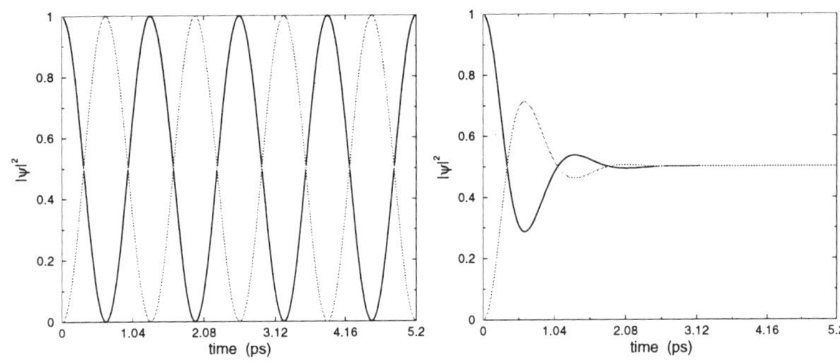

Figure 10. Electron density oscillations on the donor and its catalyst at zero degree K (Targeted transfer), without damping $\gamma_D = \gamma_C = 0$ (left) and with damping $\gamma_D = \gamma_C = 1$ (right). $\mu_D = 2$, $\chi_D = -1$, $\mu_C = 1$, $\chi_C = -1$, $\lambda_{CD} = 10^{-2}$.

4.3. *Ultrafast ET in the Trimer model*

We now test the principle of catalysis suggested by fig.6 on a trimer model with Hamiltonian

$$H_{tr} = \mu_D|\psi_D|^2 + \frac{1}{2}\chi_D|\psi_D|^4 + \mu_C|\psi_C|^2 + \frac{1}{2}\chi_C|\psi_C|^4 + \mu_A|\psi_A|^2 + \frac{1}{2}\chi_A|\psi_A|^4$$

$$+\lambda_{CD}(\psi_D^\star\psi_C + cc) + \lambda_{CA}(\psi_A^\star\psi_C + cc) + \lambda_{AD}(\psi_D^\star\psi_A + cc)$$

which now involve three molecules a well-tuned pair donor-catalyst (D) and (C) and an acceptor (A). In order to fix the ideas, assume that donor and acceptor are both soft ($\chi_D < 0, \chi_A < 0$) and that the reaction energy for the transfer of the electron from the donor to the acceptor is positive that is $\mu_A + \chi_A/2 < \mu_D + \chi_D/2$.

In the absence of catalyst ($\lambda_{CD} = \lambda_{CA} = 0$) there is a large Kramers energy barrier between donor and acceptor, which implies $\mu_D + \chi_D << \mu_A$. To be more precise, this energy barrier may be measured in eV or large fraction of eV. In that case, the direct ET between donor and acceptor would range in the normal Marcus regime. ET cannot occur at zero temperature and may only occur (significantly) at relatively large temperature since its activation energy is supposed to be large.

We now introduce the catalyst which is well tuned to the Donor (that is $\chi_C = -\chi_D$ and $\mu_C = \mu_D + \chi_D$). The initial electronic level μ_A of the acceptor should belong to the variation interval of the electronic level of the donor-catalyst system which yields $\mu_D + \chi_D < \mu_A < \mu_D$ at weak damping or $\mu_D + \chi_D/2 < \mu_A < \mu_D$ at strong damping. Fig.11 sketches $H_D(I_D)$, $H_C(I_C)$ and $H_A(I_A)$ when all these conditions are fulfilled.

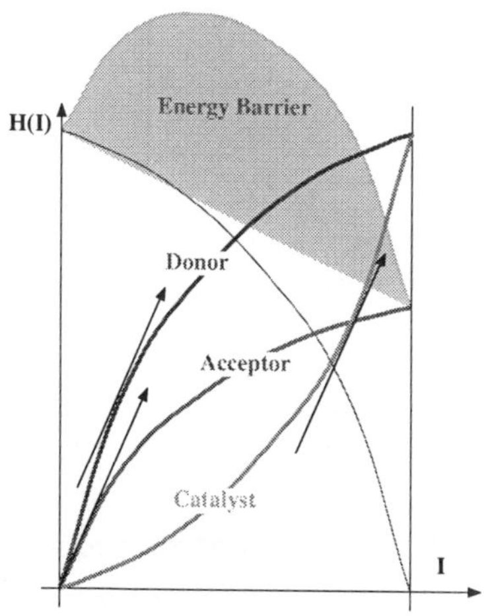

Figure 11. Energies $H_D(I)$, $H_C(I)$ and $H_A(I)$ versus electron density I for the donor, catalyst and acceptor in the situation of Ultrafast ET. The energy barrier between donor and acceptor without catalyst is plotted in gray.

Fig.12 shows that under these conditions, huge charge fluctuations suddenly appear between between weakly coupled donor acceptor and catalyst, while the donor-acceptor system alone does not exhibit any fluctuations. However, the absence of energy dissipation prevents the electron from falling on its ground-state, which is on the acceptor. The same model with damp-

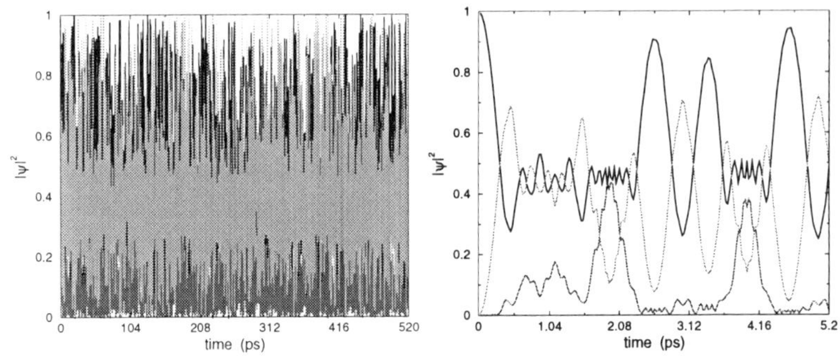

Figure 12. Electron density on donor acceptor and catalyst versus time in the trimer in the absence of damping $\gamma_D = \gamma_C = \gamma_A = 0$ at two different time scales $\mu_D = 2$, $\chi_D = -1$, $\mu_C = 1$, $\chi_C = 1$, $\mu_A = 1.5$, $\chi_A = -0.75$, $\lambda_{AD} = \lambda_{AC} = \lambda_{CD} = 10^{-2}$. The electron is initially on the donor.

ing shows that the electron finally falls on the acceptor while the catalyst has only taken transitively a fraction of the electronic charge (see 13).

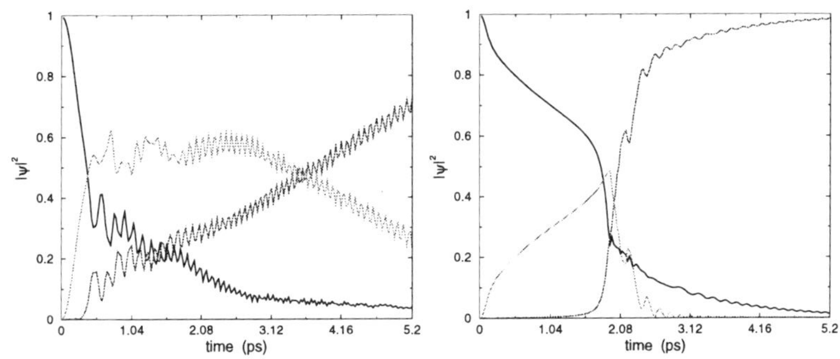

Figure 13. Same as fig.12 but with damping $\gamma_D = \gamma_C = \gamma_A = 2$ (left) or 10 (right)

This ET is highly sensitive to small perturbations of the donor-catalyst system which easily breaks the Targeted Transfer [6,7]. We have shown for example that relatively small electric fields are sufficient for blocking ET at zero degree K [4]. These principles may be extended to many-site networks of electronic levels where the electron can choose a specific path very selectively. This path can be blocked and switched under small perturbations.

Logical functions with one or few electrons could be built at the molecular level suggesting potential nanodevice applications and complex biological functions to be studied in living cells. These studies are left for further developments.

5. Ultra Fast Electron Transfer in the Photosynthetic Reaction Center as an illustration

We think that many examples of potential applications of our approach for ET can be found in biochemistry. Among them, we describe some observed features for the primary charge separation in the photosynthetic reaction center [9] which has been intensively studied during several decades with many various experimental techniques including femtosecond pulse laser experiments combined with protein mutations [12,13]. We believe that the observed features concerning this system globally agree with our predictions although more quantitative investigations should be necessary.

Photosynthesis [10] occurs in the chloroplasts of the photosynthetic cells. These chloroplasts contain stacking of many tykaloids which are small vesicles. The membrane of these tykaloids contains many complex structures of biomolecules. Among them pigment molecules (bacteriochlorophylls a and b, carotenes etc...) are organized into complexes LH1, LH2... forming in a series of rings as an onion. Their global structure operates as a kind of antenna capturing photons as electronic excitations (excitons) which systematically funnel to the center of the antenna at the Photosynthetic Reaction Center.

The excitons collected in the pigments are transferred very fast within a 100 femtosecond through the pigment molecules to the inner ring of the antenna. Then, the exciton is transferred to the the photosynthetic reaction center within about 30 picoseconds. This Reaction Center consists of many assembled polypeptides forming the scaffold of the system maintaining the cofactors (Bacteriochlorophylls, Bacteriopheophytin) and the metallic ions in appropriate positions and interactions (see fig.14). The role of the reaction center is to use the exciton energy for pumping electrons from the outside of the tykaloid membrane (the periplasm) to the inside of the tykaloid (the cytoplasm).

The exciton is captured by the strongly bonded dimer (P) of bacteriochlorophyll a which is very close to the periplasm (P) $\longrightarrow$ ($P^\star$). This excited state ($P^\star$) has the property to transfer very fast an electron to the Bacteriopheophitin BPhea (H_LA) in the middle of the tykaloid membrane. $P^* H_LA \rightarrow P^+ H_L^-$. This transfer occurs at room temperature within a half-time of 3 ps which is quite short especially when considering the distance between the electronic sites on ($P^\star$) and (H_L) which is 17Å. The free energy variation of the electron is about 0.25 eV which is still relatively small compared to the exciton energy ($P^\star$) which is 1.38 eV. Surprisingly this transfer time becomes shorter at low temperature of the order of $\approx$ 1ps at 10 K. In the absence of further transfer, the electron is stable on BPhea molecule (H_L) over few ns.

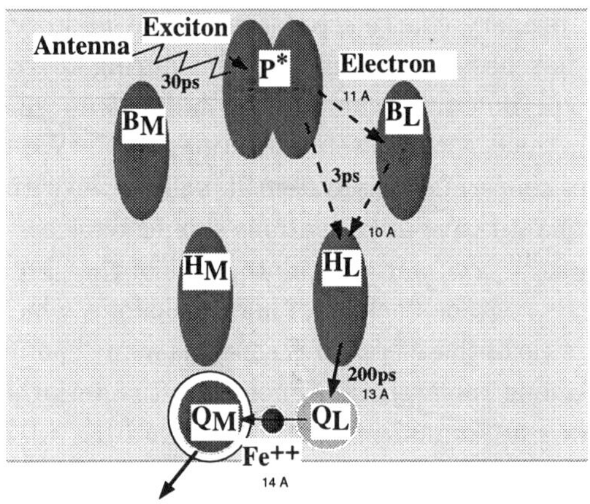

Figure 14. Scheme of the reaction center imbedded in the tykaloid membrane. Its structure consists of a scaffold of polypeptides, a special pair of Bacteriochlorophyll a forming the dimer (P) and an almost symmetric symmetric structures consisting of the two ancillary Bacteriochlorophyll a (B_L) and (B_M), the two Bacteriopheophytins (H_L) and (H_M) and the menaquinone (Q_L) and the ubiquinone(Q_M) bridged by Fe^{++}. Electron transfer between ($P^\star$) and H_L is ultrafast and occurs in 3 picosecond (room temperature) or one picosecond (low temperature).

However, it has been also demonstrate experimentally that the presence of the ancillary bacteriochlorophyll BChla (B_L) is necessary for the electron transfer to BPhea (H_L) to occur, but its exact role is controversial because no substantial accumulation of $BChla^-$ has never been observed in femtosecond experiments.

There is a subsequent series of electron transfers which becomes slower and slower till the electron completely crosses the tykaloid membrane. Although we do not focus our interest on them because they are not ultrafast, we shortly describe the following events. Still in the reaction center, the electron arrives at menaquinone (Q_L) within a half-time of 200 ps. The charge separation is now stable for about 100 ms. Next again, the electron transfers from (Q_L) to (Q_M) (ubiquinone) over the bridge Fe^{++} within a half-time of 30-100 μs, and now is stable for about 1 s. The final result is that an electron have been transferred from one side of the reaction center to the other side across the tykaloid membrane, Next, electron transfers continue elsewhere in the solvent with diffusive molecules involving cytochrome. P^+ recover its electron from the water of the periplasm, which is decomposed then releasing electrons, protons and neutral oxygen molecules The proton gradient between the two sides (cytoplasm and periplasm) of the tykaloid membrane fuels ATPase which is another complex structure in the tykaloid membrane. The flow of proton through this structure generates ATP from ADP. At this stage, a substantial part of the light energy which was harvested, has been stored in a stable form. Further chemical reactions then use this ATP energy for fixing carbon dioxide as hydrocarbons through the Calvin-Benson cycle.

We are here especially interested in understanding the ultrafast electron transfer occurring between $(P^\star)$ and (H_L) with the help of ancillary molecule (B_L). There are other puzzling phenomena associated with this transfer. It has been observed that the excited state $(P^\star)$ is associated with low frequency coherent molecule vibrations (with period 500 femtoseconds and possibly 2 picoseconds) which strongly depend on point mutations of the protein cofactor matrix [12,13]. The electron transfer is unexpectedly sensitive to minor mutations of the involved molecules and in particular to macroscopic electric fields [14] which however involves electron energy changes by at most few meV which is rather small compared to electron energy differences counted in eV. Another puzzling feature is that despite the almost symmetry of the RC, the pathway used by the electron systematically breaks this symmetry.

These observed features could be interpreted within our approach considering that the pair of Bacteriochlorophyll (P) is the donor. The two bacteriochlorophyll BChla (B) are slightly nonequivalent and only one (B_L) is well-tuned as a catalyst for on the donor for triggering ET to the nearby Bacteriopheophitin BPhea (H_L). The biomolecule (B_L) plays the role of catalyst in our trimer model for triggering ET from the donor $(P^\star)$ to the acceptor (H_L). Actually, as it can be seen above in our numerical simulations, there is only a partial and transitory ET between $(P^\star)$ and (B_L) which could explain why (B_L^-) is not observed. Moreover, ET occurs with oscillations associated with important molecular reorganization which could be observed.

Despite the almost symmetry, slight changes of the environment of (B_R) compared to those of (B_L) is sufficient to detune (B_R) from $(P^\star)$ so that ET practically does not occur to (H_R). This effect can observed numerically when considering ET in a more realistic model with 5 sites which is almost symmetric modelling the real system shown fig.14 but will not be shown here because of a lack of space.

In real living system, long evolution and natural selections has optimized the tuning of the whole system so that ET is faster at low temperature. In that case, thermal fluctuations slows down ET despite it is still very efficient at room temperature. Again, this can be observed numerically but will not be shown here.

Since ET is very well optimized, minor mutations at crucial points which affects the quality of the tuning, could sharply reduces the efficiency of ET. In particular the relatively small electric fields (in term of energy at the molecular scale) strongly affect ET by detuning (B_L) from the donor ($P^\star$). Moreover, we observed numerically that the effect of the electric field is asymmetric (as observed experimentally) because an energy barrier for the (partial) electron transfer between ($P^\star$) and (B_L) is created when the electric field has a certain sign while (ET moves toward the normal Marcus regime) when its sign is opposite, there is no barrier but only a slowing down of ET (ET moves toward the Marcus inverted regime).

6. Summary and Concluding Remarks

6.1. *Discussion on the effect of temperature*

We have not analyzed here the effect of the temperature because ultrafast ET is essentially a quantum coherent process which occurs at low temperature. Nevertheless we have shown that the effect of temperature can be simply introduced in the dynamical equations of our model as thermal fluctuations on the electronic levels. We can anticipate and make qualitative predictions on its effect.

When the pair Donor-Catalyst is well-tuned, the thermal noise will tend to detune randomly the system and consequently it will reduce the probability of transfer and thus it will increase the characteristic time for ET. On the opposite, when there is no or a very poor tuning, we return to the standard Marcus theory where thermal fluctuations generate the thermally activated electron transfer which is faster at high temperature.

In the intermediate case, where the pair Donor-Catalyst is not perfectly tuned, some small thermal fluctuation will initially favors electron transfer for overcoming small energy barriers but there will be an optimal temperature beyond which a further increase of the temperature will disfavor ET. Increasing too much the electronic level fluctuations will decreases the probability of a good tuning of the pair Donor-Catalyst and thus slow down the electron transfer. Note the similarity with the effect of stochastic resonance.

6.2. *Summary*

We presented new basic principles for a non-adiabatic theory of ET. We have shown that our new modelling recovers the Marcus theory concerning its main conclusions in the standard situations of thermally activated (and slow) electron transfer.

Next, we have shown that we can use this new modelling for opening new perspectives for understanding ultrafast ET and especially enzymatic catalysis in puzzling situations often observed for example in biochemistry where electron transfers may be not thermally activated and occurs at relatively low temperature.

In our theory, the role of the nonlinearities and damping is essential for achieving an ultrafast ET. These nonlinearities and damping originate from the nonadiabaticity of the electron transfer which is the consequence of the fact because the transfer integral between the molecules is small, the change of the electronic densities during tunnelling must be sufficiently slow so that the local environment of the electron has time to respond and changes during the tunnelling. Another important point of our theory is that it preserves the dynamics of the relative phases of the electronic wave functions. At zero temperature without thermal noise, ET is a coherent process unlike for the Marcus theory which requires thermal fluctuations and for which ET becomes a random and incoherent process.

Thus, we have shown that in a system donor acceptor which could not react directly at zero temperature because of a large activation energy, an ultrafast electron transfer could be triggered at zero temperature just by a weak coupling with a third molecule which is appropriately tuned with the donor (catalyst). Actually, this molecule does not react chemically with the donor because its covalent binding energy is negligible. It only induces slow variations of the electronic level by a partial and transitory transfer of the electron. This variation induces a resonance with the acceptor which then captures the electron. Then, the role of the damping originating from the coupling of the electron with the phonon bath, is essential to trap the electron as fast as possible in its ground state on the acceptor. For a given system, there is an optimized damping at the cross-over underdamped-overdamped (neither too large nor too small), at which ET is faster. In the underdamped regime, such ET is accompanied by both charge and phonons oscillations which concern many phonons and not only few of them.

Finally, note that more generally our theory is extendable for describing the selective transfer between large molecules of other quantum excitations than electrons, for example electronic excitons or some specific quantum bond vibration (Davydov exciton) etc.... Indeed, the set of quantum variables $\{\psi_\alpha\}$ could represent as well the complex amplitudes of the global quantum state of a system with a certain kind of excitation which could tunnel between different sites α and which is also supposed to be coupled to phonon baths. The electrostatic energies however should be different and a priori smaller than for electron transfer.

6.3. *Future Perspectives in Biochemistry*

Because of the fine tuning, this mechanism we propose for ultrafast electron transfer is highly sensitive to small perturbations which could detune the catalyst from the donor, then sharply reducing the amplitude of the electronic level oscillation which does not reach the electronic level of the acceptor and thus breaks or sharply slows down ET at zero temperature. This perturbation could be simply produced by an (directive) external electric field, it could be also molecule mutations in some crucial regions modifying the linear and (or) the nonlinear coefficient μ_α and χ_α of the catalyst or the donor even slightly. It could be also inhibited by extra molecules ("poisons") weakly interacting with the donor or the catalyst by for van der Waal, dipolar, electrostatic etc...interactions and detune the donor-catalyst pair.

Since the perturbation can be small, this effect could work at relatively large distance especially in the case of electrostatic interactions. It opens a new direction for understanding some puzzling long distance interactions between or inside biomolecules. The properties of the solvent and in particular its pH are also important because its acts on the dielectric constant of the environment which changes the capacitive nonlinear coefficient χ_α^C then detuning the pair donor-catalyst. The reverse effect could exist. Some specific molecules (activator) could activate (or reactivate) the donor-catalyst pair by restoring a fine tuning. Then, in the presence of only few of appropriate molecules, enzymes could be activated and initiate chemical reactions creating a relatively large amount of a specific product initiating new chemical reactions and changing substantially the chemical state of the cell. In fine, this is an amplification mechanism detecting the presence of the few specific molecule activating the enzyme.

Several subsequent ET could be linked as cascades in biostructures resulting in electron transport over long distance but along a selected path. Cascade of ET could occur in nonrigid structures for example in the solvent where it follows a well defined sequence of molecule species where each ET occurs when the concerned molecules which are randomly diffusing in the solvent, enter in contact and interact. The high sensitivity of ultrafast ET could be one of the mechanism explaining the highly selective behavior of enzymes allowing biochemical reactions in vivo to behave as logical systems at the molecular level which are under current investigation. Simple logical nanoscale devices based on the principles described here could be perhaps built artificially with molecules or quantum dots ans could mimic some biofunctions.

Acknowledgments

This work has been supported by the European TMR program LOCNET HPRN-CT-1999-00163.

References

1. A.M. Kuznetzov and J. Ulstrup, *Electron Transfer in Chemistry and Biology: An introduction to the theory*, Wiley series in Theoretical Chemistry (1999)
2. R.A. Marcus, Rev. Mod. Phys. **65** (1993) 599–610
3. H.A. Kramers, Physica **7** (1940) 284
4. S. Aubry and G. Kopidakis, in preparation
5. P. Senet, J.Chem.Phys. **107** (1997) 2516–2524
6. S. Aubry, G. Kopidakis, A. M. Morgante, G. Tsironis, Physica **B296** (2001) 222
7. G. Kopidakis, S. Aubry and G. Tsironis, Phys. Rev. Lett. **87** (2001) 165501
8. P. Maniadis, G. Kopidakis and S. Aubry, in preparation
9. http://photoscience.la.asu.edu/photosyn/education/antenna.html and R.E. Blankenship (2002) *Molecular Mechanisms of Photosynthesis*, Blackwell Science, in press. (B)
10. K. Schulten, *From Simplicity to Complexity and Back: Function, Architecture and Mechanism of Light Harvesting Systems in Photosynthetic Bacteria* in *Simplicity and Complexity in Proteins and Nucleic Acids* Eds. H. Frauenfelder, J. Deisenhofer and P.G. Wolynes (1999) Dahlem University Press
11. X. Hu, A. Damjanovic, T. Ritz and K. Schulten, Proc. Nat. Acad. Sci. (USA) **95** 5935 (1998).
12. M.H. Vos, J.L. Lambry, S.J. Robles, D.C. Youvan, J. Breton and J-L. Martin, Proc.Natl.Acad.Sci.USA **88** (1991) 8885-89
13. C.Rischel, D. Spiedel, J.P. Ridge, M.R. Jones, J. Breton, J-C. Lambry, J-L. Martin and M.H. Vos, Proc.Natl.Acad.Sci.USA **95** (1998) 12306-311
14. S. Tanaka and R.A. Marcus, J.Phys.Chem **B101** (1997) 5031-5045
15. D. Moses, A. Dogariu and A.J. Heeger, Phys.Rev.**B61** (2000) 9373–9379

COLLAPSE CONTROL IN AN INHOMOGENEOUS NONLINEAR SCHRÖDINGER EQUATION MODEL

P. L. CHRISTIANSEN AND YU. B. GAIDIDEI

Informatics and Mathematical Modelling, The Technical University of Denmark, DK-2800 Lyngby, Denmark
E-mail: plc@imm.dtu.dk

B. LEMESURIER

Department of Mathematics, College of Charleston, Charleston, SC 29424, USA

Collapse process in the inhomogeneous two-dimensional nonlinear Schrödinger equation is analyzed both numerically and analytically. It is shown that in the vicinity of a narrow **attractive** inhomogeneity, the collapse of beams which in the homogeneous medium would blow-up may be delayed and even arrested.

1. Introduction

The spatial contraction of wave-packets and the formation of a singularity in finite time - the wave collapse or, more generally, the blow-up of the wave-packet - is one of the basic phenomena in nonlinear physics of wave systems. Examples are self-focusing of light [1,2,3] in optics, the collapse of Langmuir waves in plasma [4], self-focusing of gravity-capillary surface waves [5], the blow-up of nonlinear electronic excitations in molecular systems [6] and the collapse in a Bose gas with negative scattering length [7,8]. Wave collapse is an efficient process of energy and/or mass localization as well as energy dissipation (see, e.g. review papers [9,11,12]). It is also very efficient mechanism of energy transduction from large scales into small (microscopic) scales and energy dissipation (see, e.g. review papers [9,11,12]).

The theory of self-focusing wave packets in optics, plasma and solid-state physics is based on the analysis of the nonlinear Schrödinger equation (in the theory of the Bose-Einstein condensation this equation is called the Gross-Pitaevskii equation [13])

$$i\partial_z\psi + \nabla^2\psi + |\psi|^{2\sigma}\psi + V(\vec{r})\psi = 0, \qquad \vec{r} \in R^D \tag{1}$$

where $\psi(\vec{r}, z)$ is the complex amplitude of the quasi-monochromatic wave train (the condensate wave function) , $\nabla^2 = \sum_{i=1}^{D} \partial_{x_i}^2$ is the D-dimensional Laplace operator, z is the propagation variable (the time variable in Bose-Einstein theory) and $\vec{r} = (x_1, ..x_D)$ is the spatial coordinate. The third term in Eq. (1) characterizes the nonlinear properties of the system : light intensity dependent refractive index in optics, effective self-interaction of Langmuir waves in plasma or the interaction between Bose-particles etc. Finally the fourth term in Eq. (1) is either an inherent space dependent refractive index of the material or an external (confining) potential. The following functionals are constants in propagation coordinate z (time):

$$N = \int |\psi|^2 \, d\vec{r}, \tag{2}$$

which is the number of excitations in plasma physics, number of atoms in Bose-Einstein condensates and the power in optics and, and the Hamiltonian of the system or energy

$$H = \int \{(\nabla\psi)^2 - \frac{1}{2}|\psi|^4 - V(\vec{r})|\psi|^2\} \, d\vec{r}. \tag{3}$$

2. Critical collapse in homogeneous media

The general properties of collapse dynamics in spatially uniform media are described in many review papers (see [9,11,14,15,16,24] *e.g.*). Therefore here we restrict ourselves to describing only the properties which are necessary for understanding pecularities of critical collapse dynamics in spatially inhomogeneous media.

Physical systems, which share the same value of the factor σD, possess many similar features such as stability properties of the stationary solutions. It was found [11] that in homogeneous systems ($V(\vec{r}) = 0$) the stationary solutions of Eq. (1) are stable when $\sigma D < 2$ and unstable when $\sigma D > 2$ with the case $\sigma D = 2$ being marginally stable. In the last case the excitation either blows up or disperses depending on whether a certain characteristic measure of the excitation is above or below a threshold value, respectively. This measure is the number of excitations N.

Collapse and dispersion of nonlinear pulses can be studied by measuring the second moment (or variance) of the shape function $|\psi(r,z)|^2$

$$\langle r^2 \rangle(z) = \frac{1}{N} \int r^2 \, |\psi(\vec{r}, z|^2 \, d\vec{r}. \tag{4}$$

From the equation (1) for $\sigma = 1$ we get the virial theorem

$$N \frac{d}{dz} \langle r^2 \rangle = 4 \left(2H - (D-2) \int |\psi|^4 \, d\vec{r} - \int |\psi|^2 \vec{r} \cdot \nabla V(\vec{r}) \, d\vec{r} \right). \tag{5}$$

In the homogeneous case ($V(\vec{r}) = 0$) and $D = 2$ the integration of Eq. (5) gives the relation

$$N\langle r^2\rangle = 4Hz^2 + C_1 z + C_2 \tag{6}$$

(C_1 and C_2 are some constants) which makes $H < 0$ a sufficient condition for a singular behavior (collapse) [17]. The behavior of the pulse near the blow-up point is rather well understood both in the supercritical ($D > 2$) case and in the critical $D = 2$ case. In the *supercritical case* $D > 2$ explicit radially symmetric self-similar solution has the form [9]

$$\psi(r,z) = (K(Z-z))^{1/2}\left(\frac{K}{Z-z}\right)^{i/K} Q\left(\frac{r}{\sqrt{K\,(Z-z)}}\right) \tag{7}$$

where $r = \sqrt{\sum_{i=1}^{D} x_i^2}$, K is a positive constant, Z is the collapse propagation distance, and the shape function $Q(r)$ is a solution of

$$-\left(\frac{d^2}{dr^2} + \frac{D-1}{r}\frac{d}{dr}\right) Q + Q - |Q|^2 Q - iK\left(1 + r\frac{d}{dr}\right) Q = 0 \tag{8}$$

with the boundary condition $\frac{d}{dr}Q(0) = 0$. In the *.critical case* $D = 2$ such solutions do not exist but there is strong non-rigorous argument [24] that the function $\psi(r,z)$ for z close to Z has the asymptotics [19,10]

$$\psi(r,z) \sim \left(\frac{\ln(\ln(1/(Z-z)))}{(Z-z)}\right)^{1/2} R\left(r\left(\frac{\ln(\ln(1/(Z-z)))}{(Z-z)}\right)^{1/2}\right) \tag{9}$$

where $R(r)$ is the so-called Townes soliton, the unique positive solution of

$$\left(\frac{d^2}{dr^2} + \frac{1}{r}\frac{d}{dr}\right) R - R + R^3 = 0, \qquad \frac{d}{dr}R(0) = 0. \tag{10}$$

The asymptotics (9) was derived by using the lens transformation [18]

$$\psi(\vec{r},z) = \frac{1}{L(z)}\Phi(\vec{\xi},\zeta)\exp\left(i\zeta + i\frac{\dot{L}}{L(z)}\frac{r^2}{4}\right) \tag{11}$$

where $r = |\vec{r}|$, overdot denotes the derivative d/dz, $L(z)$ is the beam width, new independent variables are defined as

$$\vec{\xi} = \frac{\vec{r}}{L(z)}, \qquad \dot{\zeta} = \frac{1}{L^2(z)}, \tag{12}$$

Under the lens transformation (11) the equation for the shape function, $\Phi(\vec{\xi},\zeta)$, may be represented in the form of Schrödinger equation

$$i\Phi_\zeta = -\nabla_\xi^2 \Phi + U(\vec{\xi})\Phi \tag{13}$$

where

$$U(\vec{\xi}) = 1 - \frac{1}{4}\beta\xi^2 + L^2\,V(L\vec{\xi}) - |\Phi|^2. \tag{14}$$

is an effective potential and the function $\beta(z)$ is defined by the relation

$$\ddot{L}\,L^3 = -\beta(z). \tag{15}$$

In the homogeneous case, $V(\vec{r}) = 0$, the potential energy of the inertial force , $-\frac{1}{4}\beta\xi^2$, makes for $\beta > 0$ the function $U(\xi)$ unbounded from below and as a result the motion of a particle in this potential becomes infinite. We are interested in the solutions of Eq. (13) under the boundary condition that there are only outgoing waves as $\xi \to \infty$. With this boundary condition the problem (13) is no longer self-adjoint (see a very lucid discussion of this subject for a closely related problem in [27]). The eigenvalues may have a finite imaginary part which gives the rate of radiation losses.

From Refs. [19-25] is known that in the homogeneous case ($V(\vec{r}) = 0$) the function $\beta(z)$ satisfies the equation

$$\dot{\beta} = -\frac{A}{L^2}\,\nu(\beta) \tag{16}$$

where the radiation rate ν is given by the expression

$$\nu = \theta(\beta)\,\exp\{-\frac{\pi}{\sqrt{\beta}}\} \tag{17}$$

A is a numerical constant, $\theta(\beta)$ is the Heaviside step-function. This equation can be obtained from the solvability condition for the asymptotic expansion of the self-similar shape function $\Phi(\xi, \zeta)$ [19-24] or using nonlinear eigenvalue formulation [25] or by applying a multiscales approach [26]. From Eq. (16) one can obtain (see [24] *e.g.*)

$$\beta \sim \pi^2 \left(\ln\,\ln\left(\frac{1}{Z-z}\right)\right)^{-2} \tag{18}$$

and combining (18) with (15), get Eq. (9).

Analysis, presented in [11], gives the necessary condition for the collapse $N \geq N_c \approx 11.69$ in the case $D = 2$ and the sufficient condition $H < 0$ for $D \geq 2$. Experiments show that for $D = 2$ in the case of Gaussian initial data

$$\psi(\vec{r}, 0) = h\,e^{-r^2/2}$$

collapse occurs for h above the threshold $h_c \approx 1.95$, for which the number of particles $N \approx N_c$ and the energy $H \approx 0.009$. Using a Crank-Nicholson finite difference scheme with adaptive integration step on a nonuniform grid, the propagation distance, Z, needed for a blow up to occur in the homogeneous case ($V(\vec{r}) = 0$) was calculated. For $h = 1.95$, Z is found to be 5.45 (Fig. 1).

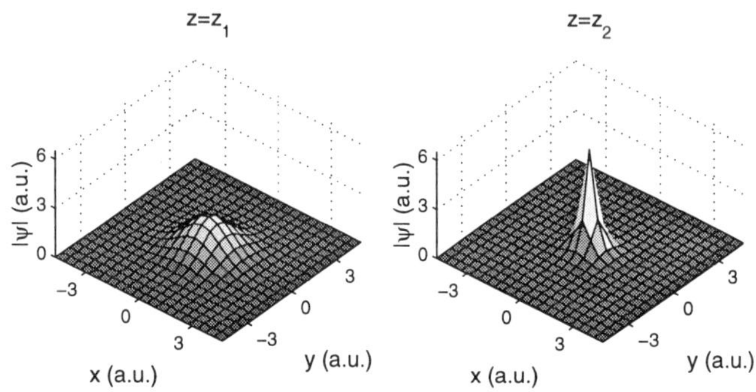

Figure 1. Gaussian collapsing beam shown at two propagation distances. $h = 1.95$, $w = 1$.

3. Collapse dynamics in the presence of attractive potential

The presence of an inhomogeneity significantly influences the dynamics of beams with arbitrary initial conditions (see [32] *e.g.*). The effects of periodic spatial modulation of the refractive index ($V(\vec{r}) = v\cos(kx)$) in the spatiotemporal evolution of pulses in nonlinear waveguides were recently investigated using a variational approach and numerical simulations [28], while exact sufficient criteria for blow-up were obtained in [29,30].

Also criteria for existence and stability of stationary solutions in inhomogeneous systems have attracted a lot of attention. In [33] a general criterion for the existence of stable stationary solutions was derived. Soliton-like excitations in the disordered two-dimensional cubic nonlinear Schrödinger equation model were investigated in Refs.[34,35]. It was shown that otherwise unstable excitations is found to be stabilized by the presence of disorder.

In physical systems, where an excitation is located in the vicinity of a smooth bell shaped inhomogeneity with a width much larger than the width of the excitation, one may model the inhomogeneity as a parabolic potential. In this case attractive potential *always* enhances self-focussing of nonlinear excitations. For example, the acceleration of the self-focusing of nonlinear excitations in molecular structures with parabolic-type inhomogeneities was found in Ref.[6] and the collapse of light beams in weakly nonlinear dispersive media with either a constant or weakly oscillating parabolic density profile was investigated in Ref. [31]. Collapse and Bose-Einstein condensation in trapped Bose gas with negative scattering length in the presence of the parabolic confining potential, $V(\vec{r}) \sim r^2$, were studied in Refs. [7,8]. However, the parabolic model breaks down when the widths of the inhomogeneity and the excitation are of comparable size. In this case the dynamical evolution of excitations strongly depends on the relation between the excitation width and the characteristic length of the inhomogeneity.

We will mainly focus on the situation, where the linear part of the operator in Eq.(1), $-\nabla^2 - V(\vec{r})$, supports a bound state. In this case stable stationary solutions with $N < N_c$ exist [33] and it is thus relevant to ask if these bound states act as attractors for certain classes of initial conditions.

3.1. *Numerical results*

We restrict ourselves to initial conditions of the form

$$\psi(\vec{r},0) = h \exp\left(-\frac{|\vec{r}-\vec{R}_0|^2}{2\,w^2} - i\vec{v}\,(\vec{r}-\vec{R}_0)\right) \tag{19}$$

where $\vec{R}_0$ is the center of the beam at $z = 0$ and $\vec{v}$ gives the initial "velocity" of the center,w characterizes the width of the beam and h gives its amplitude. We will consider the case of immobile beams and moving beams separately.

3.1.1. *Immobile beams*

In this subsection we consider the case of Gaussian trap potential

$$V_G(r) = h_p \exp\{-\frac{r^2}{2w_p^2}\} \tag{20}$$

and assume that initially the beam is immobile $\vec{v} = 0$ and the center of beam coincides with the center of the potential: $\vec{R}_0 = 0$. The initial width of the beam is chosen equal to 1. The results of our numerical simulations showed [36] that for wide attractive potentials there is only an acceleration of the collapse.

However, with potentials narrower than the initial self-induced potential of the initial data,increasing h_p often causes the eventual failure of this

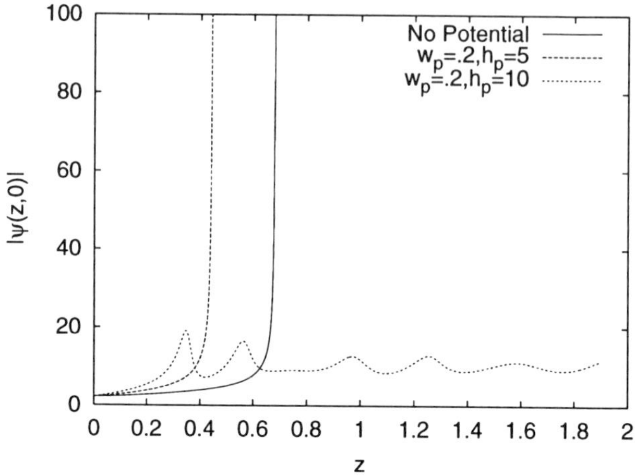

Figure 2. Beam amplitude $|\psi(0,z)|$ evolution shown for various attractive potentials. The initial beam amplitude $h = 2.2$.

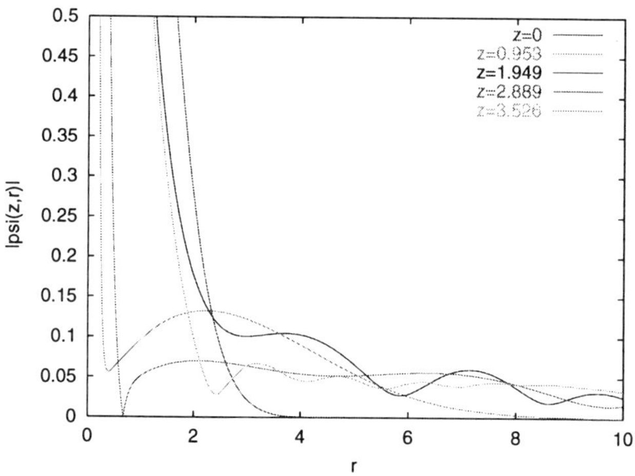

Figure 3. Spatial cross sections of the beam shown at various propagation distances. $h = 2.1$, $h_p = 3.48$ $w = 1$.

initial focussing, with the pulse repeatedly dispersing and refocussing. (see Fig. 2) Though this effect is most pronounced when the beam power is just a little above the threshold for collapse, it persists at least until the power is 30% above threshold, with the smallest inhibiting potential becoming rapidly stronger and slowly narrower as h increases[36]. Spatial cross sections of the solution amplitude at various times for $h = 2.1, h_p = 3.48, w_p = 0.3$ are presented in Fig.3.

3.1.2. *Swinging and orbiting beams*

In this subsection we extend the scope of self-focussing dynamics to include the center of the beam motion. To avoid the influence of tails of the Gaussian potential (20) for the potential, $V(\vec{r})$, we use a smoothed version of the circular step potential

$$V_H(\vec{r}) = h_p \theta(w_p - |\vec{r}|), \tag{21}$$

where θ is the Heaviside step function and h_p and $2w_p$ are the height and diameter of the potential, respectively. To monitor how the center of the beam evolves from its initial position, $\vec{R}_0$, we use the centroid

$$\vec{R}(z) = \frac{1}{N} \int_{-\infty}^{\infty} \int_{-\infty}^{\infty} \vec{r} |\psi(x, y, z)|^2 \, \mathrm{dxdy}. \tag{22}$$

We aim to illustrate how the beam may be separated into radiation and a non-collapsing core, due to the attraction towards the interior of the potential. To meet this end we divide the numerical calculations into two groups which give a broad representation of the possible scenarios of beam/potential interaction:

(1) Rectilinear motion: the beam is initially placed at nonzero distance $\left(\left|\vec{R}_0\right| > 0\right)$ from the center of the potential, $\vec{r} = 0$, with zero velocity ($\vec{v} = 0$) in the transverse plane.
(2) Orbital motion: the beam is initially placed at nonzero distance $\left(\left|\vec{R}_0\right| > 0\right)$ from the center of the potential with the velocity vector, $\vec{v}$, being perpendicular to $\vec{R}_0$.

Initial conditions belonging to the first group are 'pseudo 2D' and resembles the propagation of 1D "beams" within the framework of the 1D quintic NLSE ($\sigma = 2$) [37], whereas the second group fully exploits the two degress of freedom in the transverse plane. Fig. 4 shows the result of a numerical calculation belonging to the first category. A beam characterized by $\vec{v} = 0$ and $\left|\vec{R}_0\right| = 0.75$ is launched into a potential with $h_p = 1$ and $w_p = 1.25$. In Fig. 4a the evolution of the $y = 0$ cross-section of $|\psi|$, $|\psi(x, y = 0, z)|$, is plotted. We observe how the beam initially focuses as it is accelerated towards the

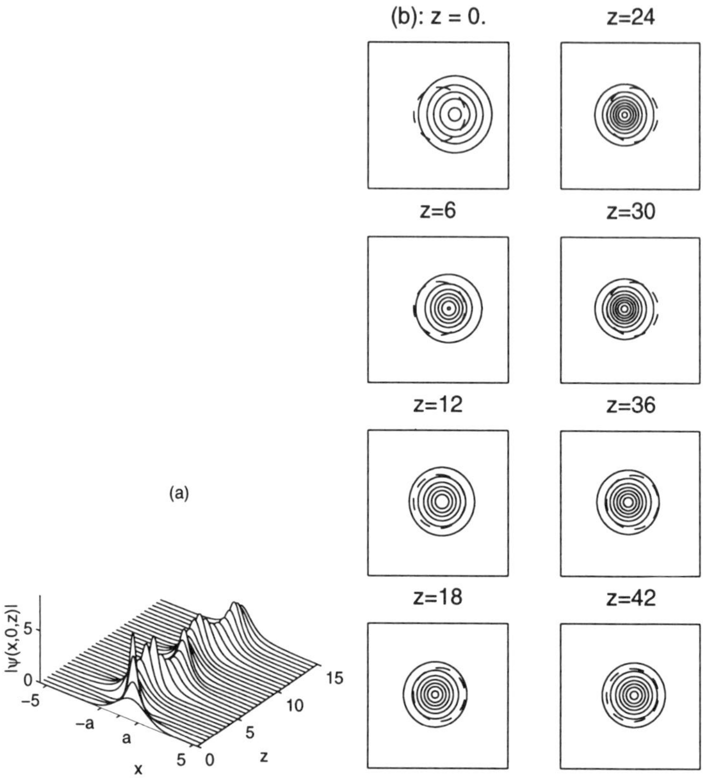

Figure 4. Rectilinear motion. Initial conditions: $\vec{v} = 0$, $\vec{R}_0 = (0.75, 0)$. In (a) the propagation of the $y = 0$ cross-section of $|\psi|$, $|\psi(x, y = 0, z)|$, is plotted. In (b) Eight contour plots depict the propagation of ψ. The following contour levels are used: $|\psi| = 0.4, 0.8, 1.1, 1.5, 1.9, 2.3, 2.7, 3.0$. The potential is indicated by a dashed circle (from Ref. [38]).

center of the potential, $\vec{r} = 0$. However, the beam amplitude ceases to increase at some point and instead undergoes moderate oscillations. In Fig. 4b the propagation of ψ is shown in contour plots for different values of z. 8 different contour levels are used to render the beam, and a dashed circle indicates the potential. Few visible contour levels thus correspond to low beam amplitudes whereas more levels are rendered for higher amplitudes. From the contour plots it is evident how the beam even at $z = 42$, almost 8 times z_0, shows no signs of approaching a collapse. Moreover, we observe how the beam profile remains almost circular through the oscillations in amplitude and width.

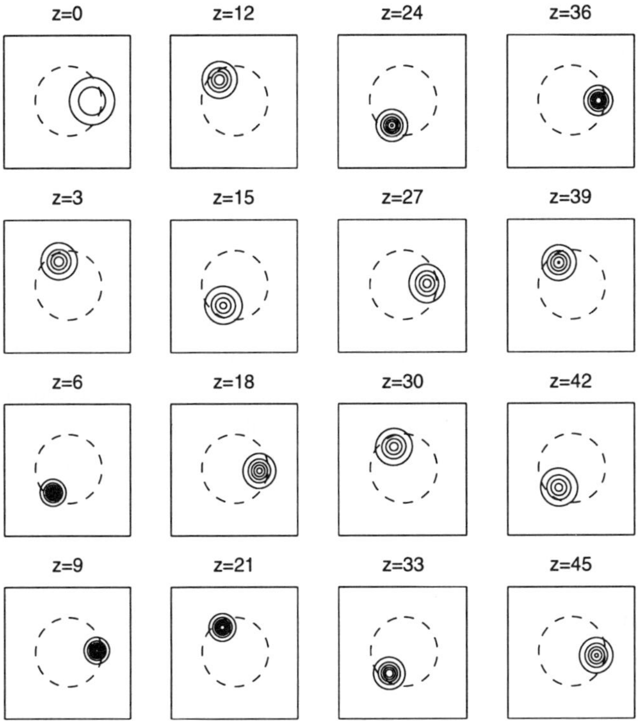

Figure 5. Orbital motion. Initial conditions: $\vec{v} = (0, -0.562)$, $\left|\vec{R}_0\right| = 1.5$. Contour levels are given by $|\psi| = 0.7, 1.4, 2.1, 2.7, 3.4, 4.1, 4.8, 5.5$. The potential with the width $w_p = 2.1$ is indicated by a dashed circle (from Ref.[38]).

When performing numerical calculations with orbital motion, we borrow concepts from celestial mechanics. For a planet moving in a circular orbit, the magnitude of the acceleration, a_{orb}, is related to the radius of the orbit, R_{orb}, and the magnitude of the velocity, v_{orb}, according to $a_{orb} = v_{orb}^2/R_{orb}$. In the present context of an orbitting beam, this relation translates into

$$\left|\ddot{\vec{R}}(z=0)\right| = \frac{\left|\dot{\vec{R}}(z=0)\right|^2}{\left|\vec{R}_0\right|}. \tag{23}$$

In Fig. 5, eight contour plots depict a beam moving in the potential well (21) with the width $w_p = 2.1$ for initial parameters given by $\left|\vec{R}_0\right| = 1.5$ and $\vec{v} = (0, -0.562)$. These initial parameters obey the relation Eq. (23). As in Fig. (4) we observe how the beam propagates without approaching a collapse. The calculation is continued until $z = 45$ and not even at this point does the beam show signs of significant distortion. The beam profile visible in the contour plots does in fact appear to be circle like through all stages of propagation. Moreover, the orbiting motion is accompanied by oscillations in the amplitude and width of the beam which decrease in strength through the propagation. More detailed results of numerical simulations on swinging and orbiting beam dynamics may be found in Ref. [38].

In conclusion, we have for a variety of initial conditions demonstrated how an initially super-critical beam may be separated into radiation and a non-collapsing core. We rely on physical arguments in concluding that the core mass must be below the critical value, $N_c = 11.69$, required for self-focusing. In the next subsection we present an analytical theory based on the results obtained in Ref.[38], which elucidates the mechanism of collapse preventing by an attractive potential.

3.2. *Analytical approach*

We assume that the inhomogeneity potential $V(\vec{r})$ is weak, narrow and bounded:

$$L\left|\frac{\nabla_{\vec{r}}V(\vec{r})}{V(\vec{r})}\right| > 1$$
$$V(\vec{r}) \to 0, \qquad r \to \infty. \tag{24}$$

We also assume that super-criticality is small: the mass of the beam, N, only slightly exceeds the critical value, N_c, *i.e.* $(N - N_c)/N_c \ll 1$. We shall restrict ourselves to the case of immobile beams assuming that the center of beam coincides with the center of the potential well (the case of moving beams is considered in [38]). Using the solvability condition for the asymptotic expansion of the self-similar shape function $\Phi(\vec{\xi}, \zeta)$ we obtain (see [38] for details) the following equation of motion for the width

$$\ddot{L} = -\frac{\Delta}{L^3} - \frac{1}{2M}\frac{\partial}{\partial L}W(L). \tag{25}$$

Here

$$\Delta = \frac{N_s - N_c}{M}, \qquad M = \frac{1}{4}\int R^2(r)\, d\vec{r} \tag{26}$$

is the excess beam mass above critical with the beam mass, N_s, defined as

$$N_s = \int_{r \le \xi_s L} |\psi(\vec{r}, z)|^2\, d\vec{r} \equiv \int_{\xi \le \xi_s} \left|\Phi(\vec{\xi}, \zeta)\right|^2 d\vec{\xi} \tag{27}$$

where the constant $\xi_s \gg 1$ characterizes the size of the beam. The function

$$W(L) = \int_{-\infty}^{\infty} V(\vec{r}L) R^2(r)\, dx dy \tag{28}$$

is an effective potential caused by the presence of the linear potential $V(\vec{r})$.

Eq (25) describes the beam dynamics (sub-critical beams for beams for $\Delta < 0$ and super-critical beams for $\Delta > 0$) in the adiabatic approximation when the mass of the beam is assumed to be constant ($\dot{\Delta} = 0$)[6]. This approximation is too crude, however, and is not sufficient in the case under consideration because as it is seen from the results of numerical simulations the beam evolution is accompanied by a radiation.

In the presence of weak and narrow potential $V(r)$ the effective potential $U(\xi)$ given by Eq. (14) does not change qualitatively its shape. It is still unbounded for $\xi \to \infty$. Due to "tunneling" through the potential barrier the beam mass, $N_s(z)$, decreases in accordance with the equation

$$\dot{N}_s = -\frac{\nu}{L^2} N_s, \tag{29}$$

or in terms of the excess mass (26)

$$\dot{\Delta} = -\frac{\nu}{L^2}\left(\frac{N_c}{M} + \Delta\right). \tag{30}$$

The raditation rate ν has the same form as in the homogeneous case (see Eq. (17)) but the parameter β relates to the excess mass Δ in the following way

$$\beta = \Delta + \frac{L^3}{2M}\frac{\partial W}{\partial L}. \tag{31}$$

Thus in terms of the excess mass and the width of beam the radiation rate ν can be represented as follows

$$\nu = \exp\left\{ -\pi \left(\Delta + \frac{L^3}{2M}\frac{\partial W}{\partial L}\right)^{-1/2} \right\}. \tag{32}$$

Eqs (30), (32) show that the width dynamics controls the tunneling rate. It is also seen in the presence *attractive* potential the radiation rate *always* increases and in this way beam gets rid more easily of an extra mass and becomes subcritical. This is the key result of our analytical approach.

The exact analytic expression for the Townes soliton is unknown. Therefore we made further simplification by replacing the function Ψ by its Gaussian approximation in the form

$$\Psi_g(\xi) = \sqrt{\frac{N_c}{\pi B^2}} \exp\{-\frac{\xi^2}{2B^2}\}. \tag{33}$$

where $B^2 =\approx 0.8$. Inserting Eq. (33) into Eq. (28), we obtain that in the case of the Gaussian well inhomogeneity potential $V(\vec{r})$ given by Eq. (20) the effective potential has the form

$$\frac{1}{M}W(L) = -\frac{N_c}{M}\frac{h_p\, w_p^2}{w_p^2 + B^2\, L^2} = -\frac{4.3\, h_p\, w_p^2}{1.25\, w_p^2 + L^2} \tag{34}$$

By using *Mathematica* we solved the set of Eqs (25), (30) and (32) with

the effective potential $W(L)$ given by Eq. (34). For $h_p = 1.95$, $w_p = 1$ and the initial conditions

$$L(0) = 1,\ \dot{L}(0) = 0,\ \Delta(0) = 0.025.$$

the results are presented in Fig. 6. There is a clear correlation between the excess mass and beam amplitude dynamics. Collapse is arrested due to fast change of the beam mass at the intial stage of evolution (at $z < 1$). The results of analytical approach are in a good agreement with the results of numerical simulations presented in Fig. 2 (see also Ref. [36] for more numerical results).

4. Summary

In summary we have in this paper shown that the presence of inhomogeneity permits the stabilization of otherwise collapsing excitations. We have demonstrated this via numerical simulations and via analysis. Analyzing the beam dynamics under the influence of attractive inhomogeneity one can conclude that

- in the case of the beam situated exactly at the center of the potential well the collapse can be delayed and even arrested if the potential is deep and narrow;
- to prevent the colapse of the beam situated away from the potential, the initial distance between the beam and the well is in a certain interval;
- the mechanism of collapse inhibition is generic: the inhomogeneity facilitates the radiation of the beam. The mass of the beam decreases and becomes less than critical. In this way the singular behavior of the beam is prevented.

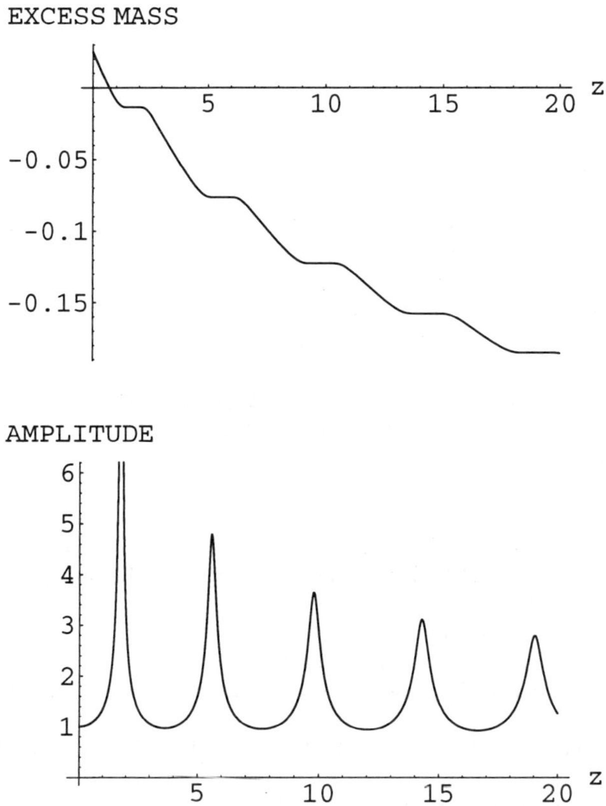

Figure 6. Analytical dependence of the excess mass (Δ) and the amplitude of beam ($\psi(0, z)$) obtained from Eqs. (25), (30) and (32).

ACKNOWLEDGMENTS

Yu.B.G. thanks MIDIT and Informatics and Mathematical Modelling, Technical University of Denmark for hospitality .

References

*. Permanent address: Bogolyubov Institute for Theoretical Physics, 252 143 Kiev, Ukraine.
1. G.A. Askar'yan, Zh. Eks.Teor.Fiz. (Sov. Phys. JETP) **42**,1568 (1962).
2. R.J. Chiao, F. Gardmire, and C.H. Townes, Phys. Rev. Lett. **13**, 479 (1964).
3. P. Kelley, Phys. Rev. Lett. **15**, 1005 (1965)
4. V.E. Zakharov, Sov. Phys. JETP **35**,908 (1972)

5. G.. Papanicolau, C. Sulem, P.L. Sulem, and X.P. Wang, Physica D **72**. 61 (1994).
6. Yu. B. Gaididei, K. Ø. Rasmussen, and P.L. Christiansen, Phys. Rev. E **52**, 2951 (1995).
7. E.V. Shuryak, Phys. rev. A **54**, 3151 (1996).
8. Yu. Kagan,A,E. Muryshev, and G.V. Shlyapnikov, Phys. Rev. Lett. **81**, 933 (1998).
9. V.E. Zakharov, in *Handbook of Plasma Physics*, edited by M.M. Rosenbluth and R.Z. Sagdeev (Elsevier,Amsterdam, 1984),p.81.
10. B.J. LeMesurier et al, Physica D **32**, 210 (1988).
11. J.Juul Rasmussen and K. Rypdal, Phys. Scr. **33**, 481 (1986); K. Rypdal and J.Juul Rasmussen, *ibid* **33**, 498 (1986).
12. E.A. Kuznetsov, Chaos **6**,381 (1996).
13. L.P. Pitaevskii,Zh. Eks.Teor.Fiz. **40**, 646 (1961) [Sov. Phys. JETP **13**,451 (1961)]; E.P. Gross, Nuovo Cimento **20**, 454 (1961).
14. E.A. Kuznetsov, A. M. Rubenchik, and V. E. Zakharov, Phys. Rep. **142**,103 (1986).
15. L. Bergé, Phys. Reports **303**, 259 (1998).
16. Yu.S. Kivshar and D. E. Pelinovsky, Phys. Reports **331**, 117 (2000).
17. S.N. Vlasov, V.A. Petrishchev, and V.A. Talanov, Radiophys. and Quant. Electr. **14**, 1062 (1971).
18. V.I. Talanov, JETP Lett. **11**, 303 (1970).
19. G.M. Fraiman, Sov. Phys. JETP **61** 228 (1985).
20. B.J. LeMesurier, G. Papanicolau, C. Sulem, and P. Sulem, Physica D **32** 210 (1988).
21. M. Landman, G. Papanicolau, C. Sulem, and P. Sulem, Phys. Rev. A **38** 3387 (1988).
22. V.M. Malkin, Phys.Lett. A **151**,285 (1990).
23. A.I. Smirnov and G.M. Fraiman, Physica D **52** 2 (1991).
24. G. Fibich and G. Papanicolaou, SIAM J. of Appl. Math. **60**, 183 (1999).
25. S. Dyachenko, A.C. Newel, A. Pushkarev, and V.E. Zakharov, Physica D **57** 96 (1992).
26. D. Pelinovsky, Physica D **119**, 301 (1998).
27. R. B. Paris and A. D. Wood, IMA J. Appl. Math. **43**, 273 (1989).
28. A.B. Acceves and C. De Angelis, Opt. Lett. **18**, 110 (1993).
29. S.K. Turitsyn, Opt. Lett. **18**, 110 (1993).
30. Yu.S. Kivshar and S.K. Turitsyn, Phys. Rev. E **49**, R2536 (1994).
31. L. Bergé, Phys. Plasmas **4**,1227 (1997).
32. A.B.Aceves, J.V.Moloney, and A.C. Newell,Phys. Rev. A **39**,1809 (1989).
33. H.A. Rose and M.I. Weinstein, Physica D **30**, 207 (1988).
34. P. L. Christiansen, Yu. B. Gaididei, M. Johansson, K. Ø. Rasmussen, D. Usero and L. Vázquez , Phys. Rev. B **56** ,14407 (1997).
35. Yu. B. Gaididei, P.L. Christiansen, Optic Lett.,**23**,1090 (1998).
36. B. LeMesurier and P. Christiansen, Regularisation and control of self-focussing in the 2D cubic Schrödinger equation by linear potential, 2001 (submitted to Physica D).
37. Yu. B. Gaididei, J. Schjødt-Eriksen, and P. L. Christiansen, Phys. Rev. E

60, 4877 (1999).
38. J. Schjødt-Eriksen, Yu. B. Gaididei, and P. L. Christiansen, Phys. Rev. E **64**, 066614 (2001)

THE DISCRETE NONLINEAR SCHRÖDINGER EQUATION — 20 YEARS ON

J. CHRIS EILBECK

Department of Mathematics,
Heriot-Watt University,
Edinburgh EH14 4AS, UK
E-mail: J.C.Eilbeck@hw.ac.uk

MAGNUS JOHANSSON

Dept of Physics and Measurement Techn.
Linköping University,
S-58183 Linköping, Sweden
E-mail: majoh@ifm.liu.se

We review work on the Discrete Nonlinear Schrödinger (DNLS) equation over the last two decades.

1. Introduction

The Discrete Nonlinear Schrödinger (DNLS) equation describes a particularly simple model for a lattice of coupled anharmonic oscillators. In one spatial dimension, the equation in its simplest form is

$$i\frac{dA_j}{dt} + \gamma|A_j|^2 A_j + \varepsilon(A_{j+1} + A_{j-1}) = 0, \tag{1}$$

where $i = \sqrt{-1}$, the index j ranges over the 1D lattice. The lattice may be infinite ($j = 0, \pm 1, \pm 2, \ldots$) or finite ($j = 1, 2, \ldots, f$). In the latter case one usually assumes periodic boundary conditions, $A_{j+f} = A_j$. The quantity $A_j = A_j(t)$ is the complex mode amplitude of the oscillator at site j, and γ is a anharmonic parameter.

The connection with the continuous Nonlinear Schrödinger (NLS) equation

$$iA_t + \gamma|A|^2 A + A_{xx} = 0 \tag{2}$$

is more clear if we write (1) in an alternative form

$$i\frac{dA_j}{dt} + \gamma|A_j|^2 A_j + \varepsilon(A_{j+1} - 2A_j + A_{j-1}) = 0. \tag{3}$$

The transformation $A_j \to A_j \exp(-2it\varepsilon)$ takes (3) into (1). With $\varepsilon = 1/(\Delta x)^2$, (3) is seen as a standard finite difference approximation to (2).

The DNLS equation (1) is a special case of a more general equation, the Discrete Self-Trapping (DST) equation[34]

$$i\frac{dA_j}{dt} + \gamma|A_j|^2 A_j + \varepsilon \sum_k m_{jk} A_k = 0. \tag{4}$$

Here $M = [m_{ij}]$ is a $f \times f$ coupling matrix. In physical applications M is real and symmetric, and clearly with M a suitably chosen constant tridiagonal matrix, we can regain (1) or (3). Physically this corresponds to the choice of nearest neighbour couplings. A more general choice for the elements of M introduce longer range couplings or different topologies to the lattice. The distinction between the DST and the DNLS equation is somewhat blurred in modern usage.

One interesting limiting special case of the DST equation is the case of the so-called *complete graph model*, when

$$m_{ij} = 1 - \delta_{ij},$$

corresponding to a lattice with each point connected directly to every other point on the lattice.

Clearly one can scale t and γ in the DNLS model to fix $\varepsilon = 1$, and this is often done in the literature. However the more general formulation (1) is useful when one wishes to consider the case $\varepsilon \to 0$, i.e. the limit of zero coupling, known nowadays as the anti-integrable or anti-continuum limit[75]. In this limit, the solution of (1) is trivial:

$$A_j = \sqrt{\frac{\omega_j}{\gamma}} e^{i(\omega_j t + \alpha_j)}, \tag{5}$$

where the frequencies ω_j and phases α_j can be chosen arbitrarily and independently at each site.

It is worth pointing out to avoid confusion that there are other possible discretizations of the NLS equation, one being the eponymous Ablowitz-Ladik (AL) model[3]

$$i\frac{dA_j}{dt} + \left(1 + \frac{1}{2}\gamma|A_j|^2\right)(A_{j+1} + A_{j-1}) = 0. \tag{6}$$

Another is a model due to Izergin and Korepin[52] which is rather lengthy to write down, see the book by Faddeev and Takhtajan[37] for details. Both the AL model and the Izergin–Korepin models have the advantage of being integrable equations[2], but it can be argued that they are less physically meaningful. The DST equation has been shown to be non-integrable when

$f > 2$ except for the rather unphysical case of a special non-symmetric interaction matrix M. It should also be mentioned that an often studied model is the Salerno equation[90], which contains a parameter interpolating between the AL model (6) with pure inter-site nonlinearity and the DNLS equation (1) with pure on-site nonlinearity. This allows e.g. the use of soliton perturbation theory[98,24] to elucidate the role of the on-site nonlinearity as a non-integrable perturbation to the integrable AL equation. The Salerno equation was extensively analyzed by Hennig and co-workers[50] and is also described in the book by Scott[92].

Another possible source of confusion is that the acronym DNLS is sometimes used for the Derivative Nonlinear Schrödinger equation[2].

The DST equation (4) can be derived from the Hamiltonian:

$$H = \sum_{j=1}^{f} \left[\omega_0^{(j)}|A_j|^2 - \frac{\gamma}{2}|A_j|^4\right] - \varepsilon \sum_{j\neq k} m_{jk} A_j^* A_k \tag{7}$$

with canonical variables: $q_j \equiv A_j$ and $p_j \equiv \mathrm{i}A_j^*$. Here $\omega_0^{(j)} \equiv \varepsilon m_{jj}$ are the harmonic frequencies of each uncoupled oscillator ('on-site energies'); with $m_{jk} = \delta_{j,k\pm1}$ the DNLS equations (1) and (3) are obtained for $\omega_0^{(j)} = 0$ and $\omega_0^{(j)} = 2\varepsilon$, respectively. There is a second conserved quantity, the *number* (or *norm*)

$$N = \sum_{j=1}^{f} |A_j|^2. \tag{8}$$

The integrability of the $f = 2$ (dimer) case follows from these two conserved quantities, and in this case all the time-dependent solutions $A_j(t), j = 1, 2$ can be expressed in terms of elliptic functions[53]. One can always scale A_j and γ so that $N = 1$, or alternatively scale A_j and N so that $\gamma = 1$.

The DNLS Hamiltonian is the starting point for a study of a quantum version of the DNLS system, see the paper by Eilbeck in these proceedings. In particular the quantum analogue of a classical discrete breather can be derived.

There are now almost 300 papers on the DNLS and DST equations, and in this short survey we can only hope to cover a small amount of available material, concentrating on our own interests. Currently a database of papers in this area is held at `http://www.ma.hw.ac.uk/~chris/dst`. For complementary aspects we recommend the review papers by Hennig and Tsironis[50] (in particular concerning map approaches with applications to stationary wave transmission) and Kevrekidis et al.[63] (in particular concerning different types of localized modes and their stability, bifurcation and interaction properties), as well as the pedagogical introduction in the textbook by Scott[92] and the general review of discrete breathers by Flach and Willis[43].

2. Stationary Solutions

Stationary solutions of the DNLS or DST equations are special solutions of the form

$$A_j(t) = \phi_j \exp(i\omega t), \tag{9}$$

where the ϕ_j are independent of time. Inserting this ansatz into the equations give an algebraic set of equations for the ϕ_j. For example, for DNLS (1), we get

$$-\omega\phi_j + \gamma|\phi_j|^2\phi_j + \varepsilon(\phi_{j+1} + \phi_{j-1}) = 0. \tag{10}$$

It is this feature that makes the DNLS a relatively simple model to work with. For small periodic lattices up to $f = 4$ it is possible to solve the resulting equations exactly and obtain all the families of stationary solutions as a function of ω and γ (for fixed N), with a fascinating bifurcation structure[34]. The complete graph model can also be solved exactly for any f[34]. For a large or infinite lattice the solutions must be found by numerical methods such as shooting methods or spectral methods. These solutions can then be investigated as a function of the parameters of the equation by numerical continuation methods (see e.g. [32] for a complete list of solutions for $f = 6$). If γ is sufficiently large, localized solutions are found which decay exponentially for large $|j|$. Two examples are shown in Fig.1. Since these solutions have a periodic time behaviour $\phi_j \exp(i\omega t)$, it seems

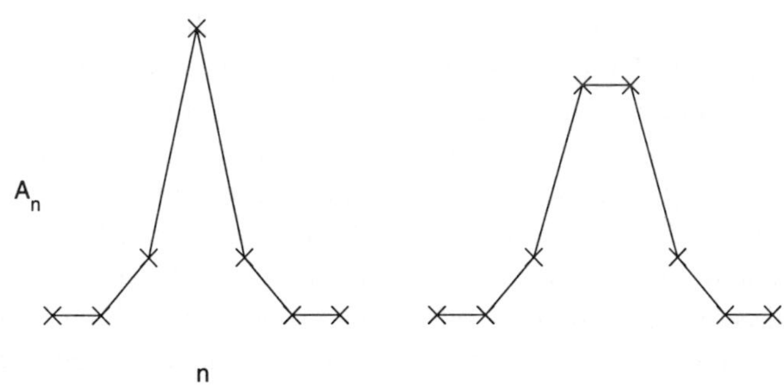

Figure 1. Example: localized stationary solutions of the DNLS model

appropriate to call them "breather" solutions. Another motivation is that the DNLS equation can be derived from the discrete Klein-Gordon equation, describing a lattice of coupled anharmonic oscillators, via a multiscale expansion in the limits of small-amplitude oscillations and weak inter-site coupling[68,28,83]. The discrete breathers of this lattice are then represented as stationary solutions to the DNLS equation. The reader should note that in the early days of DNLS studies, when breathers in discrete systems were not so well understood, these solutions were often called solitons.

The stability of such solutions in time can be investigated by looking at general perturbations *in the rotating frame* of the solutions[18]

$$\tilde{A}_j(t) = (\phi_j + \delta u_j(t)) \exp(i\omega t)$$

This reduces the linear stability problem to a study of a linear eigenvalue problem. It is perhaps to be expected that the stability of a branch of stationary solutions can change at a bifurcation point. What is surprising is that, since the eigenvalue problem is not self-adjoint, solutions can also change stability at other points on the branch. Usually the single-site peaked ('site-centred') solution shown at the l.h.s. of Fig.1 turns out to be stable, whereas the two-site peaked ('bond-centred') solution shown on the r.h.s. is not.

For the case of an infinite lattice, both solutions in Fig.1 can be smoothly continued versus coupling ε (or, equivalently by rescaling, versus ω) for all ε, without encountering any bifurcations. For $\varepsilon \to 0$, the site-centred solution will be completely localized at the central site with all other oscillator amplitudes zero, while the bond-centred solution becomes completely localized on the two central sites. For $\varepsilon \to \infty$ both solutions are smoothly transformed into the same soliton solution of the continuous NLS (2) (which explains why they are sometimes also termed 'discrete solitons'). As there are no bifurcations, the site-centred solution is stable and the bond-centred unstable for all ε in the infinite chain. Comparing the value of the Hamiltonian ('energy') of the two solutions for a fixed N, one finds that the site-centred solution always has the lowest energy. This energy difference has been proposed to act as a sort of Peierls-Nabarro potential barrier[65]. Another property of these two solutions in infinite 1D lattices is that they exist for arbitrarily small γ (or arbitrarily small N for fixed γ)

Historical note. Although the stationary DNLS equation (10) was derived already by Holstein in 1959 in his pioneering work on polarons in molecular crystals[51], the first systematic study of its single-peak breather solution as an exact solution to the fully discrete equations was performed by Scott and MacNeil in 1983[94], following Scott's interest in Davydov solitons on proteins. They investigated the family of single-peak stationary solutions using shooting methods running on a Hewlett-Packard programmable calculator. Further interest in Davydov solitons on protein molecules led to a study of a related molecule called *acetanilide.* A model of the crystalline state of this molecule was set up which was essentially four *coupled* DNLS systems. Techniques to map out the families of stationary solutions on this system were developed, including path-following from the anharmonic limit[33]. It was then realised that the *single* DNLS system was of independent interest, which led to the work described in[34].

Later and independently, Aubry, MacKay and co-workers developed a much more general approach along these lines to the breather problem in arbitrary systems of coupled oscillators[75,11]. In this context, much new attention was directed to the DNLS model and its stationary solutions. In addition, two more large bursts of interest into studies of the DNLS equation have appeared recently, following the experimental progress in the fields of nonlinear optical waveguide arrays[36] and Bose-Einstein condensates trapped in periodic potentials arising from optical standing waves[8]. These applications will be discussed briefly below. Since the DNLS equation is of general applicability and appears in completely different physical fields, new researchers drawn to its study have not always been aware of earlier results. Thus many of its properties have been independently rediscovered and appeared several times in the literature in different contexts during the last two decades.

3. Disorder

One natural generalization to the DNLS equation (1) or (3) is to consider non-constant coupling parameters ε_{jk}, equivalent to nontrivial distributions of the elements m_{jk} of the matrix M in the DST equation (4). One may also consider site dependent γ_j as well. An early application, in the large f case, was to model the dynamics of the energy distribution of modes on a globular protein. Feddersen[38] considered interactions among CO stretch oscillations in adenylate kinase, which comprises 194 amino acids ($f = 194$). Since the structure of this enzyme has been determined by x-ray analysis, the $f(f-1)/2 = 18721$ off-diagonal elements of the dispersion matrix M were calculated from Maxwell's equations. Also diagonal elements were selected from a random distribution, and the degree of localization of a particular stationary solution of the form (9) with real ϕ was defined by evaluating the quotient $\sum \phi_j^4 / \sum \phi_j^2$.

This numerical study revealed two features. Firstly, at experimentally reasonable levels of nonlinearity (γ), stable localized solutions were observed near some but not all of the lattice sites. Secondly, this anharmonic localization was observed to be distinctly different from "Anderson localization", a property of randomly interacting linear systems [9]. Thus none of the stationary states that were observed to be highly localized at large γ remained so as γ was made small. Also, none of the states that were localized at $\gamma = 0$ (i.e., Anderson localized) remained so as γ was increased to a physically reasonable level.

The transition between Anderson localized modes and breather states has more recently been extensively analysed in a series of papers by Kopidakis and Aubry for general coupled oscillator chains[69,71,70] (see also Archilla et al.[10] for a slightly different model), and has to its larger parts been understood. The generic scenario, valid also for the DNLS model, is consistent with Feddersen's observations but too complicated to describe in detail here. Briefly, there are two kinds of localized breather solutions in a disordered nonlinear lattice: 'extraband discrete breathers' (EDBs) with frequencies outside the spectrum of linear Anderson modes, and 'intraband discrete breathers' (IDBs) with frequencies inside the linear spectrum. EDBs cannot be smoothly continued versus frequency into IDBs but are lost in cascades of bifurcations. IDBs on the other hand can be continued outside the linear spectrum, but not into EDBs but only into a certain type of spatially extended multi-site breathers. The IDBs can only exist as localized solutions inside the linear spectrum provided their frequencies do not resonate with linear Anderson modes. However, for an infinite system the linear spectrum becomes dense so that the allowed frequencies for localized IDBs must constitute a (fat, i.e. of non-zero measure) Cantor set! In fact, for the DNLS case the latter result had been rigorously obtained already in 1988 by Albanese and Fröhlich[5]; see also[45,6] for other early mathematical results on the DNLS model with disorder.

It is interesting to remark that the general scenario with two types of discrete breathers, EDBs and IDBs, where the latter exist as localized single-peaked solutions only in-between resonances with linear modes, is not peculiar for random systems, but observed also in other situations when the linear spectrum is discrete with localized eigenmodes. A very recently studied example[85] is the DNLS model (1) with an added linearly varying on-site potential $\omega_0^{(j)} = \alpha j$. In this case the linear spectrum constitutes a so-called Wannier-Stark ladder (WSL) of equally spaced eigenfrequencies, with eigenstates localized around each lattice site, giving rise to Bloch oscillations (recently experimentally observed in waveguide arrays[82,84]). Then, resonances were shown to result in 'hybrid discrete solitons', interpreted as bound states of single-peaked IDBs and satellite tails corresponding to nonlinearly modified Wannier-Stark states localized some distance away from the main peak. Due to the finite (constant) distance between the linear eigenfrequencies in the WSL, IDBs remain single-peaked in frequency intervals of finite length, in contrast to the IDBs in random systems.

The fact that nonlinearity modifies localized linear modes into extended nonlinear solutions should be of some physical importance, since it provides a mechanism for transport in random systems. In fact, this aspect was considered also by Shepelyansky in 1993[95], who used the well-known Chirikov criterion of overlapping resonances to argue that, above some critical nonlinearity strength γ_c, the number of excited linear modes (and thus the spatial width) for a typical initially localized excitation in the DNLS model would spread sub-diffusively as $(\Delta n)^2 \sim t^{2/5}$ (for linear random systems, $(\Delta n)^2$ remains finite under very general conditions).

We also mention that the case with disorder residing purely in the nonlinearity strengths γ_j was studied by Molina and Tsironis[79]. In this case only partial localization of an initially single-site localized excitation could be found for large nonlinearities (dynamical self-trapping, corresponding to asymptotic approach to an exact discrete breather), while some portion was found to always escape ballistically (i.e. spreading as $(\Delta n)^2 \sim t^2$) leading to absence of complete localization. The scenario with partial self-trapping above a critical nonlinearity strength combined with asymptotic spreading through small-amplitude waves appears very generally for single-site initial conditions in the DNLS model, with or without disorder[78,15,58].

4. Mobile breathers

Since the DNLS equation is an approximation to the equations describing Davydov solitons, which are thought from numerical studies to be mobile, it is natural to ask whether the sort of breathers shown in Fig.1 can move if sufficiently perturbed. The first attempt to model this was made in a relatively obscure conference proceedings[31] in 1986. The key to getting mobility is to realise that a shape like Fig.1 will move if the figure represents $|A_j|^2$ but the phase is no longer constant and rotates through 2π as we traverse the breather. The same paper reported a very preliminary study of the interaction of the moving breather with an "impurity", or more precisely a long-range interaction due to the curved nature of the chain. There is now a growing literature on trapping of mobile breathers due to curved chains and long range-coupling (c.f.[38,46]) and on trapping due to local impurities in the lattice (c.f.[49]). Regrettably, due to space considerations, we have omitted any further discussion of this interesting area.

Feddersen[39,30] used spectral and path-following methods to make a more detailed numerical study of travelling breathers in the DNLS system using the ansatz

$$A_j(t) = u(z)\exp\{i(\omega t - \kappa j)\}, \quad z = j - ct. \tag{11}$$

Note that $c \neq \omega/\kappa$, i.e. the solution is regarded as a solitary pulse modulated by a carrier wave moving at a different velocity. His studies show a solution with this form to a high degree of numerical accuracy for a range of parameter values. However this numerical evidence cannot be regarded as a rigorous proof for the *existence* of moving breathers in the DNLS system, and this is still an outstanding question.

Much recent attention has been drawn to mobile breathers in general oscillator chains (see several other contributions to these proceedings), and many of these results can be transferred also to DNLS chains. Here, we wish to just mention particularly some results of Flach et al.[44] (see also [41]) who used an inverse approach, choosing particular given profiles of travelling waves and finding equations of motion having these as exact solutions. Generalizing the DNLS equation (1) by replacing $\gamma|A_j|^2$ with $G(|A_j|^2)$ and ε with $\varepsilon + F(|A_j|^2)$, where F and G are functions to be determined, and choosing A_j of the form (11) with *real* u, they could determine explicit expressions for the functions F and G for which the particular solution A_j exists as an exact travelling wave. In this way, they could e.g. reproduce the AL-model (6) for $G \equiv 0$ by choosing A_j to be its well-known soliton solution. Moreover, they could prove that no such travelling solution with pulse shaped u could exist for a pure on-site nonlinearity ($F \equiv 0$), and thus not for the DNLS equation (1). However, this does not prove the absence of exact moving localized DNLS breathers, for (at least) two reasons. (i) The envelope $u(z)$ could contain a non-trivial space-dependent complex phase not absorbable into the $\exp\{-i\kappa j\}$ factor. In fact, the solutions numerically found by Feddersen contained such a phase. (ii) Moving breathers could have a time-varying (e.g. periodic) shape function $u(z, t)$. This could be possible since the stationary DNLS breathers, in the regime where mobility is numerically observed, exhibit internal shape modes ('breathing modes') which can be found as localized time-periodic solutions to the linearized equations[57,67,56,63].

5. Chaotic Solutions

Since the DNLS equation for $f > 2$ is not integrable, it might be expected that it has solutions exhibiting Hamiltonian chaos, and in fact the first study of the DST equation showed chaotic-looking trajectories in the $f = 3$ case[34]. A more thorough analysis of this case was carried out by Cruzeiro-Hansson et al.[25], who estimated the region of both classical and quantum phase space occupied by chaotic states. A number of other studies have been carried out since then. For example, Hennig and coworkers[48] considered a DST trimer (4) with $m_{11} = m_{22} = m_{33} = 0$ and $m_{13} = m_{23} \ll m_{12} = m_{21}$, i.e. an (integrable) dimer interacting weakly with the third oscillator. Then, a Melnikov approach could be used to show the existence of homoclinic chaos. A similar approach for the case $f = 4$ with a dimer interacting weakly with the two other sites also demonstrated the presence of Arnold diffusion[47]. In the opposite limit of large f, homoclinic chaos has also been demonstrated and analyzed through a Melnikov analysis of a perturbed continuous NLS equation[17].

As another example of chaotic behaviour, Eilbeck[32,35] showed that on a $f = 6$ periodic lattice modelling benzene, a mobile breather could propagate which hopped around the lattice in a random way, even reversing its direction of motion at unpredictable intervals.

6. 2-dimensional DNLS lattices

As follows from the general theory of MacKay and Aubry[75], breathers exist also in higher dimensions. While we are aware of very few explicit results for the DNLS model in three dimensions[42], the two-dimensional case has been rather thoroughly studied. Some recent results are described in[63]. Instead of attempting to give a complete survey here, we will concentrate on discussing the main differences to the one-dimensional case.

In the 2D case and for a square lattice, the DNLS equation (3) with $\gamma = \varepsilon = 1$ is readily generalized to

$$i\frac{dA_{m,n}}{dt}+|A_{m,n}|^2A_{m,n}+A_{m+1,n}+A_{m-1,n}+A_{m,n+1}+A_{m,n-1}-4A_{m,n}=0, \tag{12}$$

and stationary solutions of the form (9) with frequency ω can be found analogously to the 1D case. The single-site peaked discrete soliton (breather) was first thoroughly studied in[77,72]. The following characteristics should be mentioned: (i) The solution can be smoothly continued from a single-site solution at the anti-continuum (large-amplitude) limit $\omega \to \infty$ to the so called ground state solution of the continuous 2D NLS equation[87] in the small-amplitude limit $\omega \to 0$. (ii) There is an instability-threshold at $\omega \sim 1$, so that the solution is stable for larger ω ('discrete branch') and unstable for smaller ω ('continuum-like branch'). (iii) The stability change is characterized by a change of slope in the dependence $N(\omega)$, so that $\frac{dN}{d\omega} > 0$ (< 0) on the stable (unstable) branch. (A similar criterion exists also for single-site peaked solutions to the 1D DNLS equation with on-site nonlinearities of arbitrary power[73].) (iv) The value of the excitation number N at the minimum is *nonzero*, and thus there is an *excitation threshold*[99] for its creation, in contrast to the 1D case (3) where $N \to 0$ as $\omega \to 0$ for fixed $\gamma = \epsilon = 1$. A similar scenario occurs also in 3D[42]. The effect of this excitation threshold in 2D was recently proposed to be experimentally observable in terms of a delocalizing transition of Bose-Einstein condensates in optical lattices[61]. (v) The dynamics resulting from the instability on the unstable branch is, in the initial stage, similar to the collapse of the unstable ground state solution of the continuous 2D NLS equation[87], with increased localization and blow-up of the central peak. In contrast to the continuum case, however, this process must be interrupted since the peak amplitude must remain finite, and the result is a highly localized 'pulson' state where the peak intensity $|A_{m,n}|^2$ oscillates between the central site and its four nearest neighbours[22]. This process has been termed 'quasicollapse'.[73,77] It is not known whether these pulson states represent true quasiperiodic solutions to the DNLS equation (see below).

As was shown by MacKay and Aubry[75,11] under very general conditions, two-dimensional lattices allow for a new type of localized solutions, 'vortex-breathers', with no counterpart in 1D. They can be constructed as multi-site breathers by continuation from an anti-continuum limit of a cluster of single-site breathers with identical frequencies but with uniformly spatially varying phases constituting a closed loop, such that the total phase variation around the loop ('topological charge') is a multiple of 2π. The simplest examples are three breathers in a triangle phase shifted by $2\pi/3$, or four breathers in a square phase shifted by $\pi/2$. The general existence and stability theorems[75,11] (valid also for the DNLS equation) guarantee that such solutions exist as localized solutions for weak enough coupling, and that certain configurations are linearly stable. As a consequence of the phase torsion, such solutions will carry a localized circulating current when the coupling is nonzero. For the DNLS model, vortex-breathers for a square 2D lattice were first obtained in [57]. Typically they become unstable as the coupling is increased; the mechanisms of these instabilities were described in some detail in[76,63].

Let us finally also mention a recent study[62] exploring numerically different types of breathers (including vortex-breathers) and their stability in triangular and hexagonal DNLS-lattices.

7. Quasiperiodic Breathers

A particular feature of the DNLS equation, distinguishing it from generic anharmonic lattice models, is the existence of continuous families of exact, spatially localized solutions of the form (9) but where the amplitudes ϕ_j in the rotating frame are not time-independent but time-periodic (with non-harmonic time-dependence). Such solutions are obtained by adding a term $i\dot{\phi}_j$ to the left-hand side of Eq.(10). Thus, these solutions are in general *quasiperiodic* with two incommensurate frequencies in the original amplitudes A_j (although they may also be periodic if the frequency relation is rational). At first, one may not be surprised by the existence of quasiperiodic solutions, since at least for finite-size lattices they should appear as KAM tori. However, generically (i) one would not expect them to appear in continuous families since they should be destroyed for rational frequency relations; and (ii) one would not expect them to survive as localized solutions in infinite lattices since the presence of two incommensurate frequencies in a generic anharmonic system would generate all possible linear combinations of the frequencies, i.e. a dense spectrum implying that resonance with the continuous spectrum should be unavoidable, and the breather should radiate and decay.

The key point to realize why, in spite of this, quasiperiodic breathers with two incommensurate frequencies do exist in the DNLS lattice is to note that the first frequency ω in (9) always yields harmonic oscillations, and thus no multiples of this frequency are generated. The origin to this is the *phase invariance* of the DNLS equation, i.e. invariance under transformations $A_j \to A_j e^{i\alpha}$, related to the norm conservation law (8) by Noether's theorem. A recent result[13] proves that very generally, each conservation law in addition to the Hamiltonian yields possibility for existence of quasiperiodic breathers with one additional frequency.

The existence of quasiperiodic DNLS-breathers in infinite lattices was first proposed by MacKay and Aubry[75], and later explicit proofs of existence and stability as well as numerical demonstrations for some particular examples were given[55,57] (earlier findings of quasiperiodic solutions in DNLS-related models had concerned mainly small systems[53,7] or the integrable AL model[16]). As some renewed interest has appeared on this topic[63], it is useful to comment on the differences between these two approaches. The solutions in[55,57] (see also [13]) were constructed as multi-site breathers by continuation from the anti-continuum limit $\varepsilon = 0$ of solutions with two (or more) sites oscillating with non-zero amplitude according to (5) with two different (generally incommensurate) frequencies ω_1 and ω_2. Except for some particular relations between the frequencies where resonances appear, such solutions can always be continued to some non-zero ε. On the other hand, the solutions discussed in[63] originated in internal-mode excitations (i.e. time-periodic localized solutions to the linearized equations) of a particular stationary solution, the so-called 'twisted localized mode'[27] (TLM). As for the bond-centred breather in Fig.1, the anti-continuum version of this solution has two neighbouring sites oscillating with equal $|A_j|^2$; however for the TLM these sites are oscillating in anti-phase so that the solution is spatially antisymmetric. This solution exists and is linearly stable for small ε.[55] Now, the occurrence of linear internal-mode oscillations is a very common phenomenon[67]. However, in most cases such oscillations do not yield true quasiperiodic solutions of the fully nonlinear equations since typically some harmonic will resonate with the linear continuous phonon spectrum, implying that these oscillations decay in time. This scenario appears e.g. for the single-site peaked DNLS-breather[56]. The interesting discovery by Kevrekidis and co-workers was, that for the particular case of the TLM, there are certain intervals in ε where *all* higher harmonics of the internal mode frequency are outside the continuous spectrum, and thus in these intervals the oscillating solutions of the linearized equations could be continued into truly quasiperiodic localized solutions of the nonlinear equation. As the allowed intervals are away from $\varepsilon = 0$, it is clear that this approach yields solutions which could not be obtained by direct continuation from the anti-continuum limit.

8. Wave Instabilities

Another important class of solutions in anharmonic lattices are space-time periodic *travelling waves*. For the DNLS model such solutions are very simple, since they are just rotating-wave solutions of the type (11) with constant $u = |A|$. Direct substitution into the DNLS equation (using the form (3)) yields the nonlinear dispersion relation

$$\omega = -4\varepsilon \sin^2 \frac{\kappa}{2} + \gamma |A|^2. \tag{13}$$

Linear stability analysis shows[18,68] that the travelling waves are stable if and only if $\frac{\gamma}{\epsilon}\cos\kappa < 0$. Thus, for $\frac{\gamma}{\epsilon} > 0$ only waves with $\pi/2 < |\kappa| \leq \pi$ are stable, while waves with small wave vectors $0 \leq |\kappa| \leq \pi/2$ are unstable through a *modulational instability* analogously to the continuous NLS equation. This instability destroys the homogeneous amplitude distribution of the wave, and typically[28,88,89] results in the creation of a number of small-amplitude mobile localized excitations ('breathers'), which through interaction processes (see below) may coalesce into a small number of standing large-amplitude breathers. Thus, the plane-wave modulational instability was proposed[28] generally to constitute the first step towards energy localization in nonlinear lattices (including DNLS).

Now, in a linear system one may always take linear combinations of counter-propagating waves $e^{\pm i\kappa j}$ to obtain *standing waves* (SWs) of the form $\cos(\kappa j + \beta)$. The same is of course not true in a nonlinear system due to lack of superposition principle; still however there generally exist nonlinear continuations of the linear standing waves, although they cannot be written as superpositions of counter-propagating travelling waves. Such nonlinear standing waves were investigated in detail for general coupled oscillator chains in [83] (the results for the DNLS chain were more concisely summarized in [60]). Without going into too much detail, let us state some main conclusions, referring to the DNLS form (3) with $\gamma = 1$. (i) SWs with given wave vector κ exist as stationary solutions of the form (9) for *all* values of $\frac{\omega}{\varepsilon} > -4\sin^2\frac{\kappa}{2}$. In the lower limit (corresponding to the dispersion relation (13) for a linear wave), the wave is a linear standing wave. (ii) In the anti-continuum limit $\frac{\omega}{\epsilon} \to \infty$ a SW with wave vector κ is described by a particular spatially periodic (or quasiperiodic if κ is irrational) repetition of local on-site solutions of the form (5) of oscillating and zero-amplitude solutions. The oscillating sites have the same frequency ω but generally alternating phases $\alpha = 0, \pi$. (iii) For each wave vector κ there are only two different distinct (modulo lattice translations) SW families corresponding to different spatial phases β of the linear SW $\cos(\kappa j + \beta)$. They appear as hyperbolic respectively elliptic periodic points in the map defined by the stationary DNLS equation (10). (iv) One of the SW families is stable in a regime of large $\frac{\omega}{\epsilon}$, while close to the linear limit *all nonlinear SWs with* $\kappa \neq \pi$ *are unstable* for infinite systems! The instability for the 'most stable' waves is of oscillatory type (i.e. corresponding to complex eigenvalues of the linear stability eigenvalue problem).

Investigating the long-time dynamics resulting from the SW instabilities, completely different scenarios were found[60] for $|\kappa| < \pi/2$ and $|\kappa| > \pi/2$, respectively. For the first case one finds after long times persisting large-amplitude standing breathers, while for the second case a 'normal' thermalized state is obtained. In fact, this division of the available phase space into two isolated regimes of qualitatively different asymptotic dynamics was first found by Rasmussen et al.[89], and shown to correspond to a phase transition through a discontinuity in the partition function in the Gibbsian formalism. In terms of the Hamiltonian and norm densities for a chain of f sites, the phase transition line was obtained as $\frac{H}{f} = -\gamma(\frac{N}{f})^2$, which can be seen to correspond exactly to a SW with wave vector $|\kappa| = \pi/2$. Note that the existence of the second conserved quantity N, which is peculiar for DNLS-type models, is crucial in this context.

Another interesting observation is that taking the limit $\kappa \to \pi$ for one of the nonlinear SWs generated from the anti-continuum limit as above, one obtains a solution consisting of a stable background wave with $\kappa = \pi$ having a single defect site of zero-amplitude oscillation inserted into it. This solution can be smoothly continued to the continuum limit, where it is seen to correspond to the dark-soliton solution of the *defocusing* NLS equation (note that the transformation $A_j = (-1)^j A_j$ in (1) is equivalent to reversing the sign of $\frac{\gamma}{\varepsilon}$). Also the discrete dark soliton ('dark breather') has been shown to be stable close to the anti-continuum limit, but unstable through an oscillatory instability close to the continuum limit for arbitrarily weak discreteness[59]. The typical outcome of this instability is a spontaneous motion[66]. As for the case of ordinary moving breathers, it is still an open question whether moving dark breathers exist as exact solutions, and current research is devoted to this issue. However, numerical evidence that they can exist at least to a very high numerical accuracy was given in the work of Feddersen[39,40].

Let us also mention that *asymmetric* discrete dark solitons, with different left and right background amplitudes, can exist as quasiperiodic solutions of the type described in the previous section. Such solutions were analyzed by Darmanyan et al[26] and are subject to similar instabilities.

9. Breather Interactions

In general, one cannot conclude from a linear stability analysis that a solution is fully stable, but only that small perturbations at least cannot grow exponentially in time. However, for the single-peaked DNLS-breather, a stronger result is obtained[99]: such solutions are orbitally Lyapunov stable for *norm-conserving* perturbations. This basically means that small breather perturbations will remain small (modulo a possible phase drift) for *all* times. This result is a consequence of the single-site breather being a ground state solution, in the sense that among all possible solutions at a given norm, it has the smallest value of the Hamiltonian. Thus, once again we find a property where norm conservation is crucial, and thus one should not expect that Lyapunov stability is a generic property of breathers in Hamiltonian lattices.

Still there are important issues to address concerning the fate of perturbed breathers, which cannot be predicted from stability theorems. One issue is breather-breather interactions, which correspond to large perturba-

has appeared in the last year, of which we here, quite randomly, just quote [1,61]. So far, most of the predictions are awaiting experimental verification. Some experimental confirmation that, at least to some extent, BEC in periodic potentials can be treated with DNLS models, under the condition that the inter-well potential is much larger than the chemical potential, has appeared very recently[19,20]. In these experiments the BEC was trapped in an optical lattice superimposed on a harmonic magnetic potential, and modelled by a DNLS equation with an additional quadratic on-site term $\Omega j^2 A_j$. The observed frequency of the Josephson-like coherent oscillations of the BEC centre-of-mass in the magnetic trap was shown to agree with DNLS predictions[19]. Moreover, changing the centre of the magnetic potential led to a transition from the (superfluid) regime of coherent oscillations into an insulator regime with the condensate pinned around the potential centre[20]. The onset of the transition was interpreted as the result of a discrete modulational instability, and could be estimated from the DNLS model. Many new experiments in this exciting field are awaited in the near future!

11. Conclusions

We hope the reader has enjoyed this brief introduction to this fascinating topic. We are conscious of the many details, figures and areas that we have left out, either because of space restrictions or because the topics are covered in depth elsewhere. To do the subject full justice would require a whole volume.

Acknowledgements

We thank all our colleagues, too many to mention explicitly, which in one way or another have contributed to this field. Special thanks are due to Sergej Flach and Thomas Pertsch for their helpful assistance, and to Rolf Riklund and Michael Öster for reading the manuscript. JCE would like to thank the EU for the financial support of the LOCNET programme. MJ would like to thank the Swedish Research Council for support.

References

1. F.Kh. Abdullaev, B.B. Baizakov, S.A. Darmanyan, V.V. Konotop, and M. Salerno. Nonlinear excitations in arrays of Bose-Einstein condensates. *Phys. Rev. A*, 64:043606-1–10, 2001.
2. M.J. Ablowitz and P.A. Clarkson. *Solitons, nonlinear evolution equations, and inverse scattering.* CUP, Cambridge, 1991.

3. M.J. Ablowitz and J.F. Ladik. A nonlinear difference scheme and inverse scattering. *Stud. Appl. Math.*, 55:213–229, 1976.
4. A.B. Aceves, C. De Angelis, T. Peschel, R. Muschall, F. Lederer, S. Trillo, and S. Wabnitz. Discrete self-trapping, soliton interactions, and beam steering in nonlinear waveguide arrays. *Phys. Rev. E*, 53:1172–1189, 1996.
5. C. Albanese and J. Fröhlich. Periodic solutions of some infinite-dimensional Hamiltonian systems associated with non-linear partial difference equations I. *Commun. Math. Phys.*, 116:475–502, 1988.
6. C. Albanese and J. Fröhlich. Periodic solutions of some infinite-dimensional Hamiltonian systems associated with non-linear partial difference equations. II. *Commun. Math. Phys.*, 119:677–699, 1988.
7. J.D. Andersen and V.M. Kenkre. Exact solutions for the quantum nonlinear trimer. *Phys. Stat. Sol. B*, 177:397–404, 1993.
8. B.P. Anderson and M.A. Kasevich. Macroscopic quantum interference from atomic tunnel arrays. *Science*, 282:1686–1689, 1998.
9. P.W. Anderson. Local moments and localized states. *Rev. Mod. Phys.*, 50:1191–201, 1978.
10. J.F.R. Archilla, R.S. MacKay, and J.L. Marín. Discrete breathers and Anderson localization: two faces of the same phenomenon? *Physica D*, 134:406–418, 1999.
11. S. Aubry. Breathers in nonlinear lattices: existence, linear stability and quantization. *Physica D*, 103:201–250, 1997.
12. S. Aubry and G. Kopidakis. A nonlinear dynamical model for ultrafast catalytic transfer of electrons at zero temperature. arXiv:cond-mat/0210215, 2002.
13. D. Bambusi and D. Vella. Quasi periodic breathers in Hamiltonian lattices with symmetries. *Disc. Cont. Dyn. Syst.*, 2:389–399, 2002.
14. O. Bang and P.D. Miller. Exploiting discreteness for switching in waveguide arrays. *Opt. Lett.*, 21:1105–1107, 1996.
15. L.J. Bernstein, K.W. DeLong, and N. Finlayson. Self-trapping transitions in a discrete NLS model with localized initial conditions. *Phys. Lett. A*, 181:135–141, 1993.
16. D. Cai, A.R. Bishop, and N. Grønbech-Jensen. Spatially localized, temporally quasiperiodic, discrete nonlinear excitations. *Phys.Rev.E*, 52:R5784–R5787, 1995.
17. A. Calini, N.M. Ercolani, D.W. McLaughlin, and C.M. Schober. Mel'nikov analysis of numerically induced chaos in the nonlinear Schrödinger equation. *Physica D*, 89:227–260, 1996.
18. J. Carr and J.C. Eilbeck. Stability of stationary solutions of the discrete self-trapping equation. *Phys. Lett. A*, 109:201–204, 1985.
19. F.S. Cataliotti, S. Burger, C. Fort, P. Maddaloni, F. Minardi, A. Trombettoni, A. Smerzi, and M. Inguscio. Josephson junction arrays with Bose-Einstein condensates. *Science*, 293:843–846, 2001.
20. F.S. Cataliotti, L. Fallani, F. Ferlaino, C. Fort, P. Maddaloni, M. Inguscio, A. Smerzi, A. Trombettoni, P.G. Kevrekidis, and A.R. Bishop. A novel mechanism for superfluidity breakdown in weakly coupled Bose-Einstein condensates. arXiv:cond-mat/0207139, 2002.

21. P.L. Christiansen, Yu.B. Gaididei, M. Johansson, and K.Ø. Rasmussen. Breatherlike excitations in discrete lattices with noise and nonlinear damping. *Phys. Rev. B*, 55:5759–5766, 1997.
22. P.L. Christiansen, Yu.B. Gaididei, K.Ø. Rasmussen, V.K. Mezentsev, and J. Juul Rasmussen. Dynamics in discrete two-dimensional nonlinear Schrödinger equations in the presence of point defects. *Phys. Rev. B*, 54:900–912, 1996.
23. D.N. Christodoulides and R.I. Joseph. Discrete self-focusing in nonlinear arrays of coupled waveguides. *Opt. Lett.*, 13:794–796, 1988.
24. Ch. Claude, Yu.S. Kivshar, O. Kluth, and K.H. Spatschek. Moving localized modes in nonlinear lattices. *Phys. Rev. B*, 47:14228–14232, 1993.
25. L. Cruzeiro-Hansson, H. Feddersen, R. Flesch, P.L. Christiansen, M. Salerno, and A.C. Scott. Classical and quantum analysis of chaos in the discrete self-trapping equation. *Phys. Rev. B*, 42:522–526, 1990.
26. S. Darmanyan, A. Kobyakov, and F. Lederer. Asymmetric dark solitons in nonlinear lattices. *JETP*, 93:429–434, 2001.
27. S. Darmanyan, A. Kobyakov, and F. Lederer. Stability of strongly localized excitations in discrete media with cubic nonlinearity. *JETP*, 86:682–686, 1998.
28. I. Daumont, T. Dauxois, and M. Peyrard. Modulational instability: first step towards energy localization in nonlinear lattices. *Phys. Rev. A*, 46:3198–3205, 1996.
29. A.S. Davydov. The theory of contraction of proteins under their excitation. *J. Theor. Biol.*, 38:559–569, 1973.
30. D.B. Duncan, J.C. Eilbeck, H. Feddersen, and J.A.D. Wattis. Solitons on lattices. *Physica D*, 68:1–11, 1993.
31. J.C. Eilbeck. Numerical simulations of the dynamics of polypeptide chains and proteins. In C. Kawabata and A.R. Bishop, editors, *Computer Analysis for Life Science – Progress and Challenges in Biological and Synthetic Polymer Research*, pages 12–21, Tokyo, 1986. Ohmsha.
32. J.C. Eilbeck. Nonlinear vibrational modes in a hexagonal molecule. In A. Davydov, editor, *Physics of Many-Particle Systems*, volume 12, pages 41–51, Kiev, 1987.
33. J.C. Eilbeck, P.S. Lomdahl, and A.C. Scott. Soliton structure in crystalline acetanilide. *Phys. Rev. B*, 30:4703–4712, 1984.
34. J.C. Eilbeck, P.S. Lomdahl, and A.C. Scott. The discrete self-trapping equation. *Physica D*, 16:318–338, 1985.
35. J.C. Eilbeck and A.C. Scott. Theory and applications of the discrete self-trapping equation. In P.L. Christiansen and R.D. Parmentier, editors, *Structure, Coherence, and Chaos in Dynamical Systems*, pages 139–159. Manchester University Press, 1989.
36. H.S. Eisenberg, Y. Silberberg, R. Morandotti, A.R. Boyd, and J.S. Aitchison. Discrete spatial optical solitons in waveguide arrays. *Phys. Rev. Lett.*, 81:3383–3386, 1998.
37. L.D. Faddeev and L.A. Takhtajan. *Hamiltonian methods in the theory of solitons.* Springer-Verlag, Berlin, 1987.
38. H. Feddersen. Localization of vibrational energy in globular protein.

Phys. Lett. A, 154:391-395, 1991.

39. H. Feddersen. Solitary wave solutions to the discrete nonlinear Schrödinger equation. In M. Remoissenet and M. Peyrard, editors, *Nonlinear Coherent structures in Physics and Biology, v. 393 of Lecture Notes in Physics*, pages 159–167. Springer, 1991.
40. H. Feddersen. Numerical calculations of solitary waves in Davydov's equations. *Physica Scripta*, 47:481–483, 1993.
41. S. Flach and K. Kladko. Moving discrete breathers? *Physica D*, 127:61–72, 1999.
42. S. Flach, K. Kladko, and R.S. MacKay. Energy thresholds for discrete breathers in one-, two-, and three-dimensional lattices. *Phys. Rev. Lett.*, 78:1207–1210, 1997.
43. S. Flach and C.R. Willis. Discrete breathers. *Phys. Rep.*, 295:182–264, 1998.
44. S. Flach, Y. Zolotaryuk, and K. Kladko. Moving lattice kinks and pulses: An inverse method. *Phys. Rev. E*, 59:6105–6115, 1999.
45. J. Fröhlich, T. Spencer, and C.E. Wayne. Localization in disordered, nonlinear dynamical systems. *J. Stat. Phys.*, 42:247–274, 1986.
46. Yu.B. Gaididei, S.F. Mingaleev, and P.L. Christiansen. Curvature-induced symmetry breaking in nonlinear Schrödinger models. *Phys, Rev. E*, 62:R53–R56, 2000.
47. D. Hennig and H. Gabriel. The four-element discrete nonlinear Schrödinger equation – non-integrability and Arnold diffusion. *J. Phys. A*, 28:3749–3756, 1995.
48. D. Hennig, H. Gabriel, M.F. Jørgensen, P.L. Christiansen, and C.B. Clausen. Homoclinic chaos in the discrete self-trapping trimer. *Phys. Rev. E*, 51:2870–2876, 1995.
49. D. Hennig, K.Ø. Rasmussen, G.P. Tsironis, and H. Gabriel. Breatherlike impurity modes in discrete nonlinear lattices. *Phys. Rev. E*, 52:R4628–R4631, 1995.
50. D. Hennig and G.P. Tsironis. Wave transmission in nonlinear lattices. *Phys. Rep.*, 307:334–432, 1999.
51. T. Holstein. Studies of polaron motion. *Ann. Phys.*, 8:325–389, 1959.
52. A.G. Izergin and V.E. Korepin. A lattice model related to the nonlinear Schrödinger equation. *Sov. Phys. Dokl.*, 26:653–654, 1981.
53. S.M. Jensen. The nonlinear coherent coupler. *IEEE J. Quantum Electron.*, QE-18:1580–1583, 1982.
54. M. Johansson. Decay of discrete nonlinear Schrödinger breathers through inelastic multiphonon scattering. *Phys. Rev. E*, 63:037601-1–4, 2001.
55. M. Johansson and S. Aubry. Existence and stability of quasiperiodic breathers in the discrete nonlinear Schrödinger equation. *Nonlinearity*, 10:1151–1178, 1997.
56. M. Johansson and S. Aubry. Growth and decay of discrete nonlinear Schrödinger breathers interacting with internal modes or standing-wave phonons. *Phys. Rev. E*, 61:5864–5879, 2000.
57. M. Johansson, S. Aubry, Yu.B. Gaididei, P.L. Christiansen, and K.Ø. Rasmussen. Dynamics of breathers in discrete nonlinear Schrödinger models. *Physica D*, 119:115–124, 1998.

58. M. Johansson, M. Hörnquist, and R. Riklund. Effects of nonlinearity on the time evolution of single-site localized states in periodic and aperiodic discrete systems. *Phys. Rev. B*, 52:231–240, 1995.
59. M. Johansson and Yu.S. Kivshar. Discreteness-induced oscillatory instabilities of dark solitons. *Phys. Rev. Lett.*, 82:85–88, 1999.
60. M. Johansson, A.M. Morgante, S. Aubry, and G. Kopidakis. Standing wave instabilities, breather formation and thermalization in a Hamiltonian anharmonic lattice. *Eur. Phys. J. B*, 29:279–283, 2002.
61. G. Kalosakas, K.Ø. Rasmussen, and A.R. Bishop. Delocalizing transition of Bose-Einstein condensates in optical lattices. *Phys. Rev. Lett.*, 89:030402-1–4, 2002.
62. P.G. Kevrekidis, B.A. Malomed, and Yu.B. Gaididei. Solitons in triangular and honeycomb dynamical lattices with the cubic nonlinearity. *Phys. Rev. E*, 66:016609-1–10, 2002.
63. P.G. Kevrekidis, K.Ø. Rasmussen, and A.R. Bishop. The discrete nonlinear Schrödinger equation: a survey of recent results. *Int. J. Mod. Phys. B*, 15:2833–2900, 2001.
64. S.W. Kim and S. Kim. The structure of eigenmodes and phonon scattering by discrete breathers in the discrete nonlinear Schrödinger chain. *Physica D*, 141:91–103, 2000.
65. Yu.S. Kivshar and D.K. Campbell. Peierls-Nabarro potential barrier for highly localized nonlinear modes. *Phys. Rev. E*, 48:3077–3081, 1993.
66. Yu.S. Kivshar, W. Królikowski, and O.A. Chubykalo. Dark solitons in discrete lattices. *Phys. Rev. E*, 50:5020–5032, 1994.
67. Yu.S. Kivshar, D.E. Pelinovsky, T. Cretegny, and M. Peyrard. Internal modes of solitary waves. *Phys. Rev. Lett.*, 80:5032–5035, 1998.
68. Yu.S. Kivshar and M. Peyrard. Modulational instabilities in discrete lattices. *Phys. Rev. A*, 46:3198–3205, 1992.
69. G. Kopidakis and S. Aubry. Intraband discrete breathers in disordered nonlinear systems. I. Delocalization. *Physica D*, 130:155–186, 1999.
70. G. Kopidakis and S. Aubry. Discrete breathers and delocalization in nonlinear disordered systems. *Phys. Rev. Lett.*, 84:3236–3239, 2000.
71. G. Kopidakis and S. Aubry. Intraband discrete breathers in disordered nonlinear systems. II. Localization. *Physica D*, 139:247–275, 2000.
72. E.W. Laedke, K.H. Spatschek, V.K. Mezentsev, S.L. Musher, I.V. Ryzhenkova, and S.K. Turitsyn. Instability of two-dimensional solitons in discrete systems. *JETP Lett.*, 62:677–684, 1995.
73. E.W. Laedke, K.H. Spatschek, and S.K. Turitsyn. Stability of discrete solitons and quasicollapse to intrinsically localized modes. *Phys. Rev. Lett.*, 73:1055–1059, 1994.
74. S.-S. Lee and S. Kim. Phonon scattering by breathers in the discrete nonlinear Schrödinger chain. *Int. J. Mod. Phys. B*, 14:1903–1914, 2000.
75. R.S. MacKay and S. Aubry. Proof of existence of breathers for time-reversible or Hamiltonian networks of weakly coupled oscillators. *Nonlinearity*, 7:1623–1643, 1994.
76. B.A. Malomed and P.G. Kevrekidis. Discrete vortex solitons. *Phys. Rev. E*, 64:026601-1–6, 2001.

77. V.K. Mezentsev, S.L. Musher, I.V. Ryzhenkova, and S.K. Turitsyn. Two-dimensional solitons in discrete systems. *JETP Lett.*, 60:829–835, 1994.
78. M.I. Molina and G.P. Tsironis. Dynamics of self-trapping in the discrete nonlinear Schrödinger equation. *Physica D*, 65:267–273, 1993.
79. M.I. Molina and G.P. Tsironis. Absence of localization in a nonlinear random binary alloy. *Phys. Rev. Lett.*, 73:464–467, 1994.
80. R. Morandotti, H.S. Eisenberg, Y. Silberberg, M. Sorel, and J.S. Aitchison. Self-focusing and defocusing in waveguide arrays. *Phys. Rev. Lett.*, 86:3296–3299, 2001.
81. R. Morandotti, U. Peschel, J.S. Aitchison, H.S. Eisenberg, and Y. Silberberg. Dynamics of discrete solitons in optical waveguide arrays. *Phys. Rev. Lett.*, 83:2726–2729, 1999.
82. R. Morandotti, U. Peschel, J.S. Aitchison, H.S. Eisenberg, and Y. Silberberg. Experimental observation of linear and nonlinear optical Bloch oscillations. *Phys. Rev. Lett.*, 83:4756–4759, 1999.
83. A.M. Morgante, M. Johansson, G. Kopidakis, and S. Aubry. Standing wave instabilities in a chain of nonlinear coupled oscillators. *Physica D*, 162:53–94, 2002.
84. T. Pertsch, P. Damberg, W. Elflein, A. Bräuer, and F. Lederer. Optical Bloch oscillations in temperature tuned waveguide arrays. *Phys. Rev. Lett.*, 83:4752–4755, 1999.
85. T. Pertsch, U. Peschel, and F. Lederer. Hybrid discrete solitons. *Phys. Rev. E.*, 66:Scheduled 1 December, 2002.
86. M. Peyrard. The pathway to energy localization in nonlinear lattices. *Physica D*, 119:184–199, 1998.
87. J. Juul Rasmussen and K. Rypdal. Blow-up in Nonlinear Schroedinger equations. *Physica Scripta.*, 33:481–504, 1986.
88. K.Ø. Rasmussen, S. Aubry, A.R. Bishop, and G.P. Tsironis. Discrete nonlinear Schrödinger breathers in a phonon bath. *Eur. Phys. J. B*, 15:169–175, 2000.
89. K.Ø. Rasmussen, T. Cretegny, P.G. Kevrekidis, and N. Grønbech-Jensen. Statistical mechanics of a discrete nonlinear system. *Phys. Rev. Lett.*, 84:3740–3743, 2000.
90. M. Salerno. Quantum deformations of the discrete nonlinear Schrödinger equation. *Phys. Rev. A*, 46:6856–6859, 1992.
91. A.C. Scott. Davydov's soliton. *Phys. Rep.*, 217:1–67, 1992.
92. A.C. Scott. *Nonlinear Science.* OUP, Oxford, 1999.
93. A.C. Scott and J.C. Eilbeck. On the CH stretch overtones in benzene. *Chem. Phys. Lett.*, 132:23–28, 1986.
94. A.C. Scott and L. MacNeil. Binding energy versus nonlinearity for a "small" stationary soliton. *Phys. Lett. A*, 98:87–88, 1983.
95. D.L. Shepelyansky. Delocalization of quantum chaos by weak nonlinearity. *Phys. Rev. Lett.*, 70:1787–1790, 1993.
96. A. Smerzi, S. Fantoni, S. Giovanazzi, and S.R. Shenoy. Quantum coherent atomic tunneling between two trapped Bose-Einstein condensates. *Phys. Rev. Lett.*, 79:4950–4953, 1997.
97. A. Trombettoni and A. Smerzi. Discrete solitons and breathers with dilute

Bose-Einstein condensates. *Phys. Rev. Lett.*, 86:2353–2356, 2001.

98. A.A. Vakhnenko and Yu.B. Gaididei. On the motion of solitons in discrete molecular chains. *Theor. Math. Phys.*, 68:873–880, 1987.
99. M.I. Weinstein. Excitation thresholds for nonlinear localized modes on lattices. *Nonlinearity*, 12:673–691, 1999.

EXPERIMENTAL STUDIES AND THEORY OF NONLINEAR ROTATIONAL DYNAMICS IN THE QUANTUM REGIME: THE INTERPLAY OF STRUCTURE, DYNAMICS AND LOCALIZATION IN CRYSTALS

FRANÇOIS FILLAUX AND BÉATRICE NICOLAÏ

LADIR-CNRS and Université Pierre et Marie Curie
2 rue H. Dunant, 94320 Thiais, France
E-mail: fillaux@glvt-cnrs.fr

ALAIN COUSSON

Laboratoire Léon Brillouin (CEA and CNRS)
CEA Saclay, 91 191, Gif sur Yvette cedex, France
E-mail: cousson@llb.saclay.cea.fr

Experimental and theoretical studies of quantum rotation of methyl groups in crystals are presented. Tunnelling spectroscopy with the inelastic neutron scattering technique and determination of angular probability densities of methyl groups in their rotational planes with the single crystal neutron diffraction technique are introduced. The importance of temperature and deuteration effects is emphasized. Three examples are presented. i) In the manganesediacetate tetrahydrate crystal, $Mn(CH_3COO)_2;4H_2O$, methyl groups are in three different environments, their rotational axes have different orientations and they are well separated from each other. They can be regarded as isolated single rotors and tunnelling excitations are localized. ii) In the lithiumacetate dihydrate crystal, $CH_3COOLi;2H_2O$, all methyl groups are equivalent. Close-contact pairs of face-to-face methyl groups are distributed in a nearly hexagonal planar structure with rather short methyl-methyl distances. The face-to-face methyl groups twisted by 60° perform combined rotation represented with symmetry adapted coordinates. Ordering of the methyl groups at a low temperature upon deuteration reveals significant interaction between pairs. However, in the 2D structure arising from methyl-methyl interaction there is no evidence for collective rotation of the pairs. Tunnelling excitations are localized within pairs. iii) In the 4-methylpyridine (C_6H_7N) crystal the distances between methyl groups are quite similar to those in lithiumacetate. However, in the tetragonal structure, the face-to-face methyl groups twisted by 90° to each other cannot perform combined rotation. The corresponding effective potential is virtually a constant. There is no phase transition upon methyl deuteration. Methyl groups form orthogonal infinite chains parallel to *a* or *b* and virtually isolated from each other. The collective rotation in 1D is represented in the displacive regime with the quantum sine-Gordon theory. Tunnelling transitions are represented with extended states in an energy-band structure. Additional transitions are attributed to travelling states of a dimensionless pseudo particle: the quantum breather mode. All excitations are totally delocalized.

1. Introduction

Light particles moving in a potential with topological degeneracy manifest their quantum nature *via* tunnelling, one of the evidences of the profound difference between classical and quantal worlds. Owing to the spatial extension of the wave function, the degenerate ground state in the classical regime splits into sublevels. In the condensed matter, the observation of tunnelling sheds light onto fundamental problems of quantum mechanics in a complex environment.

For light quantum rotors like methyl groups, the topological degeneracy arises from the intrinsic periodicity of the angular coordinate and rotational tunnelling is observed in many crystals at a low temperature.[1] The upper limit of the tunnel splitting is the rotational constant $B = \hbar^2/2I_r$, where I_r is the moment of inertia of the rotor. For an isolated and rigid CH_3 group rotating around its axis of inertia supposed to be fixed, B ranges from 0.650 to 0.700meV and from 0.325 to 0.350meV for the deuterated analogue CD_3. The magnitude of the tunnel splitting depends on the particle mass and potential shape (distances between identical sites, barrier height...). Deuteration is of dramatic consequence to the tunnelling frequency.

In many molecular crystals the frequency range for tunnelling is well separate from the density-of-states for optical phonons and dynamical interaction with the lattice is very weak. Tunnelling transitions are specific to the rotational coordinates and can be represented with a rather simple Hamiltonian. Therefore, rotational tunnelling is unique to observing nonlinear dynamics in the quantum regime. For example, quantum rotation of isolated rotors can be regarded as the solution of the Mathieu equation at the molecular level (see below section 2).[2] Energy localization is characteristic of these tunnelling transitions.[3,4] For some systems, methyl-methyl interaction may compete with localization. For example, the rotational dynamics in the lithiumacetate dihydrate crystal are solutions of the nonlinear and nonintegrable Hamiltonian for coupled pairs of rotors (see below 3).[5,6,8,7,9] Tunnelling excitations remain largely localized in pairs. Finally, collective rotational dynamics in the 4-methylpyridine crystal are solutions of the sine-Gordon Hamiltonian in the quantum regime (see below section section 4).[10,11,12,13,14,16,15] Tunnelling excitations are totally delocalized

For methyl groups, the great specificity of rotational tunnelling can be fully exploited with the inelastic neutron scattering (INS) technique, because the cross-section of H atoms for incoherent neutron scattering is much greater than for any other atom, by about one order of magnitude.[17] Any intensity arising from the crystal density-of-states (primarily acoustic phonons in the tunnelling energy range) can be ignored.

Nowadays, transitions above $\approx 1\mu$eV are observed with advanced spectrometers available at various neutron sources. However, the observation of tunnelling transitions for deuterated analogues is hampered by the much weaker cross section of D atoms and by the dramatic decrease of the splitting upon deuteration. This information is often missing and this is a limitation for modelling rotational dynamics.

Pioneering INS works with rather modest resolution have first revealed single tunnelling transitions and dynamical models for single rotors were sufficient to interpreting the spectra.[3,4] With better spectrometer resolution, multi-component spectra due to dynamical coupling were observed and different models were proposed to account for local or collective rotational dynamics. The diversity of models is partly due to the lack of information and this is a long lasting source of polemics that keep the field lively. Unfortunately, the hope that advanced methods of quantum chemistry and molecular dynamics simulation could provide realistic modelling of the effective potentials experienced by methyl rotors is not yet realized.[18,19] As long as tunnelling frequencies cannot be calculated with good accuracy (say a few μeV), experiments remain the best source of knowledge of these quantum dynamics. This is a strong incentive to undertaking new experiments in order to remove ambiguities, as much as possible. Studies of various isotope derivatives and mixtures, tunnelling spectra measured on single crystals, determination of the angular probability density and of the kinetic momentum distribution are among the salient progresses which have shed a new light on quantum rotation in solids, during the last decade, or so.

This paper is meant to be an introduction to tunnelling spectroscopy with neutrons and to the analysis of model Hamiltonians for quantum rotation. We present three prototypical examples of quantum rotors with totally different dynamics. The impact of advanced techniques and data analysis to theoretical developments is emphasized. We hope to convince the reader that rotational dynamics of methyl groups is a unique benchmark for quantum mechanics in the solid states. Moreover, this field of investigation is a lively forum for experimental chemists/physicists on the one hand and theoreticians/mathematicians, on the other. To launch fruitful collaborations between such experts from different fields is a major goal of the LOCNET network.

2. The single methyl rotor

The simplest model for methyl rotation is the single rigid rotor in a threefold potential. The Hamiltonian

$$H_1 = -B\frac{\partial^2}{\partial\theta^2} + \frac{V_{03}}{2}(1 - \cos 3\theta) \tag{1}$$

can be transformed into a Mathieu equation.[2] Eigen states $E_{n\sigma}$ and wave functions $\Psi_{n\sigma}(\theta)$ depend on two quantum numbers: the principal torsional quantum number n for the oscillator limit with full degeneracy and a sublevel index σ, which gives the symmetry of the wave function. As there is no analytical solution for the eigen values, numerical calculations using the basis set of the free rotor are carried out with the variational method (Figure 1).[20] In this simple model, there is only one tunnelling transition and it is straightforward to estimate threefold potential barriers from measurements. As the tunnel splitting varies practically exponentially with the potential barrier, tunnelling is a very sensitive probe of the local potential experienced by methyl groups in crystals. Experimental values can be comparer with those estimated with various models (for example quantum chemistry methods).[18,19]

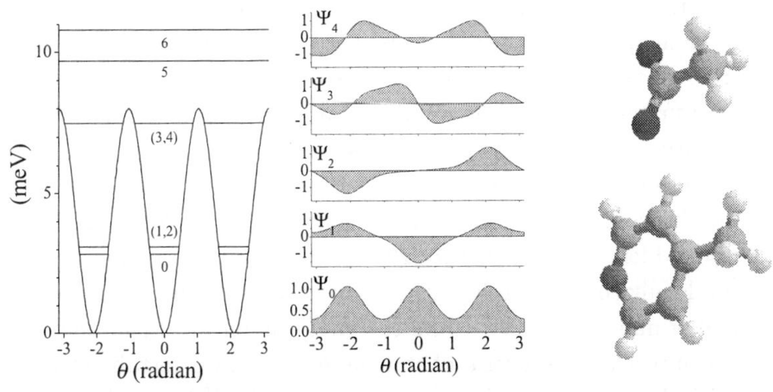

Figure 1. Schematic representation of the eigen states and eigen functions for a methyl rotor in a threefold potential and molecular models for the acetate entity (upper) and 4-methylpyridine (lower).

The effective potential can be decomposed into an internal potential that is determined by the molecular frame bearing the methyl group and an external contribution arising from the environment. As internal barriers are negligible for the acetate entity and for the 4-methylpyridine molecule (see Figure 1), the observed potential barriers are entirely due to the environment and can be tentatively related to the crystal structure.

A tutorial example is the crystal of manganesediacetate tetrahydrate, $Mn(CH_3COO)_2$;$4H_2O$, which contains 3 crystallographically inequivalent methyl groups.[21] The INS spectrum of a powdered sample at 1.5 K reveals 3 tunnelling transitions at 1.2, 50 and 137 μeV, with equal intensities.[22] The corresponding potential barriers (49.0, 17.5 and 11.5 meV, respectively) emphasize the great sensitivity of the tunnelling frequency to the local potential. However, effective potentials arising from the crystal environment at each site are sums of many contributions (atom...atom, electrostatic, multipolar...) that are poorly known and cannot be calculated with good accuracy. Unavoidable uncertainty precludes a firm assignment scheme for the tunnelling frequency at each rotor site.

2.1. *INS spectra of oriented single crystals*

The diffraction techniques are used to determine crystal structures with great accuracy. For ordered crystal rather sharp spots of intensity (Bragg-peaks) are observed for well defined orientations of the crystal and detector with respect to the incident beam. The diffraction pattern defines the reciprocal lattice that is a Fourier transformation of the direct lattice. The intensity of the peaks is determined by the lattice symmetry and the scattering cross-section of the atoms at each site.

X-ray diffraction is the most used technique available in many laboratories as a routine facility. The intensity scattered by atoms increases with the number of electrons. Consequently, the Bragg-peak intensities are largely dominated by contributions from heavy atoms whilst hydrogen atoms can be totally hidden. As a general rule, hydrogen atoms of methyl groups cannot be localized accurately with X-ray. However, the crystal structure is a source of information regarding the environment of the methyl groups.

For example, the structure of manganesediacetate tetrahydrate determined with X-ray reveals the orientations of the rotational axes of the methyl groups (see Figure 2).[23,24,25] Then, with properly oriented single crystals, the INS intensity of the tunnelling transitions can be probed as a function of the orientation of the momentum transfer vector $\mathbf{Q}$ with respect to the rotational axes. ($\mathbf{Q} = \mathbf{k_0} - \mathbf{k_f}$ with $|\mathbf{k_0}| = 2\pi/\lambda_0$ and $|\mathbf{k_f}| = 2\pi/\lambda_f$, where λ_0 and λ_f are the incident and scattered wavelengths, respectively.) The intensity is a maximum when $\mathbf{Q}$ is perpendicular to the axis of rotation and it vanishes if $\mathbf{Q}$ is parallel. According to such measurements, the tunnelling frequencies at 1.2, 50 and 137 μeV are attributed to sites C, B and A, respectively.

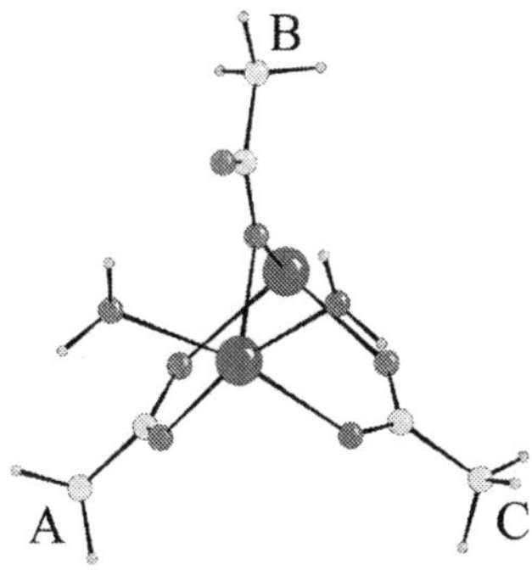

Figure 2. Schematic representation of the arrangement of the three inequivalent methyl groups in the manganesediacetate tetrahydrate crystal at 14 K, after ref. 21

2.2. *The density probability of methyl rotors*

The neutron diffraction technique is complementary to X-ray diffraction. The scattering cross section of electrons is negligible and Bragg-peaks arise from scattering by nuclei. This technique is unique to determine the position of H-atoms because the nuclear cross-sections are on the same order of magnitude for all nuclei. For example, the orientation of the methyl groups and of the water molecules were determined for manganesediacetate tetrahydrate (see Figure 2).[21]

With the Fourier difference method, data obtained from single-crystal neutron diffraction provide a full view of the probability density of the H(D)-atoms. For this purpose, once the crystal structure has been determined, Bragg-peak intensities can be calculated for an ideal crystal in which the scattering cross section of the H(D)-atoms of the methyl groups is set to zero. The difference from the original pattern contains specific information on the methyl H(D)-atoms. Further Fourier back transformation gives the probability density distribution in direct space (for example, see Figure 3).

Because diffraction arises from coherent scattering by a large number of atoms in a regular lattice, disorder appears as a perturbation of the lattice periodicity averaged over space and time. It is impossible to distinguish statistical and dynamical disorder. However, the dynamical orientational disorder of methyl group is so large compared to disorder arising from other vibrations that the angular probability density is quite representative of the wave function and can be somehow related to the potential.

2.2.1. *Methyl rotors at low temperature*

The Fourier difference maps for the three methyl groups in the manganese-diacetate tetrahydrate crystal at 14 K are presented in Figure 3. The

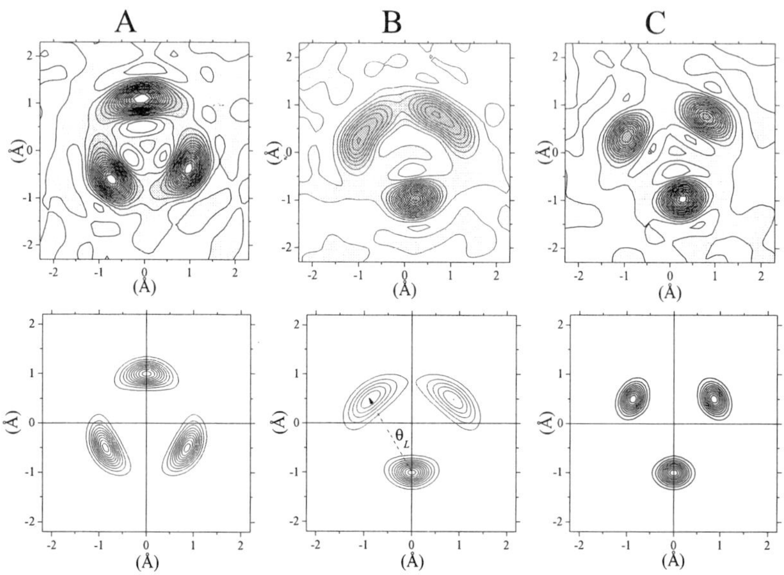

Figure 3. Top: measured probability density distributions for the three inequivalent methyl groups in the manganese diacetate tetrahydrate crystal measured at 14 K, after ref. 21. Labels A, B and C refer to those in Figure 2. Bottom: calculated maps (see text). A: $V_3 = 11.5$ meV, $u_r^2 = 0.05$ Å^2. B: $V_3 = 17.6$ meV, $u_r^2 = 0.05$ Å^2, $\theta_L^2 = 0.06$ rad^2. C: $V_3 = 49.0$ meV, $u_r^2 = 0.05$ Å^2.

different dynamics are immediately recognized by visual examination. The highest potential barrier (lowest tunnelling frequency) occurs for the methyl group C whose protons are quite localized. The more delocalized protons for the methyl group A correspond to the lowest potential barrier (highest tunnelling frequency).

The map for the methyl group B is much more intriguing. Whereas the distributions for A and C have clearly the threefold symmetry anticipated for rigid methyl groups with fixed axes, the probability densities at the three proton sites are quite different for methyl B. Of course, as each proton is totally delocalized the total probability is the same for each site. The probability density is actually a comprise of several contributions arising from both the methyl groups and the crystal lattice. It is necessary to analyze these maps more thoroughly in order to highlight the rotational dynamics.

One can distinguish three contributions to the probability density: (*i*) the rotational dynamics; (*ii*) the deformation of the methyl groups due to internal vibrations; (*iii*) the lattice dynamics and/or disorder. Because the tunnelling occurs at a very low frequency, lattice dynamics and internal vibrations can be represented with time/space-averaged distributions. It is convenient to distinguish isotropic and anisotropic contributions. The former, as opposed to the latter, does not disturb the threefold symmetry.

i) From the tunnelling frequency and Eq. (1) the temperature dependent angular probability density is determined by the Boltzmann distribution:

$$P_d\left(\theta, T\right) = \frac{\sum\limits_{n\sigma} \Psi_{n\sigma}^2\left(\theta\right) \exp\left(-E_{n\sigma}/kT\right)}{\sum\limits_{n\sigma} \exp\left(-E_{n\sigma}/kT\right)} \tag{2}$$

ii) Methyl groups are not perfectly rigid. The 8 degrees of freedom for internal motions give rise to complex vibrations that can be represented with a static distribution for the H-atoms. The projection onto the rotational plane gives an isotropic distribution with a Gaussian like profile. The variance (mean-square amplitude) of ≈ 0.01 Å^2 is largely temperature independent.

iii) Similarly, lattice vibrations give rise to an isotropic Gaussian distribution for the C-atom with a variance of ≈ 0.02 Å^2 at a low temperature. This value may increase significantly as the temperature is closer to the melting point. The mean square displacements in the rotational plane are further multiplied by a factor of ≈ 2 due to the distance between the rotational plane and the C-atom.

All together, the isotropic mean square displacement due to vibrations $u_{iso}^2(T)$ is ≈ 0.05 Å^2 at a low temperature and the observed angular distribution is a convolution:

$$P\left(\theta, T\right) = P_d\left(\theta, T\right) \otimes \rho_\varphi\left(\varphi - \theta, T\right); \tag{3}$$

with

$$\rho_\varphi\left(\varphi, T\right) = \left(\frac{1}{\pi \varphi_{iso}^2(T)}\right)^{\frac{1}{2}} \exp\left[-\frac{\left(\varphi - \theta\right)^2}{\varphi_{iso}^2(T)}\right] \tag{4}$$

where $\varphi_{iso}^2(T) = u_{iso}^2(T)/r^2$. The rotational radius for protons r is ≈ 1 Å. The radial distribution is gaussian in shape and can be represented as:

$$\rho_r\left(r, T\right) = \left(\frac{1}{\pi u_{iso}^2(T)}\right)^{\frac{1}{2}} \exp\left[-\frac{\left(r - r_0\right)^2}{u_{iso}^2(T)}\right]. \tag{5}$$

With this rather simple model, it transpires that the probability densities measured for the A and C rotors are directly related to the square of the rotational wave functions, with minor contribution from other internal and lattice dynamics (see Figure 3).

For the B rotor, the dynamical probability density is convoluted with an anisotropic distribution of the angular coordinate θ_L representing the orientational disorder of the methyl axis with respect to one of the potential minima. The physical picture emerging from this map is quite unforeseen. In the effective potential $V(\theta_L)$, the slow methyl rotational is quenched, one of the minima is much deeper at one proton site than at the others. Indeed, this is not in conflict with the threefold symmetry of the effective rotational potential $V_{03}(\theta)$ averaged over the fast lattice coordinate θ_L.

2.2.2. *Methyl rotors at room temperature*

Tunnelling bands are well observed only at very low temperature, usually below 50 K. At higher temperatures, bandwidths increase and transitions merge into the huge peak for elastic scattering. The prevailing interpretation is that the quantum dynamics switches to classical diffusion at high temperature. Nowadays, density distributions obtained at any temperature compatible with the stability of the crystal shed light onto the rotational dynamics at high temperature.

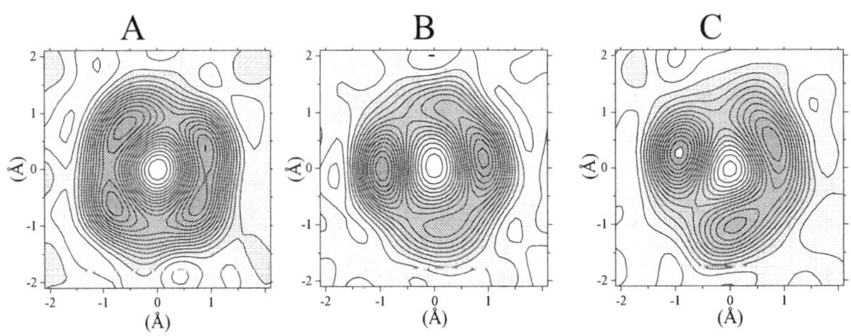

Figure 4. Measured probability density distributions for the three inequivalent methyl groups in the manganese diacetate tetrahydrate crystal at 300 K after ref. 21.Labels A, B and C refer to those in Figures 2 and 3.

At room temperature, the proton distributions for rotors A and B are almost isotropic (see Figure 4). Nearly free rotation arises mainly from thermal population of rotational levels close to, or above, the top of the potential barrier (see Figure 1). (The rather modest anisotropic features may arise from convolution with lattice displacements.) For the C rotors, partial localization of the protons due to hindered rotation is still observed and the effective potential barrier still exists at 300 K. In addition, a marked anisotropic contribution from the lattice dynamics is observed. The quantum dynamics is likely to survive even at room temperature and mixing of higher rotational states with phonons may take place.

3. Quantum rotational dynamics for pairs of coupled rotors

In the lithiumacetate dihydrate ($CH_3COOLi,2H_2O$ or Liac$-h_7$) crystal the rotational dynamics is better represented with pairs of coupled methyl groups.

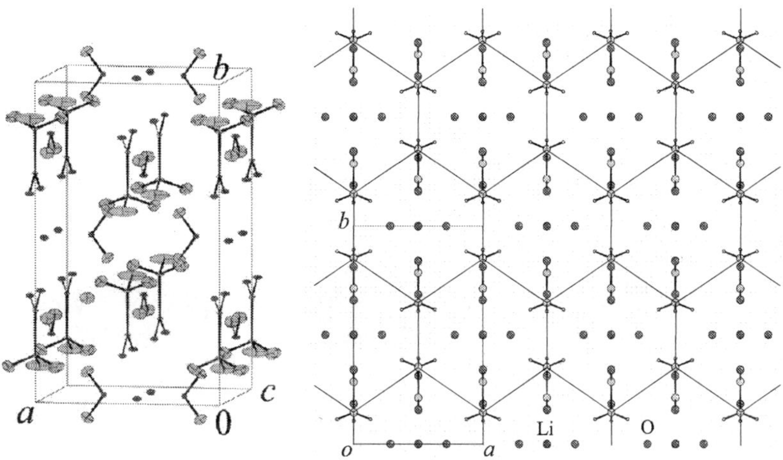

Figure 5. Crystal structure of the methyl deuterated lithium acetate dihydrate after ref. 26. Left: view of the unit cell with thermal ellipsoids. Right: projection onto the (a, b) plane. The lines along nearest neighbor interactions are guides for eyes. For the sake of clarity, the hydrogen atoms of the water molecules are hidden.

3.1. *A honey comb network in 2D of quantum rotors*

For the fully hydrogenated derivative the crystal symmetry ($Cmmm$) remains unchanged upon cooling from room temperature down to 1 K.[27,19] The methyl groups are arranged in infinite chains, along the a crystal axis, of face-to-face coaxial pairs parallel to b (see Figure 5). The distances between consecutive axes of rotation ($a/2 \approx 3.4$Å) and between the methyl-carbon atoms within pairs ($0.3b \approx 3.3$Å) are significantly shorter than the van der Waals radii and significant interactions along the chains and within pairs are schematically represented as a honey comb like structure in 2D, parallel to the (a, b) planes. The inter-layer distances of ≈ 6.56Å between (a, b) planes are much greater and interactions between methyl groups along the c axis is negligible.

3.2. *A phase transition induced by methyl deuteration*

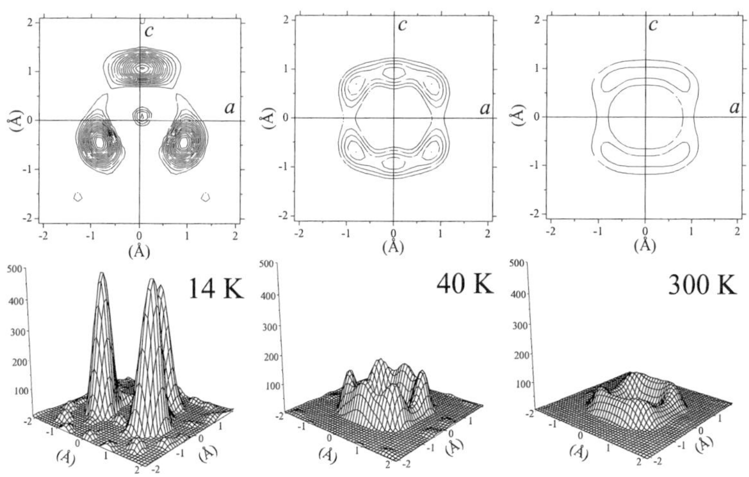

Figure 6. Probability density distributions of the deuterium atoms in the methyl deuterated lithiumacetate dihydrate ($CD_3COOLi,2H_2O$) after ref. 26: at 14 K in the ordered phase *Pman*, at 40 K in the *Cmmm* phase, slightly above the transition, and, in the same phase, at room temperature.

For Liac$-h_7$ the methyl groups are totally disordered at any temperature and the angular probability density is totally isotropic, even at ≈ 1 K.[19] Quite surprisingly, the methyl deuterated derivative ($CD_3COOLi,2H_2O$ or Liac$-d_3h_4$) undergoes a phase transition at 17.5 K, from *Cmmm* to *Pman* symmetry. The ordering of the centrosymmetric pairs of methyl groups is clearly observed with the neutron diffraction technique (Figure 5).[28,26] As all other atomic coordinates remain virtually unchanged, there is no significant variation of the effective potential arising from heavy atoms and the phase transition is entirely due to mass effects on the quantum dynamics.

In the ordered phase at 14 K the localization of the probability density distribution (Figure 6) is consistent with a rather high potential barrier. The two methyl groups in centrosymmetric pairs are indistinguishable and their images are superimposed in the Fourier difference. (The three peaks do not mean that the two methyl groups are in the eclipsed conformation.) At 300 K, the disorder is fully established. At 40 K, there is a mixture of ordered and disordered methyl groups which means that the potential barrier does not vanish at the phase transition. The density is a superposition of angular distributions for delocalized (≈50%) and localized deuterium atoms (≈50%). The six maxima do not mean that the rotational potential changes progressively from threefold to sixfold symmetry and then to free rotation upon increasing the temperature. At 40 K deuterium atoms are still partially ordered but face-to-face methyl groups are no longer indistinguishable in the *Cmmm* phase.

3.3. *The tunnelling spectrum*

Early measurements of the proton spin-lattice relaxation have revealed almost free rotation of the methyl groups.[29] However, the tunnelling bands observed in the 0.3 meV range (see Figure 7) are quite below the frequency anticipated for almost free rotation. Moreover, since all methyl groups in the crystal experience the same effective potential the rather complex spectrum must be interpreted in terms of dynamical correlation between indistinguishable quantum rotors.

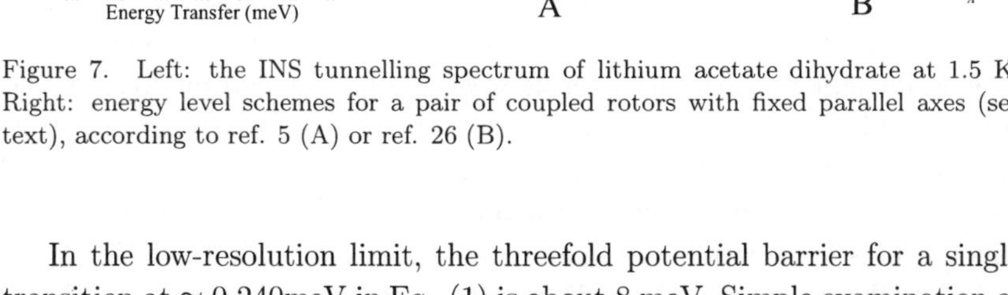

Figure 7. Left: the INS tunnelling spectrum of lithium acetate dihydrate at 1.5 K. Right: energy level schemes for a pair of coupled rotors with fixed parallel axes (see text), according to ref. 5 (A) or ref. 26 (B).

In the low-resolution limit, the threefold potential barrier for a single transition at ≈ 0.240meV in Eq. (1) is about 8 meV. Simple examination of Figure 1 reveals that the proton density should be quite localized at a low temperature. This is in conflict with the full delocalization experimentally observed.

For the CD_3 derivative the tunnelling frequency calculated at 33μeV with the same potential barrier is actually observed at 12.5μeV.[6] Apparently, there is a significant increase of the effective barrier to $\approx$ 11.6meV. This may contribute to the ordering of the CD_3 at low temperature (see below section 3.7).

3.4. *The coupled pair model*

The single rotor is clearly unable to account for the complexity of the observed spectrum. Coupled rotation of pairs of methyl groups with parallel axes suggested by the crystal structure was first represented with the Hamiltonian depending on the angular coordinates θ_1 and θ_2:[7,6]

$$H_2 = -B\left(\frac{\partial^2}{\partial\theta_1^2} + \frac{\partial^2}{\partial\theta_2^2}\right) + \frac{V_{03}}{2}\cos 3\theta_1 + \frac{V_{03}}{2}\cos 3\theta_2 + \frac{W_{12}}{2}\cos 3\left(\theta_1 - \theta_2\right) \quad (6)$$

Each top experiences the same on-site potential V_{03} and the coupling depends only on the phase difference of the two rotors. In contrast to the textbook case of coupled harmonic oscillators, this Hamiltonian cannot be diagonalized by simple transformation into normal coordinates.[9] On the other hand, numerical calculation of the eigen states and eigen functions requires very large basis sets with dimension N^2, where N is the size of the basis set for each rotor. Early calculations carried out with a very limited basis sets ($N = 9$)[7] were certainly lacking of accuracy and deserve some reservations. Furthermore, as symmetric and antisymmetric states are no longer separate, the interpretation of the energy level scheme is quite cumbersome.

In order to analyze Eq. (6) the potential terms were ignored ($V_{03} = W_{12} = 0$) and energy levels were distinguished according to their symmetry: the ground state (A_1A_2 at $E = 0$), first excited states corresponding to the transfer of one quantum to one of the two tops ($E_1^{\pm}A_2$ and $A_1E_2^{\pm}$ at $E = B$) and higher states corresponding to the transfer of one quantum to each top ($E_1^{\pm}E_2^{\pm}$ and $E_1^{\pm}E_2^{\mp}$ at $E = 2B$).[7] Then, the energy level scheme for non-zero potential terms was obtained by continuation of the zero-potential limit. It was concluded that $AE^{\pm}$ states should occur in all cases (see Figure 7A). However, according to basic quantum mechanics for two coupled and indistinguishable rotors, $AE^{\pm}$ states are not eigen states. This speculation is certainly an error and leads to misinterpretation of the spectrum. According to the energy level scheme in Figure 7A, the INS spectrum was decomposed into four components at 0.283, 0.250, 0.214 and 0.030 meV. The estimated values (namely $V_{03} \approx 7.6$ meV and $W_{12} \approx -17.0$ meV)[6] are totally in conflict with the crystal structure.

Firstly, the on-site potential barrier quite similar to that for the single rotor must be rejected. Secondly, the negative value for W_{12} favors the eclipsed configuration ($\theta_1 = \theta_2$), whereas the crystal structure shows the staggered configuration for face-to-face methyl groups at low temperature (see Figure 5). (It is unlikely that the eclipsed configuration would be favored only for the hydrogenated methyl groups.) Thirdly, we are not able to confirm these calculations. Finally, among the 4 transitions predicted in Figure 7A, three of them are hot transitions and their intensities should decrease at a low temperature. At 1.5 K the relative populations of the $|AE\rangle$ level at ≈ 0.250meV and $|EE\rangle$ levels at ≈ 0.500meV should be $\approx 710^{-2}$ and $\approx 510^{-3}$ respectively. Therefore, the spectrum should be largely dominated by the single $|AA\rangle \rightarrow |AE\rangle$ transition. However, this is not confirmed by experiments.[8]

3.5. *Symmetry adapted coordinates*

In order to propose a more rigorous analysis of the dynamics of coupled pairs, it is necessary to pay attention to the crystal structure in which the face-to-face methyl groups are indistinguishable and form centrosymmetric pairs. Consequently, the dynamics are better represented with symmetry-adapted coordinates corresponding to in-phase $\theta_i = (\theta_1 + \theta_2)/2$ and anti-phase $\theta_a = (\theta_1 - \theta_2)/2$ rotation. In contrast, again, to the harmonic case, the choice of the symmetry adapted coordinates is much more constrained. It is imposed by the periodicity of the rotors. The coordinates must be compatible with the threefold and sixfold periodicity for in-phase and anti-phase rotation. Any change of the periodicity would give a totally different energy level scheme.

In the strong coupling regime $\{|W_{12}| \gg |V_{03}|$ in Eq. (6)$\}$ the equilibrium position of one rotor with respect to the other is well defined and the two dynamics are separable. Then, the Hamiltonian is rewritten as:

$$H_{2ia} = H_{2i}(\theta_i) + H_{2a}(\theta_a) \tag{7}$$

with

$$\begin{aligned} H_{2i}(\theta_i) &= -\frac{B}{2}\frac{\partial^2}{\partial\theta_i^2} + \frac{V_{3i}}{2}\cos 3\theta_i \\ H_{2a}(\theta_a) &= -\frac{B}{2}\frac{\partial^2}{\partial\theta_a^2} + \frac{V_{3a}}{2}\cos 3\theta_a + \frac{W_{12a}}{2}\cos 6\theta_a \end{aligned} \tag{8}$$

For in-phase rotation the coupling term vanishes and the dynamics is that of a single rotor with moment of inertia $2I_r$ experiencing a threefold potential. If the coupling potential is strongly attractive the pair is in the eclipsed conformation and the effective potential is $V_{3i} \approx 2V_{03}$. Alternatively, if the staggered configuration is favored by a repulsive coupling term, the effective sixfold potential should be very weak and give rise to nearly free rotation. In that case, V_{03} in Eq. (6) cannot be determined experimentally. (This is the price for the separation of the dynamics but this is not really a penalty since free rotation is always due to a delicate balance between many contributions). For anti-phase rotation, the effective potential comprises a contribution from both V_{3a} and W_{12}.

An important consequence of the symmetry adapted coordinate is that the rotational constant is divided by a factor 2 compared to the single rotor. The maximum frequency for free rotation is now in the range 0.325 – 0.350meV for a pair of CH_3 and 0.162 – 0.175meV for a pair of CD_3. Therefore, the transitions observed around 0.270meV may correspond to pairs rotating almost freely, in accordance with measurements of the spin-lattice relaxation.[29]

The energy level scheme for tunnelling presented in Figure 7B is composed of one tunnelling state for in-phase rotation in a three-fold potential ($|0_{E\pm}\rangle_i$) and three states for anti-phase rotation in a six-fold potential ($|0_{E\pm}\rangle_a$, $|1_{E\pm}\rangle_a$ and $|1_A\rangle_a$). There are four fundamental transitions ($|0_A\rangle \rightarrow |0_{E\pm}\rangle_i$, $|0_A\rangle \rightarrow |0_{E\pm}\rangle_a$, $|0_A\rangle \rightarrow |1_{E\pm}\rangle_a$ and $|0_A\rangle \rightarrow |1_A\rangle_a$ and two "hot" transitions ($|0_{E\pm}\rangle_a \rightarrow |1_{E\pm}\rangle_a$ and $|0_{E\pm}\rangle_a \rightarrow |1_A\rangle_a$). Even at 1.5 K, the $|0_{E\pm}\rangle_a$ level at $\approx$ 0.030meV is largely populated (80%) and temperature effects on the band intensity is marginal. However, if hot transitions were not resolved from their parent bands, the attribution of the four transitions previously distinguished in the spectrum[7,6] should be straightforward.

3.6. *Isotope mixtures*

A more focussed view of the rotational dynamics was tentatively sought for with mixtures of CH_3 and CD_3 derivatives.[6] The physical and chemical differences for the two analogues are negligible compared to intermolecular interaction in the crystal, and mixtures are statistical distributions of pairs. If x and $(1-x)$ are the relative concentrations of CD_3 and CH_3, respectively, then, the relative concentrations of pairs CH_3-CH_3, CH_3-CD_3 and CD_3-CD_3 are $(1-x)^2$, $2x(1-x)$ and x^2, respectively.

According to the energy level scheme in Figure 7A, the totally decoupled mixed pairs have four tunnelling levels: $|A_H A_D\rangle$, $|A_H E_D\rangle$, $|E_H A_D\rangle$ and $|E_H E_D\rangle$. Since $|E_H\rangle$ and $|E_D\rangle$ states are quite separate the splitting arising from the relative rotation of the two tops ($|E_H^{\pm} E_D^{\pm}\rangle$ and $|E_H^{\pm} E_D^{\mp}\rangle$) can be ignored. This decoupling is the normal consequence of the removal of the indistinguishability for mixed pairs. Therefore, the spectrum is composed of degenerate transitions $|A_H\rangle \rightarrow |E_H\rangle$ and $|A_D\rangle \rightarrow |E_D\rangle$ for decoupled CH_3 and CD_3 tops, respectively. If there were no change of the effective potential upon isotope substitution, the frequency for a decoupled rotor should be at the barycenter of the spectral profile for the coupled pair.

The tunnelling spectra of mixtures containing small concentrations of CH_3 (from 5 to 20%) in CD_3 environment were decomposed into a main peak at 0.130meV and a weak shoulder at 0.180meV attributed to CH_3-CD_3 and CH_3-CH_3 pairs, respectively.[6] For an increasing concentration of CH_3 rotors, the intensity at 0.130meV decreases, that at 0.180meV increases to a maximum value for a concentration of 66% and then decreases whilst new features centered at 0.250meV appear. Therefore, there are at least 3 bands instead of the 2 anticipated with the model. The extra band was attributed to CH_3-CH_3 pairs surrounded by deuterated rotors. The effective potential barrier is increased and the band at 0.250meV is shifted downward to 0.180meV. However, each pair has four nearest neighbors methyl groups to which should correspond four different environments and effective potentials in isotope mixtures. If the different environments are not resolved, a smooth frequency shift should be observed, rather than a change of relative intensities for well separate transitions.

The interpretation of the spectra is hampered by the different crystal structures for the CH_3 and CD_3 derivatives. The phase transitions may take place at various temperatures depending on the isotope concentration and it is difficult to distinguish dynamical and structural effects. In this context, isotope mixtures of CH_3 and CH_2D derivatives are more appropriate. There is no phase transition (this point deserves further confirmation) and the perturbation of the effective potential, if any, is minimized. Moreover, a great advantage compared to CD_3 is that CH_2D rotors have significant scattering cross section. Therefore, tunnelling arising from each pair CH_3-CH_3, CH_3-CH_2D and CH_2D-CH_2D can be observed. However, owing to the different pair potentials for H...H, H...D and D...D atoms,[13] anti-phase tunnelling is cancelled for CH_2D-CH_2D pairs.

The spectra presented in Figure 8 provide more detailed information than those previously reported.[6] They were decomposed into rather narrow components with relative intensities depending on the isotope concentration. The relative frequency shift for each component is less than 10%, which suggests quite moderate changes of the effective potential. For the sake of continuity, the spectra of the mixtures containing 80% and 100% of CH_3 were decomposed into 4 components between 0.150 and 0.300meV, instead of three reported previously.[7,8,6] Therefore, the symmetry adapted coordinates in Eqs (7) and (8) and the energy level scheme in Figure 7 are the most realistic.

At a low concentration in CH_3 (20%) the most intense band at 0.166meV

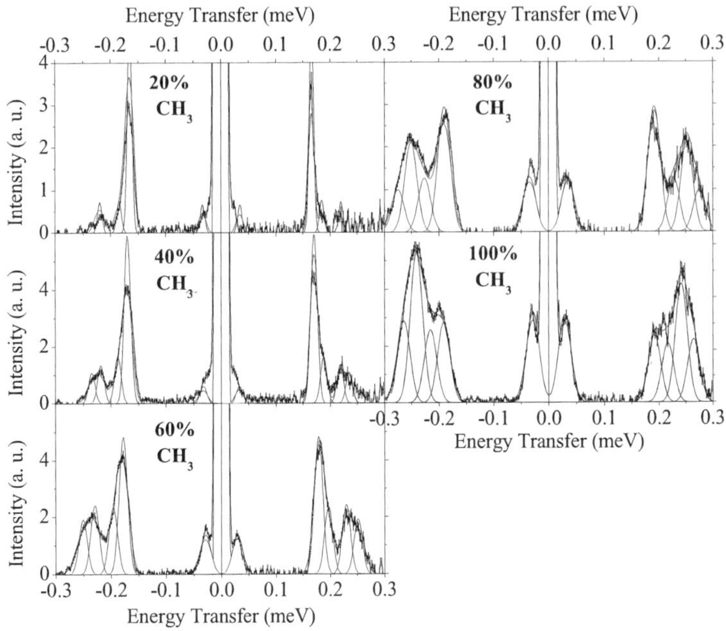

Figure 8. INS spectra and tentative band decomposition of isotope mixtures of $CH_3COOLi,2H_2O$ and $CH_2DCOOLi,2H_2O$ at 1.7 K, after ref. 15.

and the weaker band at 0.185meV can be attributed to in-phase rotation of CH_2D-CH_2D (B = 0.246meV) and CH_3-CH_2D (B = 0.281meV), respectively. The weak transition at 0.220meV is tentatively assigned to anti-phase tunnelling of CH_3-CH_2D.

For increasing concentration in CH_3 the band intensity around 0.170meV decreases whilst the frequency shifts smoothly upwards to 0.191meV for 100% of CH_3 rotors. This behavior suggests unresolved transitions due to CH_2D-CH_2D and CH_3-CH_3 pairs, respectively. We extrapolate the frequency for in-phase tunnelling of CH_2D–CH_2D pairs to be 0.164meV. The band intensity at $\approx$ 0.030meV due to out-of-phase rotation of symmetric CH_3–CH_3 pairs and the intensities of the three bands at high frequency increase simultaneously with the concentration in CH_3.

With the unambiguous assignment of the in-phase tunnelling for CH_3-CD_3 (0.130meV) and CH_2D-CH_2D (0.164meV), we assign the transition at 0.241meV in Liac$-h_7$ to in-phase tunnelling of CH_3-CH_3 entities. The average of the estimated potential values for pairs containing H atoms is $V_{3i} = (1.82 \pm 0.15)$meV. The rather modest dispersion of $\approx$10% may arise from experimental errors, band shape analysis and variations of the effective potential with the mass.

For anti-phase rotation we have virtually no information for the deuterated species. For $CH_3{-}CH_3$ pairs, we assign the doublet at 0.191 and 0.216meV to $|0_{E\pm}\rangle_a \rightarrow |1_{E\pm}\rangle_a$ and $|0_A\rangle \rightarrow |1_{E\pm}\rangle_a$, respectively. The splitting of 0.025meV compares to the $|0_A\rangle \rightarrow |0_{E\pm}\rangle_a$ transition observed at 0.029meV. Similarly, we assign the component at 0.264meV to $|0_A\rangle \rightarrow |1_A\rangle_a$ and we suppose that the companion transition $|0_{E\pm}\rangle_a \rightarrow |1_A\rangle_a$ anticipated at 0.235meV is a part of the main band at 0.241meV. Here we suppose that all transitions have virtually the same intensity. The best fit is obtained with the effective potential terms in Eq. (8) $V_{3a} = 0.19$ and $W_{12a} = 31.64$, in meV units. As anticipated for staggered methyl groups, the on-site potential is almost perfectly cancelled. The condition $W_{12a} \gg V_{3a}$ is for well separate in-phase and anti-phase dynamics is largely satisfied.

3.7. *Coupling between pairs and phase transition mechanism*

The phase transition is related to long-range ordering of the methyl groups in the honeycomb-like arrangement of acetate entities in the (a,b) plane (see Figure 5). Owing to the great stability of the positions of heavy atoms that remain virtually unchanged through the phase transition, the local potential arising from the mean crystal field is not changed. Therefore, the key parameter for the effective coupling between adjacent pairs of rotors is the angular mean-amplitude depending on mass and temperature $\bar{\theta}(T) = \langle \theta^2(T) \rangle^{1/2}$ with:

$$\langle \theta^2(T) \rangle = \int_{-\pi/3}^{\pi/3} d\theta P(\theta, T)\, \theta^2. \tag{9}$$

For anti-phase dynamics coupling between adjacent pairs is negligible compared to the high sixfold potential barrier. For in-phase rotation, the rather weak effective on-site potential can be attributed to interaction between pairs. The Hamiltonian including coupling with the four nearest-neighbors of a pair (see Figure 9) can be written as:

$$H_{ci} = -\frac{B}{2}\frac{\partial^2}{\partial\theta_i^2} + \frac{V_{03i}}{2}(1 - \cos 3\theta_i) + \sum_{n=1}^{4}\frac{V_{ci}}{2}\left[1 - \cos 3(\theta_{in} - \theta_i)\right]. \tag{10}$$

In the weak-coupling (between pairs) regime, there is no correlation between the angular coordinate of the pair (θ_i) and those of the neighbors (θ_{in}). The effective potential can be thus expanded as:

$$\begin{aligned}\frac{V_{3i}}{2}(1-\cos 3\theta_i) &= \frac{V_{03i}}{2}(1-\cos 3\theta_i)\\ &\quad+\sum_{n=1}^{4}\frac{V_{ci}}{2}(1-\cos 3\theta_i\cos 3\theta_{in}-\sin\theta_i\sin\theta_{in})\,.\end{aligned} \tag{11}$$

Averaging over the distribution of phase differences gives then the effective potential:

$$V_{3i}^{eff} \approx V_{03i} + 4V_{ci}\cos\left[3\bar{\theta}\,(T)\right], \tag{12}$$

With the potential functions determined at low temperature for in-phase rotation, $\bar{\theta}_{iH} \approx 21°$ for CH_3 pairs and $\bar{\theta}_{iD} \approx 11°$ for CD_3 analogues. Then, if the increase of the effective potential barrier from ≈ 1.9 to ≈ 4.5 meV upon methyl deuteration is due to the change of $\bar{\theta}\,(T)$ we obtain $4V_{ci} \approx$ 9.2meV and $V_{03i} \approx -3.7$meV. The opposite signs mean that the methyl orientation favored by the on-site potential is opposite to the orientation favored by interaction with next nearest neighbors. The effective potential in Eq. (12) vanishes for $\bar{\theta}_0 \approx 22°$. This is the key parameter to account for the phase transition upon deuteration.

for CH_3 pairs, $V_{3i}^{eff} \approx 1.8$meV. There is no long-range correlation for methyl orientation at any temperature. The density probability derived from the rotational wave functions is markedly localized. Therefore, the delocalized angular density distribution observed with diffraction means that methyl group are statistically distributed among several orientations. In this case, dynamical and statistical probability densities can be distinguished. This was largely overlooked in previous calculations with quantum chemistry methods.[19]

For CD_3, the long-range interaction is dominant at 14K and the orientation of the CD_3 groups along the a axis is imposed by the interaction between methyl groups. In the absence of this methyl-methyl interaction along a, the orientation of the pairs would be rotated by $\pi/3$. As the temperature is increased, the rotational levels at ≈ 2 meV are populated, the mean angular amplitude increases and the long-range ordering decreases. The phase transition takes place when the thermal bath energy is similar to ≈ 2 meV or ≈ 22 K. This value compares favorably to the temperature of 17.5 K for the phase transition.

4. Quantum rotational dynamics for infinite chains of coupled rotors

In the 4MP crystal the rotation of methyl-groups is nearly free[30,31,32] but the dynamics are quite different from those in lithiumacetate dihydrate and this is a consequence of the different crystal structures for the two systems (compare Figures 9 and 5).

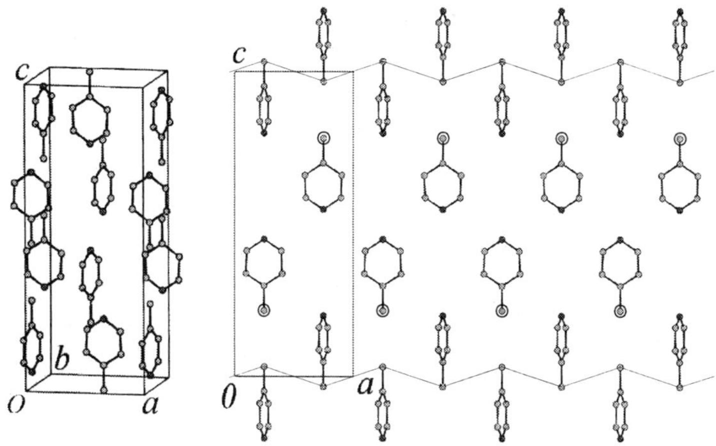

Figure 9. Schematic view of the structure of the 4-methylpyridine crystal ar 10 K, after ref. 33. Left: view of the unit cell. Right: projection onto the (a, c) plane showing the infinite chains parallel to a (along the zigzag line) or parallel to b (circles). For the sake of clarity, all H-atoms are hidden.

In the tetragonal structure ($I4_1/a$) all molecules are equivalent.[34] The C−C rotational axes are parallel to the c axis. The C_2 site symmetry impose disordering of the methyl groups and even at 10 K the angular probability density is virtually isotropic.[33] The structure remains unchanged until the first order phase transition at 254 K, close to the melting point at 276.8K. There is no further phase transition for the deuterated crystal.[35]

The shortest intermolecular distances of ≈ 3.46 Å between face-to-face methyl groups parallel to the c axis and the next shortest methyl-methyl distances of ≈ 4.0 Å perpendicular to the c axis, are quite similar to distances in lithiumacetate dihydrate (≈ 3.3 and 3.4 Å, respectively, see section 3 and Figure 5). However, in contrast to the honey-comb like arrangement, the equidistant methyl groups in 4-methylpyridine form two equivalent sets of orthogonal infinite chains parallel to the a and b crystal axes. The zigzag lines in Figure 9 correspond to chains parallel to a and circles represent intersections with the (a, c) plane of chains parallel to b. There is only one close-contact pair in common for nearest orthogonal chains.

Within the tetragonal symmetry, to any particular orientation of a methyl group corresponds 4 indistinguishable orientations obtained by symmetry with respect to the molecular plane and by $\pm\pi/2$ rotations. The twelve equivalent proton sites for a pair give a quasi-isotropic ring for the density distribution. Each methyl group experiences a virtually isotropic mean-field corresponding to the pair potential averaged over all orientations. The effective intra-pair potential is a constant and there is no coupling between collective excitations along a or b. Collective rotational dynamics occur exclusively in 1D.

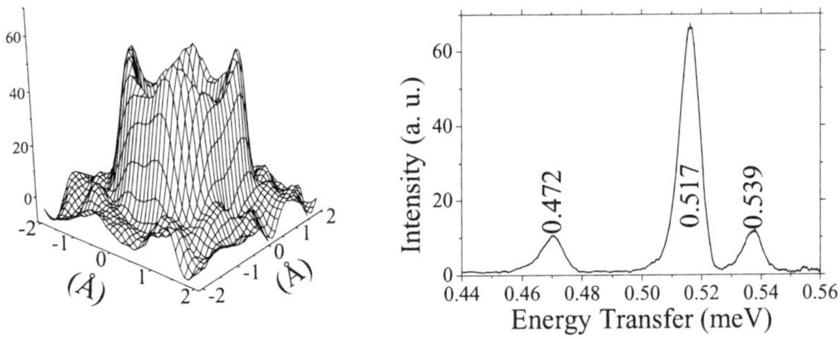

Figure 10. Left: probability density distribution of protons in the rotational plane of 4-methylpyridine at 10 K, after ref. 33. Right: the INS tunnelling spectrum of 4-methylpyridine at 0.5 K after ref. 13

These conclusions emphasize that a superficial examination of the intra-pair distances can be misleading. For example, Ohms and co-workers[34] concluded that the crystal structure implies strong correlation between the mutual orientations of adjacent methyl-groups which would have to be twisted by 60° with respect to each other and perform combined hindered rotations. Similarly, it is dangerous to utilize calculations of the effective potential, even with advanced quantum chemistry methods, for molecular-dynamics simulations without caution. Quantum chemistry gives logically a large coupling potential term for the close-contact pairs if the local symmetry is relaxed. Then, molecular dynamics simulation within the framework of classical mechanics give a false view of the methyl rotation.[36]

The tunnelling spectrum of 4-methylpyridine is totally different from that of lithium acetate. Early INS measurements with rather modest resolution have revealed a tunnelling transition at $\approx$ 0.5meV and this was regarded as a manifestation of nearly free rotation.[30] This was the first observation of rotational tunnelling with INS and this remains the highest tunnelling frequency ever reported for a methyl group.

Further INS experiments with better resolution have revealed a complex spectrum composed of an intense band at 0.517(4)meV and weaker bands at 0.472(4)meV and 0.539(4)meV (see Figure 10).[12] The splitting arises from dynamical correlation and a model based on coupled pairs of rotors, with some similarity with lithiumacetate dihydrate, was first proposed.[37] However, this is unlikely regarding the crystal structure.

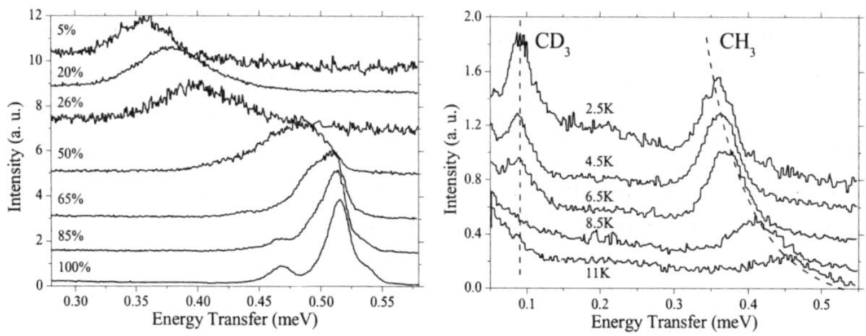

Figure 11. Tunnelling spectra of isotopic mixtures of hydrogenated and deuterated molecules, after ref. 12. Left: various concentrations of 4MP$-h_7$ at 2.5 K. Right: temperature effect for the mixture containing 5% of 4MP$-h_7$ and 95% of 4MP$-d_7$. Dash lines are guides for eyes.

The INS spectra of isotopic mixtures containing hydrogenated and deuterated molecules (4MP$-h_7$ and 4MP$-d_7$, respectively) show dramatic frequency shifts (up to $\Delta\nu/\nu \approx 30\%$), depending on concentration and temperature (see Figure 11)[12] which cannot be explained with any coupled pair model.[38] These experiments establish the collective nature of the methyl group dynamics and this was represented with the quantum sine-Gordon theory.[12]

4.1. *The quantum sine-Gordon*

This section is a summary of the relevant equations. More details were presented in refs 12,13,16. The Hamiltonian for an infinite chain of coupled rotors can be written as:

$$H = \sum_j -\frac{\hbar^2}{2I_r}\frac{\partial^2}{\partial\theta_j^2} + \frac{V_0}{2}\left(1 - \cos 3i\theta_j\right) + \frac{V_c}{2}\left[1 - \cos 3i\left(\theta_{j+1} - \theta_j\right)\right], \quad (13)$$

where θ_j is the angular coordinate of the jth rotor in the one-dimensional chain with parameter L. V_0 is the on-site potential which does not depend on lattice position, and V_c is the coupling ("strain" energy) between neighboring rotors. The index i = 1, 2 ... determines the potential periodicity compatible with the C_{3v} symmetry of the methyl-groups. In the strong coupling (or displacive) limit, when $\theta_{j+1} - \theta_j$ is sufficiently small, Eq. (13) is equivalent to the sine-Gordon equation:

$$H \approx \sum_j -\frac{\hbar^2}{2I_r}\frac{\partial^2}{\partial\theta_j^2} + \frac{V_0}{2}\left(1 - \cos 3i\theta_j\right) + \frac{(3i)^2 V_c}{4}\left(\theta_{j+1} - \theta_j\right)^2. \quad (14)$$

If variations of θ from site to site are small, then the site index can be replaced by a continuous position variable x (see Figure 12). Kinks, rotons (phonons) and breathers are the elementary excitations.[39,40,41,42,43,44,45,46,47,48,49,50,51] Since the kink density vanishes at low temperature and phonons are beyond the spectral range under investigation, these excitations can be ignored.

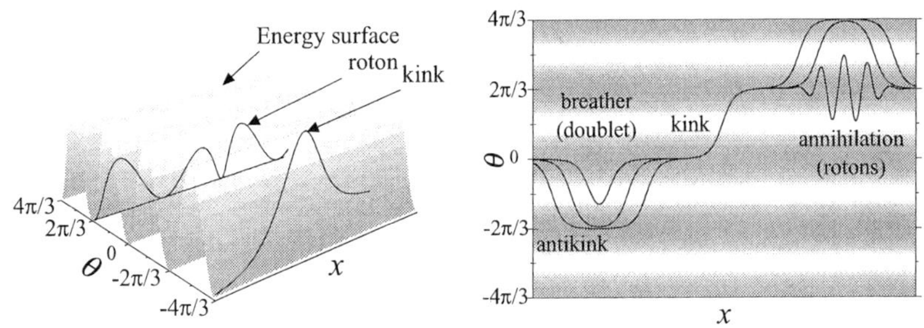

Figure 12. Artistic view of the sine-Gordon excitations.

The breather or doublet is a well-defined elementary nonlinear excitation with long life-time and as such behaves very much like a particle. In the classical regime, the waveform is:

$$\varphi_B^v(x,t) = 4\mathrm{Arct}\left[\frac{\sqrt{\frac{\omega_0^2}{\omega_B^2}-1}\,\sin\frac{\omega_B\left(t-vx/c_0^2\right)}{\sqrt{1-v^2/c_0^2}}}{\cosh\frac{(x-vt)\sqrt{1-\omega_B^2/\omega_0^2}}{d\sqrt{1-v^2/c_0^2}}}\right] \quad \text{with} \begin{cases} \omega_0^2 = \frac{V_0(3i)^2}{2I_r}\,; \\ c_0^2 = \frac{V_c L^2 (3i)^2}{2I_r} = \omega_c^2 L^2\,; \\ d = \frac{c_0}{\omega_0}. \end{cases} \quad (15)$$

The breather envelope travelling at velocity v is oscillating harmonically at frequency ω_B which may vary continuously from 0 to ω_0. Breathers can be viewed as kink-antikink bound states (Figure 12).

In the quantum regime [40,41] the classical breather yields a discrete spectrum of mass at rest ${}^{q}M^{0}_{B,l}$ depending on the quantum number l:

$$ {}^{q}M^{0}_{B,l} = 2\ {}^{q}M^{0}_{K\pm} \sin \frac{l\,(3i)^2}{16\left[1-\frac{(3i)^2}{8\pi}\right]} \text{ with } l = 1, 2, ... < \frac{8\pi}{(3i)^2} - 1 \quad ; \quad (16) $$

${}^{q}M^{0}_{K\pm}$ is the renormalized mass at rest of the classical kink:

$$ {}^{q}M^{0}_{K\pm} = \frac{8I_r}{(3iL)^2}\left[1-\frac{(3i)^2}{8\pi}\right] \text{ and } 3\text{i} < \sqrt{8\pi}. \quad (17) $$

In the case of methyl-groups, the breather-mode of the quantum sine-Gordon theory exists only for the threefold potential. Then, there is only one mass state (l =1) and this is the fundamental state. For energy transfer below the dissociation threshold corresponding to the creation of a kink-antikink pair, only kinetic energy can be transferred to the breather. This defines the dispersion law as:

$$ {}^{q}E_{B,l}\,(p_B) = \sqrt{{}^{q}E^{02}_{B,l} + p^2_B c^2_0}; \quad (18) $$

where ${}^{q}E^{0}_{B,l} = {}^{q}M_{B,l,0}c^2_0$ is the energy at rest of the breather mode and p_B is the kinetic momentum. Because the Hamiltonian in Eqs (13) and (14) possess translational invariance, the ground state can be represented with breathers at rest totally delocalized along the chain. The de Broglie wave length associated to the pseudo-particle is $\lambda = \infty$. The breather can propagate along the chains if, and only if, the de Broglie wavelength is an integral fraction of the lattice parameter:[13]

$$ \lambda = \frac{h}{p_B} = \frac{L}{n}, \quad n = 0, \pm 1, \pm 2, \ldots \quad (19) $$

Then, the kinetic energy spectrum is:

$$ {}^{q}E_{B,l,n} = \sqrt{{}^{q}E^{2}_{B,l,0} + n^2\hbar^2\omega^2_c}. \quad (20) $$

Eq. (20) is in accord with the INS [12,13,16], infrared and Raman [10,11] spectra of the 4-methylpyridine crystal. The band at 0.517meV in pure 4MP-h_7 (Figure 10) corresponds to the $0 \rightarrow 1$ transition between the travelling-states.

In Eq. (14) the periodical nature of the methyl-group rotation was discarded by the expansion of the coupling term. Consequently, collective tunnelling is not included in the sine-Gordon Hamiltonian. In the full Hamiltonian, Eq. (13), methyl tunnelling can be regarded as a one-dimensional band-structure problem. The dispersion law can be computed numerically. The tunnel splitting in the ground state varies continuously between two extremes located at the zone center, where the methyl groups are tunnelling in-phase, and at the zone boundary, where tunnelling occurs out-of-phase:

$$\begin{aligned} H_{ip} &= -\tfrac{\hbar^2}{2I_r}\tfrac{\partial^2}{\partial\theta_j^2} + \tfrac{V_0}{2}\left(1-\cos 3i\theta\right) \\ H_{op} &= -\tfrac{\hbar^2}{2I_r}\tfrac{\partial^2}{\partial\theta_j^2} + \tfrac{V_0}{2}\left(1-\cos 3i\theta\right) + \tfrac{V_c}{2}\left(1-\cos 6i\theta\right). \end{aligned} \tag{21}$$

These extremes correspond to the maxima of the density-of-states observed with the INS technique. Consequently, the bands at 0.537 and 0.470meV in pure 4MP-h_7 (see Figure 10) were attributed to in-phase and out-of-phase tunnelling transitions of the chain, respectively.

In the sine-Gordon theory three transitions (namely, two tunnelling and one breather-travelling mode) are determined with two parameters: V_0 and V_c. The potential barriers in Eq. (21) were first determined to fit the observed tunnelling transitions at 0.539 (in-phase) and 0.472meV (out-of-phase). Then, the calculated energy at rest of the breather mode and the first travelling state correspond to observation with better than 10% accuracy. This is remarkable since the sine-Gordon equation is an approximation of the Hamiltonian for the chain, which is already an approximation of the real system under consideration. The full width at half maximum of the breather waveform is about $5L$ (see below Figure 14) and numerical calculations confirm that the pinning potential in the discrete lattice is virtually zero. It is thus confirmed that the continuous limit approximation is relevant.

4.2. *Breather dynamics in isotopic mixtures*

In isotopic mixtures 4MP$-h_7$ and 4MP$-d_7$ molecules are distributed randomly among the crystal sites. They form clusters of various sizes, s_h and s_d, respectively. The rotational constant of the deuterated methyl-group being divided by a factor of 2 with respect to the hydrogenated analogue, the travelling-states are quite different. The translational invariance is broken and bound-states are observed for breathers. The frequency shifts observed at a very low temperature can be explained with reflective walls at cluster boundary.[12] The lowest travelling state in a cluster is determined by the size (sL) of the box, such as: $\lambda = sL$. Transitions between travelling states are then:

$$\begin{aligned} \nu_{sn} &= \sqrt{{}^{q}E_{B,l,0}^2 + n^2\hbar^2\omega_c^2} - \sqrt{{}^{q}E_{B,l,0}^2 + \hbar^2\tfrac{\omega_c^2}{s^2}} \quad ; \\ n &= \pm 1, \pm 2, \cdots \quad ; \quad s = 1, 2, \cdots \end{aligned} \tag{22}$$

As the cluster size decreases, the energy of the ground state increases with respect to that of the infinite chain and the frequency of the transition decreases. This equation accounts for the frequency shift of the main band from 0.517 to 0.360meV (Figure 10). This view provides also a straightforward mechanism to account for the frequency shift with temperature (see Figure 11). As the temperature increases, a continuum of breather-roton states is populated and can bridge the gap between adjacent clusters. Then, the trapped breather becomes progressively free and the travelling transition shifts continuously from 0.360 backwards to 0.500meV.

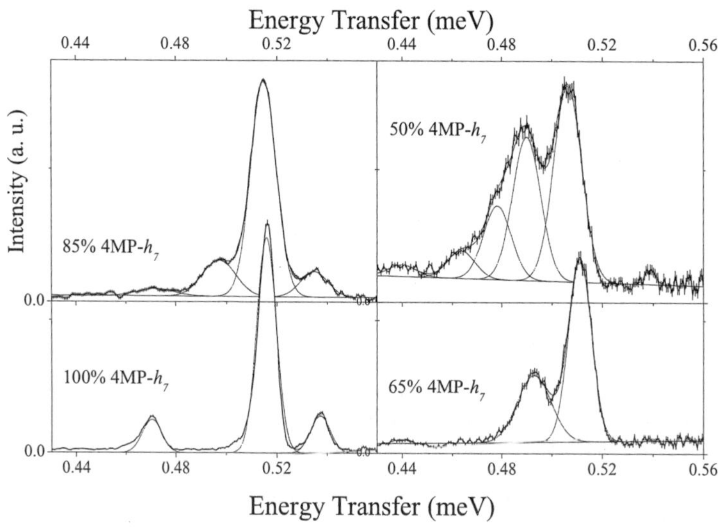

Figure 13. Inelastic neutron scattering spectra of isotopic mixtures of 4MP$-h_7$ and 4MP$-d_7$. The temperature was 0.5 K for 100% 4MP-h7 and 1.8 K for the mixtures, after ref. 16.

For modest concentrations of deuterated molecules the INS spectra obtained with better resolution reveal a more complex behavior (see Figure 13). The previously unresolved broad bands in Figure 10 are now resolved into several components. The breather band at 0.517meV in pure 4MP$-h_7$ shifts downwards slightly as the concentration of deuterated molecules increases ($\approx$ 0.506meV for 50% of 4MP$-d_7$). The weak tunnelling bands disappear rapidly whilst new bands appear with increasing intensities with the concentration of deuterated species. These spectra are well explained with the quantum sine-Gordon theory.

At a low concentration, the deuterated molecules are largely isolated. For a single chain the mean size of hydrogenated clusters is then proportional to the inverse of the concentration of deuterated molecules: $\bar{s}_h \approx 2/c_d$. However, the pairs of indistinguishable molecules along the c axis give rise to a twofold degeneracy. Consequently, the mean cluster size for a chain is $\bar{s}_h \approx 4/c_d$. Expansion of Eq. (21) for $c_d \ll 1$ gives the mean frequency for the travelling transition of the breather:

$$\bar{\nu}_{01} \approx \frac{\hbar^2 \omega_c^2}{2^q E_{B,l,0}^2} \left(1 - \frac{c_d^2}{16}\right) \tag{23}$$

This equation is rather close to the observation (Figure 14 Left). It may

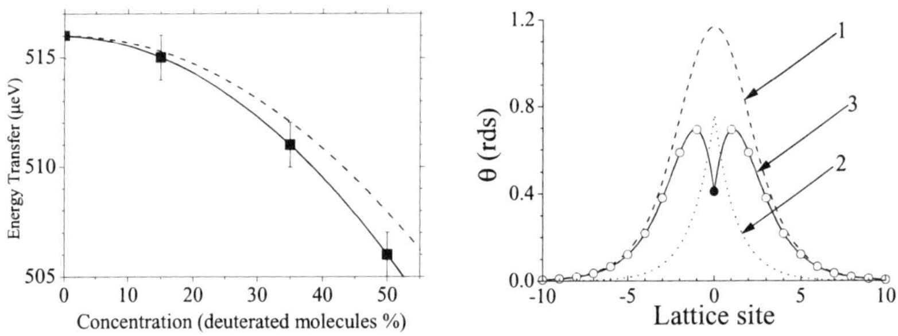

Figure 14. Left: variation of the main band frequency with the concentration of deuterated molecules. Data points: solid squares. Best fit: solid line with $\nu = 516\ 1 - 0.074c_d^2$. Dash line: theory according to Eq. (23). Right: wave form of the breather (1, dash), of the plasma wave (2, short dash) and of the breather (3, solid line) interacting with a single deuterated impurity (solid circle) surrounded by hydrogenated molecules (circles). after ref. 16.

be concluded that the observed frequency shift is essentially due to the collective nature of the dynamics. The twofold degeneracy confirms that the chain dynamics are essentially 1D in nature, with no significant contribution of the methyl-methyl interaction within the close-contact pairs. Dynamics along the a or b axis are totally uncorrelated and degenerate.

The band at ≈ 0.490meV in isotopic mixtures (see Figure 13) cannot be rationalized with the cluster size statistics presented above. The frequency corresponds roughly to that anticipated for clusters with s_h between 4 and 5. However, the intensity is virtually proportional to the amount of deuterated molecules whereas the amount of small clusters should be proportional to the square of the concentration. Therefore, this band was attributed to breathers trapped by isolated deuterated molecules.

The Hamiltonian representing an infinite chain of hydrogenated molecules containing a single deuterated molecule at the site $j = 0$ can be written as:[16]

$$H = \sum_j -\frac{\hbar^2}{2I_r}\frac{\partial^2}{\partial\theta_j^2} + \frac{V_0}{2}\left(1 - \cos 3i\theta_j\right) + \frac{V_c}{2}\left[1 - \cos 3i\left(\theta_{j+1} - \theta_j\right)\right] + \frac{\hbar^2}{4I_r}\frac{\partial^2}{\partial\theta_0^2} \tag{24}$$

The impurity is equivalent to a driving force at frequency $\omega_B/2\pi$, below the lowest phonon frequency at $\omega_0/2\pi$. A solution of the approximate Hamiltonian linearized with respect to the potential minimum is the standing plasma-wave centered at site $j = 0$ whose amplitude decreases exponentially on both sides (see Figure 14).[51] The waveform of a breather centered at x_B interacting with the impurity is :

$$\Theta\left(x,t\right) = \Phi_B\left(x - x_B, t\right) - \Phi_B\left(-x_B, t\right)\frac{\exp\left(-\sqrt{1 - \omega_B^2/\omega_0^2}\,|x|\right)}{2\sqrt{1 - \omega_B^2/\omega_0^2}} \tag{25}$$

The waveform is represented for $x_B = 0$ in Figure 14. The breather is attracted by the impurity and the effective potential averaged over the fast internal oscillation is:[51]

$$U_{eff}\left(x\right) = -4F\cot\mu\cosh\left(x\sin\mu\right)\left[1 + \cot^2\mu\cosh^2\left(x\sin\mu\right)\right]^{-3/2} \tag{26}$$

with $\mu = \frac{(3i)^2}{16\left[1-(3i)^2/8\pi\right]}$. The shape of the effective potential resembles the form of the breather-wave. The width at half minimum ($\Delta U_{eff} \approx$ 4–5 lattice sites) gives the amplitude of the breather center-of-mass oscillations around the impurity. Thus, the travelling transition is anticipated between 0.485 and 0.500meV,[12] in accord with the observation. The band at 0.490meV is thus an experimental confirmation that the width of the breather wave is $\approx$ 4–5 lattice sites. The effective potential tends to zero exponentially as x goes to infinity and impurities separated by more than $\approx$ 5 lattice sites are virtually isolated.

According to the energy at half minimum of the effective potential ($\approx$ 0.5 meV/6 K), the breather is trapped only at a low temperature.[12] The frequency of the travelling state increases with temperature, up to the frequency of the unperturbed breather.

4.3. *Discussion*

To the best of our knowledge the 4-methylpyridine crystal is a unique example of sine-Gordon dynamics in the quantum regime. The quantum sine-Gordon theory for infinite chains of coupled rotors accounts for a long list of experiments (four transitions –INS and Raman–, isotope mixtures, temperature effects, partial deuteration...) with only two parameters V_0 and V_c. Experiments have confirmed theoretical predictions for the quantization of the breather internal energy[40,41] or interaction with local impurities.[51] Some observations are directly related to the width of the breather waveform and can be regarded as direct evidence of the existence of this soliton. Conversely, experiments emphasize the dimensionless-pseudoparticle/planar-wave character of the breather mode and quantization of the travelling mode is a natural consequence of the Bragg rule. The crystal lattice "tolerates" the breather mode, although an exact solution only in the continuous limit, but, owing to discreteness, it "restricts" the travelling velocity and kinetic energy to a series of discrete extended states, *via* diffraction. This has been overlooked in theoretical works.[54] Measurements currently in progress confirm the anisotropy anticipated for particles travelling along chains parallel to the a and b crystal axes. The discrete spectrum in momentum space is markedly different from a phonon band structure and the main consequence of nonlinearity turns out to be localization in momentum rather than in position.

Alternative theoretical approaches proposed during the last decade (all of them have not been published yet) were not able to fit the data with reasonable accuracy and/or with such a limited number of parameters.[36,37,38,52,53,54,55]

Voll[38] has extended the coupled pair of Clough[7] to a quantum mechanical treatment of loops containing many coupled rotors, with the motivation of extrapolating to the infinite chain limit. He concluded that the tunnelling spectra should be hardly distinguishable from that of the single rotor but the interpretation in terms of single rotor potential would be erroneous. This conclusion is quite pertinent for 4-methylpyridine. However, calculations are so cumbersome that they were limited to the maximum number of four coupled rotors in a loop. Further developments are necessary in order to achieve reasonable modelling of the 4-methylpyridine crystal.

Neumann[36] has calculated potential terms in the 4-methylpyridine crystal with quantum chemistry methods. He has considered coupling between four nearest-neighbor methyl-groups. In addition, he introduced precession-rotation, as Schiebel and co-workers[19] did for lithiumacetate, which amounts to consider coupling with phonons. Finally, the complex potential function for methyl rotation depends on a very large number of parameters (approximately 8, but this was not clearly stated) in addition to rescaling the rotational constant, as a consequence of precession. With so many parameters the model is not unique, and the contact with real physics is questionable. The calculated spectrum of 4MP$-h_7$ looks very nice but the continuous frequency shift in isotope mixtures is not accounted for.

The ambiguity of the term "breather" in nonlinear sciences has been a source of more or less artificial/semantical controversies. Breathers are often regarded as manifestations of intrinsic energy localization in nonlinear lattices upon energy transfer to sufficiently high excited states.[56,54] In the classical regime, sine-Gordon breathers can be regarded as nonlinear excitations with a continuum of internal amplitude to which energy can be transferred. However, the internal frequencies of the breathers are always markedly below the roton density-of-states. In the quantum regime, the sine-Gordon breather is totally different and may appear rather paradoxical, as usually the case in the quantal word. The continuum of internal energy turns into a discrete spectrum depending on a specific quantum number. Moreover, in the particular case of threefold periodicity for the on-site potential there is only one mass state and this is a part of the ground state. This breather cannot be annihilated or created and it cannot be regarded any more as an "excitation". Although this property of the quantum sine-Gordon dynamics was emphasized by Dashen and co-workers,[40] this view has been contended by MacKay[55] who claimed that this is misunderstanding theoretical works. In order to avoid any further polemical discussions, it is necessary to reconsider the Lagrangian as formally written by Dashen and co-workers:[40]

$$\mathcal{L} = \frac{1}{2}\left(\partial_\mu \phi^2\right) + \frac{m^4}{\lambda}\left[\cos\left(\frac{\sqrt{\lambda}}{m}\phi\right) - 1\right]. \tag{27}$$

The mass states for the breather are $M_n = (16m/\gamma')\sin(n\gamma'/16)$ with $\gamma' = \left(\lambda/m^2\right)\left[1 - \lambda/(8\pi m^2)\right]^{-1}$. Coleman[42] has emphasized the meaning of $\sqrt{\lambda}/m$ in the quantization of the sine-Gordon Hamiltonian. In classical mechanics, changing this factor amounts to multiplying the Hamiltonian by a constant and this has no effect other than redefining the energy-scale. In the quantum regime, the energy scale is determined by $\hbar\lambda/m^2$ and a change of $\sqrt{\lambda}/m$ is equivalent to rescaling the Planck's constant h. Therefore, the periodicity of the on-site potential for methyl rotors imposes $\sqrt{\lambda}/m = 3$. This is a non-adjustable value and all parameters in Eq. (14) were determined accordingly.[12] (Much the same, the choice of symmetry adapted angular coordinates for coupled rotors, in section 3.5, must be compatible with the potential periodicity.) The alternative value of 1.46 proposed by MacKay is obviously irrelevant for methyl rotation. This author also questioned the quantization of the travelling mode, arguing that this does not apply to electrons in metals. It is quite difficult to follow this line of reasoning. Is it possible to ignore that electrons, and other particles, can be diffracted by crystal lattices according to the Bragg law? Even more surprising, the same author suggested that the disappearance of the breather transition on a time-scale of ≈ 70 hours[14] supports the decay of the breather mode itself at low temperature. However, this is a dramatic confusion between up and down scattering! The band that disappears with time corresponds to the $|1\rangle \rightarrow |0\rangle$ transition from the excited to the ground state of the travelling mode. This is merely the decay of thermally populated travelling states upon cooling down the sample. Simultaneously, the band corresponding to the $|0\rangle \rightarrow |1\rangle$ transition is still observed, in accordance with the existence of breathers as parts of the ground state. All these unfounded criticisms reveal a rather superficial examination of experimental data and a lack of realism in applying the quantum sine-Gordon theory to the 4-methylpyridine crystal.

In spite of spectacular successes, the theory proposed for infinite chains of coupled rotors was built with different pieces that may not fit to each other. On the one hand the breather mode is an analytical solution of the sine-Gordon Hamiltonian in Eq. (14). On the other, tunnelling was represented with extended states forming a band structure, according to Eq. (13). However, the tunnelling transitions disappear slowly but undoubtedly with time[14] and this is in conflict with the band structure. These transitions should arise from unstable or metastable species that were not included in the original theory. It is now timely to pursue theoretical models to account for these observations.

5. Conclusions

Tunnelling spectroscopy is unique to observing quantum nonlinear dynamics in crystals and advanced neutron diffraction techniques provide graphic views of the angular probability densities. In addition to these techniques, temperature effects and selective deuteration are necessary to fully determine rotational dynamics. This is a prerequisite to establishing realistic theoretical models. Comparison of methyl rotation in various systems emphasizes the interplay of structures and dynamics and prompt us to elicit existing theories.

In the manganesediacetate tetrahydrate crystal, methyl groups are in three different environments, their rotational axes have different orientations and they are well separated from each other. Methyl groups can be regarded as isolated single rotors.

In the lithiumacetate dihydrate, all methyl groups are equivalent. Close-contact pairs of face-to-face methyl groups with their axis parallel to b

are distributed in (a, b) planes in a nearly hexagonal structure with rather short methyl-methyl distances. The crystal symmetry allows the face-to-face methyl groups to be twisted by 60° and to perform combined rotation. The dynamics of centrosymmetric close contact pairs is well represented with symmetry adapted coordinates. As each methyl group of a pair is surrounded by methyl groups from different pairs their is no collective rotation of the pairs. Methyl-methyl interactions between pairs give rise to an effective potential that depend on the mean angular amplitude of the rotor. For CH_3 rotors, the coupling is rather weak and methyl groups are disordered at any accessible temperature. Only in the deuterated derivative coupling between pairs is strong enough to impose the ordering of the methyl group orientation at low temperature. This is a remarkable example of interplay between dynamics and structure. The symmetry imposes combined rotation of pairs whilst the rotor mass imposes ordering/disordering of the rotors and thus different crystal structures.

In the 4-methylpyridine crystal the distances between methyl groups are quite similar to those in lithiumacetate. However, in the tetragonal structure, there is a conflict between the C_2 local symmetry and the periodicity of the rotors. The face-to-face methyl groups must be twisted by 90° to each other and they cannot perform combined rotation with periodicity less than twelvefold. The corresponding effective potential is virtually a constant. Methyl groups form infinite chains parallel to a or b quite isolated from each other. The collective rotation in 1D is represented with the quantum sine-Gordon theory. There is no phase transition upon methyl deuteration.

References

1. M. Prager and A. Heidemann, Rotational Tunnelling and Neutron Spectroscopy: A Compilation (1995).
2. J. R. Durig, S. M. Craven, and W. C. Harris, *Vibrational Spectra and Structure*, vol. 1 p.73 (Marcel Dekker, New York, 1972).
3. W. Press and A. Kollmar, Solid State Commun. **17**, 405 (1975).
4. W. Press, *Single particle rotation motion in molecular crystals, Springer tracts in modern physics*, vol. 92 (Springer, Berlin, 1981).
5. S. Clough, A. Heidemann, and M. N. J. Paley, J. Phys. C: Solid State Phys. **13**, 4009 (1980).
6. A. Heidemann, H. Fredrich, E. Günther, and W. Häusler, Z. Phys. B **76**, 335 (1989).
7. S. Clough, A. Heidemann, A. H. Horsewill, and M. N. J. Paley, Z. Phys. B **55**, 1 (1984).
8. A. Heidemann, K. J. Abed, C. J. Barker, and S. Clough, Z. Phys. B **11**, 355 (1987).

9. J. Dorignac and S. Flach, Phys. Rev. B **65**, 214305 (2002).
10. L. Soulard, F. Fillaux, G. Braathen, N. L. Calvé, and B. Pasquier, Chem. Phys. Letters **125**, 41 (1986).
11. N. LeCalvé, B. Pasquier, G. Braathen, L. Soulard, and F. Fillaux, J. Phys. C: Solid State Phys. **19**, 6695 (1986).
12. F. Fillaux and C. J. Carlile, Phys. Rev. B **42**, 5990 (1990).
13. F. Fillaux, C. J. Carlile, and G. J. Kearley, Phys. Rev. B **44**, 12280 (1991).
14. F. Fillaux, C. J. Carlile, J. Cook, A. Heidemann, G. J. Kearley, S. Ikeda, and A. Inaba, Physica B **213&214**, 646 (1995).
15. F. Fillaux, G. J. Kearley, and C. J. Carlile, Physica B **226**, 241 (1996).
16. F. Fillaux, C. J. Carlile, and G. J. Kearley, Phys. Rev. B **58**, 11416 (1998).
17. A.-J. Dianoux and G. Lander, *Neutron Data Booklet* (ILL neutrons for science, 2002).
18. M. Neumann and M. R. Johnson, Chem. Phys. **215**, 253 (1997).
19. P. Schiebel, G. J. Kearley, and M. R. Johnson, J. Chem. Phys. **108**, 2375 (1998).
20. J. D. Lewis, T. B. MalloyJr, T. H. Chao, and J. Laane, J. Mol. Structure **12**, 472 (1972).
21. B. Nicolaï, G. J. Kearley, A. Cousson, W. Paulus, F. Fillaux, F. Gentner, L. Schröder, and D. Watkin, Acta Cryst. B **57**, 3644 (2001).
22. A. Heidemann, S. Clough, P. J. McDonald, A. J. Horsewill, and K. Neumaier, Z. Phys. B **58**, 141 (1985).
23. J. N. van Niekerk and F. R. L. Schoening, Acta Cryst. **6**, 227 (1953).
24. G. M. Brown and R. Chidambaram, Acta Cryst. B **29**, 2393 (1973).
25. E. F. Bertaut, D. T. Qui, P. Burlet, and J. M. Moreau, Acta Cryst. B **30**, 2234 (1974).
26. B. Nicolaï, A. Cousson, and F. Fillaux, unpublished.
27. J. L. Galigné, M. Mouvet, and J. Falgueirettes, Acta Cryst. B **26**, 368 (1970).
28. G. J. Kearley, B. Nicolaï, P. G. Radaelli, and F. Fillaux, Solid State Chem. **126**, 184 (1996).
29. P. S. Allen and P. Branson, J. Phys. C: Solid State Phys. **11**, L121 (1978).
30. B. Alefeld, A. Kollmar, and B. A. Dasannacharya, J. Chem. Phys. **63**, 4415 (1975).
31. K. J. Abed, S. Clough, C. J. Carlile, B. Rosi, and R. C. Ward, Chem. Phys. Letters **141**, 215 (1987).
32. N. L. Calvé, D. Cavagnat, and F. Fillaux, Chem. Phys. Letters **146**, 549 (1988).
33. E. K. Morris, Ph.D. thesis, Université d'Orsay (1997).
34. U. Ohms, H. Guth, W. Treutmann, H. Dannohl, A. Schweig, and G. Heger, J. Chem. Phys. **83**, 273 (1985).
35. C. J. Carlile, B. T. M. Willis, R. M. Ibberson, and F. Fillaux, Z. Kristallogr. **193**, 243 (1990).
36. M. A. Neumann, Ph.D. thesis, Université Joseph Fourier, Grenoble, France (1999).
37. C. J. Carlile, S. Clough, A. J. Horsewill, and A. Smith, Chem. Phys. **134**, 437 (1989).
38. G. Voll, Z. Phys. B **90**, 455 (1993).

39. A. Scott, F.Chu, and D. McLaughlin, IEEE **61**, 1443 (1973).
40. R. Dashen, B. Hasslacher, and A. Neveu, Phys. Rev. D **10**, 4130 (1974).
41. R. Dashen, B. Hasslacher, and A. Neveu, Phys. Rev. D **11**, 3424 (1975).
42. S. Coleman, Phs. Rev. D **11**, 2088 (1975).
43. M. J. Rice, A. R. Bishop, J. A. Krumhansl, and S. E. Trullinger, Phys. Rev. Letters **36**, 432 (1976).
44. R. Jackiw, Rev. Modern Phys. **49**, 681 (1977).
45. M. B. Fogel, S. E. Trullinger, A. R. Bishop, and J. A. Krumhansl, Phys. Rev. B **15**, 1578 (1977).
46. J. F. Currie, S. E. Trullinger, A. R. Bishop, and J. A. Krumhansl, Phys. Rev. B **15**, 5567 (1977).
47. S. E. Trullinger, Solid State Commun. **29**, 27 (1979).
48. E. Stoll, T. Schneider, and A. R. Bishop, Phys. Rev. Letters **42**, 937 (1979).
49. J. F. Currie, J. A. Krumhansl, A. R. Bishop, and S. E. Trullinger, Phys. Rev. B **22**, 477 (1980).
50. R. Rajaraman, *Solitons and instantons. An introduction to solitons and instantons in quantum field theory* (North-Holland, Amsterdam, 1989).
51. Y. S. Kivshar and B. A. Malomed, Rev. Modern Phys. **61**, 763 (1989).
52. J. A. D. Wattis, Phd, Mathematics Department, Heriot-Watt University (1993).
53. J. A. D. Wattis, Physica D **82**, 333 (1995).
54. A. Scott, *Nonlinear Science. Emergence & Dynamics of Coherent Structures* (Oxford University Press, 1999).
55. R. S. MacKay, *in Nonlinear dynamics and chaos: where do we go from here? Eds J. Hogan, A. Champneys, B. Krauskopf, M. di Bernardo, M. Homer, E. Wilson and H. Osinga* (IOOP, 2002), chap. Many-body quantum mechanics, pp. 100–134.
56. R. S. MacKay and S. Aubry, Nonlinearity **7**, 1623 (1994).

ENERGY BARRIERS IN COUPLED OSCILLATORS: FROM DISCRETE KINKS TO DISCRETE BREATHERS

J-A SEPULCHRE
Institut Non Linéaire de Nice
1361 route des Lucioles, 06560 Valbonne, FRANCE
E-mail: jacques-alexandre.sepulchre@inln.cnrs.fr

This tutorial paper aims to review some elementary concepts of the theory of "localisation by nonlinearity and discreteness, and energy transfer" in coupled oscillators. For didactic purpose, this work focusses on the concept of energy barriers in coupled pendula, rather than on the related and more general theme of "effective dynamics" in coupled oscillators, although an introduction to the latter is necessary to understand the former. Further studies concerning the material of these notes are found mainly in refs.[1,9,26,28].

1. Introduction

The "Peierls-Nabarro barrier" (PN) is an important concept for spatially discrete systems. It is discussed or used in several papers about discrete breathers (e.g. [1,6,10,12,14,16,19,22,28] for a non exhaustive list). However it is not a trivial concept, especially for newcomers to the field of discrete breathers. This might be reinforced by the fact that in the last years there have been several papers which tackled this subject but sometimes with diverging points of view. In the case of the lattice nonlinear Schrodinger equations, a clear definition of the PN barrier was early demonstrated in[14,22], by using the conserved quantity associated with the complex-phase symmetry of this model. For more general systems, however, it was suggested that there should be only a weak analogy between the notion of PN barrier developed in the theory of kinks and that which should concern DBs [6,10,19]. In fact, since 1993 R. Mackay has proposed what should be the appropriate way to define the PN barrier for DB, but his idea was not worked out until relatively recently [1,28]. Still, several papers, including some which were presented at the El Escorial conference, use different definitions of PN barriers, but with no rigorous justification. Therefore the purpose of this paper is to explain again this concept in a didactic way and to place it under the theme of energy barriers in coupled oscillators. To be pedagogical, this topic, as well as basic concepts of the theory like that of discrete breather, will be introduced with a single (section 2), or a pair (section 4), or a chain of pendula (section 3 and 5). Section 2 and 3 review results related to stationary states, or equilibria of coupled oscillators, whereas section 4 and 5 show how these results can be extended to periodic solutions. There is a picture which will recur in this paper, and which is the following: an energy barrier is associated with a separatrix belonging to critical points of some effective potential. In fact, the knowledge of the effective potential, or better, of the effective dynamics, gives a more comprehensive view of the dynamics than just energy barriers. However the latter are simpler to figure out, and it will be illustrated in the last section that the simple knowledge of an energy barrier may already be useful to make some prediction about the dynamics.

2. Energy barrier in a single oscillator

In this section we briefly review concepts referring to the dynamics of a single conservative oscillator. A good example is the pendulum.

2.1. *A single pendulum*

A plane pendulum is a rigid body formed by a mass m attached to a rod of length l which can rotate without friction around a fixed axis. The motion of the pendulum can be described by a single degree of freedom, e.g. the angle θ between the rod and the vertical axis. Neglecting the mass of the rod, the Hamiltonian of a single pendulum submitted to gravity g can be written as[a]:

$$H(p,\theta) = \frac{p^2}{2ml^2} + mgl(1-\cos\theta) \tag{1}$$

The pendulum has equilibria, $\theta = 0$ or $\theta = \pm\pi$, which are critical points of $H(p,\theta)$, i.e. $\nabla H = 0$ at these points. The stability properties of these equilibria can be studied by computing the eigenvalues of the Hessian matrix D^2H at these points. An alternative way is to look at the curves $H(p,\theta) = E$ in the (p,θ) phase plane (Figure 1). These curves are also described by the equations:

$$p^2 = A(E) + B\cos\theta \tag{2}$$

[a]In these notes the Hamiltonian formalism will be used extensively. For readers who are not familiar with it, a brief description of this point of view, as well as of some of the motivations to use it, is recalled in the appendix.

with $A(E) = 2ml^2(E - mgl)$ and $B = 2m^2gl^3$. Depending on the value of E, three cases occur. The two first cases describe periodic motions of the pendulum.

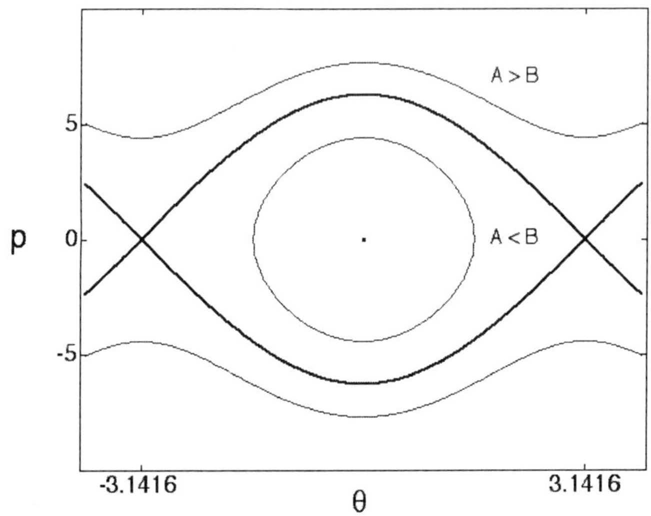

Figure 1. Phase portrait of a single pendulum, $m = 1$, $l = 1$ (see text).

- $|A| < B$: this is the vibration (called also the libration). In this case p^2 becomes zero for some angles $\pm\theta_m = \pm arcos(A/B)$ inside $[-\pi, \pi]$. At these points p changes sign and so does $\dot{\theta}$. The phase curves in this case are closed, encircling the equilibrium $\theta_m = 0$ corresponding to the limiting case $A = -B$. So $\theta_m = 0$ is a stable equilibrium.
- $|A| > B$: this is the rotation. Suppose $A > 0$, the other case is analoguous. Then, as $p(\theta) > 0$, θ keeps increasing in time. In the limiting case $A/B \gg 1$, the speed of the rotation is quasi-constant.

The third case is somehow physically unrealistic, but is the most important to understand the overall pendulum dynamics.

2.2. *The separatrix*

The case $A = B$ is in fact degenerate. It comprises the equilibria ($p = 0, \theta = \pm\pi$) (corresponding obviously to the same state for the pendulum). It comprises also the curve $p_+(\theta) = A(1 + \cos\theta)$, $-\pi < \theta < \pi$ (and also the symmetric curve $p_-(\theta) = -p_+(\theta)$). This trajectory is called a *separatrix*, as it separates the two regimes "vibration" and "rotation". It is also called a *heteroclinic* trajectory, linking asymptotically the unstable equilibria $p = 0, \theta = \pm\pi$. Contrary to the previous cases, this motion is not periodic.

Physically, a separatrix is associated with an energy barrier. For the pendulum this is computed as the difference in energy between the unstable and the stable equilibria, that is $\Delta E = 2mgl$. Indeed, it can be interpreted as the minimum (or infimum) energy to be transfered to a pendulum at rest to make it rotate. This concept will be generalised for coupled pendula, and more generally for coupled oscillators, in the next sections. Prior to this, the following paragraph discusses another important feature of the dynamics of nonlinear oscillators.

2.3. *Anharmonicity*

Conservative oscillators exhibit periodic solutions in families, which can be parametrised in general by the conserved quantity, e.g. the energy. A property of nonlinear oscillators which has deep consequences in nonlinear phenomena is that the period (or the frequency) of their periodic motions varies with the energy. This property should be called non-isochronicity, but it is commonly named *anharmonicity*. To describe this point in a simple way, it is convenient to consider the weakly nonlinear regime, where the pendulum Hamiltonian is approximated by its limited Taylor expansion:

$$H(p,\theta) \simeq \frac{p^2}{2ml^2} + mgl(\frac{\theta^2}{2} - \frac{\theta^4}{24}) \tag{3}$$

Then it turns out that it is quite usefull to change variables and to represent (p, θ) by a complex number:

$$\psi = \frac{1}{\sqrt{2}}(\frac{p}{\alpha} + i\alpha\theta) \tag{4}$$

with $\alpha = (m^2gl^3)^{1/4}$. The change of coordinates $(p,\theta) \to (i\overline{\psi}, \psi)$ can be easily checked to be canonical (see Appendix). So the new Hamiltonian is obtained by substitution of (4) into eq. (3), which readily gives:

$$H(i\overline{\psi}, \psi) = \omega_0|\psi|^2 + \frac{\chi}{12}(\psi - \overline{\psi})^4 \tag{5}$$

with $\omega_0 = \sqrt{g/l}$ and $\chi = -\frac{1}{8}(1/ml^2)$. The equation of motion for ψ is simply $i\dot{\psi} = \frac{\partial H}{\partial \overline{\psi}}$.

In fact, by a succession of changes of variables, H can be put in the form[3,25]:

$$H(i\overline{\psi}, \psi) = \sum_{n=0}^{\infty} c_n |\psi|^{2n} \tag{6}$$

which means that H can be expressed as a function of $|\psi|^2$ only, say

$$H = h(|\psi|^2) \tag{7}$$

This can be proved in several ways[b] and the final result is called the Birkhoff normal form of H. Moreover, if one is interested only up to the quartic order ($n = 2$) of this form, there is no need to compute any change of variables ! It suffices to drop out the terms which are not of the form $|\psi|^{2n}$ in (5), which gives:

$$H(i\overline{\psi}, \psi) = \omega_0 |\psi|^2 + \frac{\chi}{2} |\psi|^4 \tag{8}$$

Therefore the dynamics for ψ becomes:

$$i\dot{\psi} = \omega_0 \psi + \chi |\psi|^2 \psi \tag{9}$$

which generalises to $i\dot{\zeta} = h'(|\zeta|^2)\zeta$ for the full Birkhoff normal form. Equation (9) admits readily the solution

$$\psi(t) = \psi_0 e^{i(\omega_0 + \chi \psi_0^2)t} \tag{10}$$

because it is easily proved that $|\psi_0|^2$ is a conserved quantity (check $\frac{d}{dt}(\psi\overline{\psi}) = 0$). Consequently, the frequency of $\psi(t)$ depends on its amplitude, or equivalently it depends on the initial energy, as was claimed before.

When χ is negative, as for the pendulum, the frequency decreases with amplitude and the oscillator is called 'soft', whereas it is called 'hard' when $\chi > 0$ with frequency increasing with the amplitude (or with the energy).

Let us end this section by introducing the well-known action-angle variables. In this context, it is simply written by expressing ψ in polar coordinates:

$$\psi = \sqrt{I} e^{i\phi} \tag{11}$$

---

[b]e.g. by averaging. For the first change of variables, start with $\psi = Ae^{-i\omega_0 t}$, assuming the evolution of A is slow compared with the period $2\pi/\omega_0$ and then consider the averaged Hamiltonian. Other ways to proceed are multiple time-scale asymptotic series, Lie transforms, etc...[25,33]

It is easy to show that the new variables are canonical. Then the Hamiltonian becomes simply $H = h(I)$. This is the simplest way to describe a single oscillator. Notice however that these coordinates are not appropriate when $I = 0$, as ϕ becomes undetermined in (11).

3. Peierls-Naborro potential in a chain of oscillators

In the previous section the dynamics of a single pendulum was reviewed. The important conclusion was that the energy difference between the stable and the unstable equilibria can be interpreted as an energy barrier to be overcome to make the pendulum rotate. In this section this kind of energy barrier is generalised to equilibria of a chain of identical pendula linearly coupled. The latter is modeled by the following Hamiltonian:

$$H = \sum_{n\in\mathbb{Z}} \left[\frac{p_n^2}{2ml^2} + mgl(1 - \cos\theta_n) + \frac{\gamma}{2}(\theta_{n+1} - \theta_n)^2 \right] \tag{12}$$

Before analysing some of the equilibria of H, let us motivate further the study of this model. Indeed, whereas it has obvious limitations compared with realistic physical systems (e.g. no damping, only one-dimensional, identical units, etc...) nevertheless this model is very rich and has opened many paths in condensed matter physics. We recall some of its basic features. First, from the theoretical point of view, this chain of pendula is related to the sine-Gordon equation (SG):

$$\frac{\partial^2 u}{\partial \tau^2} + \sin u = \frac{\partial^2 u}{\partial x^2} \tag{13}$$

Indeed, when the space is discretised ($x = na$, $u(na) = \theta_n$) and by an appropriate scaling of time and space ($\tau = \omega_0^2 t$, $a^2 = mgl/\gamma$), the SG equation is seen as a continuum limit of the dynamics deduced from (12). Note that there are other continuum approximations of the chain of pendula, which take into account dispersion effects[11]. Now the interesting property of SG is that it is integrable, and possesses a lot of soliton solutions[36], which is usefull to analyse some behaviors of the chain (12), although the latter is not an integrable system.

A second reason which makes the chain of pendula a nice model to study is that is is a particular case of the Frenkel-Kontorova model[11c]. This model has played a leading role in the (low-dimensional) condensed matter physics

[c]See also contributions on the FK model in this volume

and is defined by the following Hamiltonian:

$$H = \sum_{n\in\mathbb{Z}} \left[\frac{y_n^2}{2} + (1 - \cos(\frac{2\pi x_n}{b}) + \frac{\lambda}{2}(x_{n+1} - x_n - a)^2 \right] \tag{14}$$

Here x_n represents the position of the nth (unit) mass which is submitted to a substrate periodic potential of period b and to first-neighbour interaction with equilibrium distance a. Then the chain of pendula is recovered by choosing $a = b = 2\pi$ and changing variables to $x_n = na + \theta_n$ (and with an appropriate scaling of momentum and energy). Notice that the case $a \neq b$ and in particular a/b irrational is also very interesting, but nothing of this theory will be considered here[4].

To cite only a few domains where the FK model applies, mention the theory of dislocations (first historically[23]) and the study of domain walls in ferroelectrics or in ferromagnetism. More recently, and closely related to topics of this volume, similar models have been used in biomolecules (e.g. DNA dynamics) and in the Josephson Junction arrays (dissipation is taken into account). For a recent review of the Frenkel-Kontorova model and some of its applications, see ref [11].

In these applications, the concept of *kink* is central. This object can be defined simply in the continuum case. A kink is a stationary solution $K(x)$ of the SG equation which obeys the boundary conditions: $K(x \to -\infty) = 2\pi$, $K(x \to \infty) = 0$. An anti-kink is defined with the reverse boundary condition. So function $K(x)$ satisfies:

$$\sin K(x) = K''(x) \tag{15}$$

It can be checked that a solution of this problem is given by $K(x) = 4\,\mathrm{atan}\,(e^{-x})$. In fact, as eq. (15) describes an inverted pendulum, $K(x)$ is nothing but the (π-shifted θ component of the) separatrix depicted on Fig. 1.

Let us note, however, that the notion of kink extends to other equations than Sine-Gordon. For general Klein-Gordon equations (replace the function sin by U') it exists as soon as the corresponding potential admits at least two stable equilibria. Notice also that if $K(x)$ is a kink, so is $K(x-x_0)$ and thus there is a one-parameter family of kinks in space-continuous systems. This is no longer true in the discrete case, e.g. for a weakly coupled chain of pendula. Here, the equation of the stationary states writes:

$$\sin\theta_n = \lambda(\theta_{n+1} + \theta_{n-1} - 2\theta_n) \tag{16}$$

The solutions of this equation thus stationarise the energy (12) , $\nabla H = 0$, but only the stable equilibria are minima of H. A discrete kink is a stationary state of H, such that θ_n is decreasing between the boundary conditions $\theta_n \to 2\pi$ if $n \to -\infty$ and $\theta \to 0$ if $n \to \infty$. In the framework of the FK model, it corresponds to a locally compressed state. An antikink is a local expansion and has the reverse boundary conditions. In fact, there can be many solutions which satisfy these conditions. To analyse this problem the usual method is to look at (16) as a time-discrete dynamical system defined by the map:

$$\begin{aligned} \rho_{n+1} &= \theta_n \\ \theta_{n+1} &= 2\theta_n + \frac{1}{\lambda}\sin\theta_n - \rho_n \end{aligned} \tag{17}$$

kinks are then seen as orbits $(\rho_n, \theta_n)_{k\in\mathbb{Z}}$ which asymptoticaly connect $(2\pi, 2\pi)$ to $(0, 0)$, i.e. heteroclinics orbits. (Antikinks exist due to a "time"-reversing symmetry of this system). The theory of system (17) -called the *standard map*- is well documented, see e.g. [5]. Here let us retain only that there are two types of kinks. The first type, denoted in what follows by $\theta_n^{(0)}$, is an unstable configuration whose centroid is sitting exactly at one particle; the second type, denoted by $\theta_n^{(\frac{1}{2})}$ is a stable kink, thus a configuration of minimal energy. One shows that its centroid sits in the middle of two particles (see Fig. 2).

What about traveling kinks ? In the continuum case, these are easily shown to exist, by mean of the so-called Lorenz boost[11,36]. In the discrete case this property is absent and exact traveling kinks do not seem to exist[11,36,30]. However transient traveling kinks are easily observed in numerical simulations. This theory has already a long history, but again let us recall the main points in order to make links with traveling discrete breathers. The two types of kinks mentioned above have not the same energy since $\theta_n^{(\frac{1}{2})}$ is a state of minimum energy, and $\theta_n^{(0)}$ is not. Define the Peierls-Nabarro barrier E_{PN} as their difference in energy . The goal of the next paragraph is to interpret E_{PN} as the minimum energy necessary to make a stable discrete kink traveling, at least transiently. There is an analogy with the energy barrier for the pendulum discussed above, and this can be made explicit by considering a collective coordinate description of the kinks, as seen in next section.

3.1. *The Peierls-Nabarro potential for kinks*

The idea of the collective coordinate method is to make a good change of variables in order to highlight the most relevant degrees of freedom of the dynamics (w.r.t some given problem). There are several ways to proceed

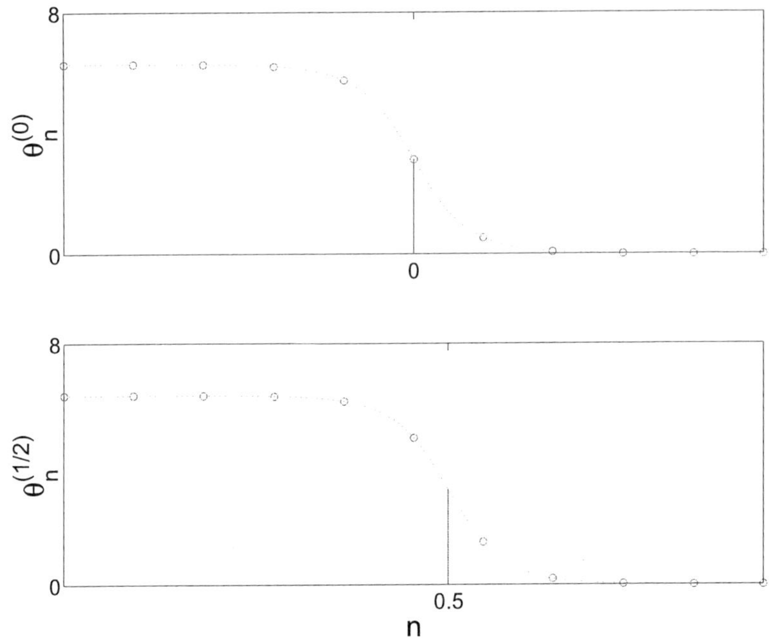

Figure 2. Representation of the two discrete kinks $\theta_n^{(0)}$ and $\theta_n^{(\frac{1}{2})}$. (see text).

and several levels of refinement. Perhaps the most sophisticated level is performed by Willis et al.[38] because they do not eliminitate any DOF in their approach. Here we consider succintly a rougher theory which is sufficient for the connection with discrete breathers later on. The scheme[d] is to consider an approximate family of discrete kinks interpolating between the two exact kinks $\theta_n^{(0)}$ and $\theta_n^{(\frac{1}{2})}$. Consider for example:

$$\theta_n^{(Q)} = K(n-Q) + \kappa_n(Q) \tag{18}$$

[d] This scheme can be motivated by MacKay's theory of Hamilonian slow manifolds[26]. Consider a manifold of quasi-stationary states of some Hamiltonian systems, assuming that it is quasi-invariant. Then a Hamiltonian slow dynamics confined near this so-called slow manifold can be constructed.

where $\kappa_n(Q)$ is a correction which takes into account the difference between the discrete sampling of the continuum kink and the exact discrete kink[e]. The continuum kink is said to be "dressed" by the interaction with the lattice and it turns out that its shape is stiffer than in the continuum case[38].

An interpretation of the PN barrier can already be given in the simplest case where $\kappa_n(Q)$ is neglected. So assume now:

$$\theta_n^{(Q)} = K(n-Q) \tag{19}$$

and define the momentum associated to Q by $P = M(Q)\dot{Q}$, where $M(Q)$ is defined by considering the kinetic energy:

$$\frac{1}{2}\sum_n\left(\frac{d\theta_n^{(Q)}}{dt}\right)^2 = \frac{1}{2}\sum_n K'(n-Q)^2\dot{Q}^2 \equiv \frac{1}{2}M(Q)\dot{Q}^2 \tag{20}$$

Next an effective Hamiltonian for P and Q can be written by substitution of (19) in (12), which gives:

$$H(P,Q) = \frac{P^2}{2M(Q)} + V_{\text{PN}}(Q) \tag{21}$$

where $V_{\text{PN}}(Q)$ is called the Peierls-Nabarro potential. The latter is written explicitly as:

$$V_{\text{PN}}(Q) = \sum_n \Big[mgl(1-\cos K(n-Q)) + \frac{\gamma}{2}(K(n+1-Q) - K(n-Q))^2\Big] \tag{22}$$

which is a periodic function of Q by construction, $V_{\text{PN}}(Q+1) = V_{\text{PN}}(Q)$. Its Fourier series can be analytically estimated in the weakly discrete case by using Poisson summation formulae (e.g. appendix of [38] or [28]) and typically it can be truncated as its coefficients decreases exponentially:

$$V_{\text{PN}}(Q) \simeq B_0 + B_1\cos(2\pi Q) \tag{23}$$

Likewise, $M(Q) \simeq M_0 + M_1\cos(2\pi Q)$. In the case of the SG kink one can compute explicitly M_0, M_1, B_0 and B_1 [11]. No need to reproduce these results here but let us note that these coefficients are positive. Therefore the effective dynamics for Q derived from H can be written as:

$$M(Q)\ddot{Q} = -\frac{M'(Q)}{2}\dot{Q}^2 + 2\pi B_1\sin(2\pi Q) \tag{24}$$

---

[e]E.g., the linear interpolation: $\kappa_n(Q) = 2[(\theta_n^{(0)} - K(n-Q))(\frac{1}{2}-Q) + (\theta_n^{(\frac{1}{2})} - K(n-Q))Q]$ for Q mod $\frac{1}{2}$. This could be improved with a nonlinear interpolation.

Neglecting $M'(Q)$, which can be justified given the values of A_1 and B_1, the collective coordinate Q of the kink obeys an effective (inverted) pendulum dynamics. Then the phase portrait discussed in Fig. 1) can be revisited in the following suggestive way: The closed trajectories are interpreted as kinks whose mean position oscillates around stable equilibria (half-integer values of Q). These vibrating kinks are said to be *pinned* or trapped by the PN potential. On the other hand the curves referring to the "rotation" of the pendulum are interpreted as traveling kinks, since it describes an unbounded motion for Q ($|\dot{Q}| > 0$). Intermediate to these motions is the separatrix which is associated with the PN barrier. The latter is thus a *depinning* energy, i.e. the minimum energy to turn the pinned mode into a traveling state. This can be computed as the difference in energy between the unstable and the stable kinks. This energy barrier can be estimated as $2B_1$. This estimate has been refined by several authors, either in the weak or in the strong discrete case[11]. The best estimates are obtained by considering the framework of the heteroclinic orbits of the standard map as mentioned earlier.

Finally, let us note that the actual dynamics of the traveling kinks is more complicated. Numerical simulations show that traveling kinks lose energy by radiating phonons, therefore they slow down and finally reach a pinned mode. This process called *radiative damping*, can be described accurately by a thorough study of the linear stability of the kink configurations[21].

4. Energy barriers in a pair of pendula

In this section, we step back from a chain of pendula to a pair of coupled pendula. The goal is to introduce other types of energy barriers which exist when the pendula are no longer in equilibria but in oscillating states.

4.1. *Periodic solutions of two coupled oscillators*

This section recalls properties concerning periodic solutions of coupled oscillators. Consider a system formed by two identical pendula, linearly coupled. The Hamiltonian can be written as following:

$$H = \frac{1}{2ml^2}(p_1^2 + p_2^2) + mgl((1 - \cos\theta_1) + (1 - \cos\theta_2)) + \frac{\gamma}{2}(\theta_1 - \theta_2)^2 \quad (25)$$

As for the single pendulum, the analysis is greatly simplified by considering the truncated Birkhoff normal form applying to the weak amplitude case. It will be argued in next section, however, that the present discussion extends to large amplitude (or strongly nonlinear) case.

So, assuming the Taylor expansion of $\cos\theta$ in (25), and using the complex variables introduced in (4), the Hamiltonian becomes:

$$H = \omega_0(|\psi_1|^2+|\psi_2|^2)+\frac{\chi}{12}((\psi_1-\overline{\psi}_1)^4+(\psi_2-\overline{\psi}_2)^4)-\frac{\gamma}{4\alpha^2}(\psi_1-\overline{\psi}_1-\psi_2+\overline{\psi}_2)^2 \tag{26}$$

The lowest order Birkhoff normal form consists to keep only terms of the form $|\psi_1|^{2k}|\psi_2|^{2l}$, with integers $l+k \leq 2$ [3,25]:

$$H = \tilde{\omega}_0(|\psi_1|^2+|\psi_2|^2)+\frac{\chi}{2}(|\psi_1|^4+|\psi_2|^4)-\tilde{\gamma}\mathrm{Re}\,(\psi_1\overline{\psi}_2) \tag{27}$$

with $\tilde{\omega}_0 = \omega_0(1+\frac{\gamma}{2mgl})$ and $\tilde{\gamma} = \frac{\gamma}{\alpha^2}$, and α has been defined above in (4). Notice that in order that the last term be of the same order than the $O(4)$ term, $\tilde{\gamma}$ should be $O(2)$, so the coupling is weak.

4.2. *The 2-oscillator sphere*

Recall that $|\psi|^2$ was conserved for the single oscillator. Here, for the same reason,

$$A = |\psi_1|^2+|\psi_2|^2 \tag{28}$$

is a conserved quantity[f]. This fact is a source of great simplification. Indeed, the original phase space $\mathbb{C}^2$ containing all the (ψ_1,ψ_2) can be reduced not only by fixing A, which makes a S^3 sphere in $\mathbb{C}^2$, but also by identifying solutions which differs only by a global phase $(\tilde{\psi_1},\tilde{\psi_2}) \sim e^{i\Phi}(\psi_1,\psi_2)$. This reduction is standard in the theory of symmetric Hamiltonian systems[29], and the result can be expressed like $S^3/S^1 \simeq \mathbb{C}P^1 \simeq S^2$. As this reduction is really useful to understand some global dynamics of two weakly coupled oscillators, let us explain how to find a parametrisation of the resulting S^2 sphere without assuming any knowledge of the reader concerning this symmetry reduction.

The goal is to represent a state of two weakly coupled oscillators (up to a global phase) by a unique point on a sphere S^2. The latter will be named the 2-*oscillator sphere*. First show that (ψ_1,ψ_2) can be represented by only one complex number $z=\psi_1/\psi_2$ if $\psi_2 \neq 0$. Indeed, given z, one can define:

$$(\tilde{\psi_1},\tilde{\psi_2}) = (\frac{z}{\sqrt{1+z^2}}\sqrt{A},\frac{1}{\sqrt{1+z^2}}\sqrt{A}) \tag{29}$$

---

[f] H is invariant under the global phase shift $\psi \to e^{i\phi}\psi$, which is a continuous symmetry parametrised by ϕ, thus giving a conserved quantity by Noether theorem, which is found to be A.

which satisfies eq. (28), and such that $\tilde{\psi_1}/\tilde{\psi_2} = z$. So, eq. (29) is equivalent to (ψ_1, ψ_2) up to a complex phase. If $\psi_2 = 0$, the state $(\psi_1, 0)$ is represented equivalently by $(\sqrt{A}, 0)$, and the latter is thought as the point $z = \infty$. Now consider the Riemann sphere on which the extended complex plane $\mathbb{C}\cup\{\infty\}$ can be mapped using the projection depicted on Fig. 3. Then, using action-angle coordinates by writing ψ_n $(n = 1, 2\ )$ in polar coordinates like in eq. (11), the mapping of $z = \sqrt{I_1/I_2}e^{i(\phi_1-\phi_2)}$ on this Riemann sphere gives a point with the following cylindrical coordinates:

$$\zeta = \frac{1}{2}\frac{I_1 - I_2}{I_1 + I_2}$$
$$\phi = \phi_1 - \phi_2 \tag{30}$$

The latter equation defines the relative phase ϕ. Define also the relative

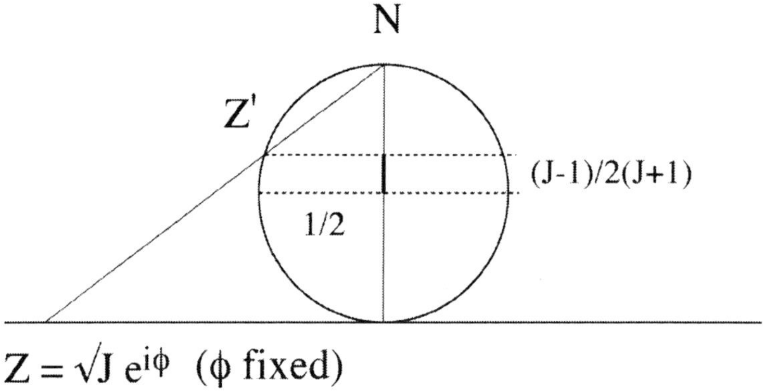

Figure 3. All $z = \sqrt{J}e^{i\phi}$ in $\mathbb{C}$ can be mapped on a sphere of diameter 1 -the *Riemann sphere*- such that the projected point z' has cylindrical coordinates $(\frac{1}{2}\frac{J-1}{J+1}, \phi)$. A point at infinity in $\mathbb{C}$ is mapped on the north pole N, $(\frac{1}{2}, \phi$ undetermined) .

variable $I = (I_1 - I_2)/2$, and so $\zeta = I/A$. In conclusion, the state (ψ_1, ψ_2) of a 2-oscillator system with $A \neq 0$ can be represented on a sphere by mean of the relative coordinates (30). This was already stated without demonstration in[9]. This representation is quite usefull, especially to visualise the following states of the 2-oscillator system: the North and South poles describe situations where one of both oscillators is at rest, respectively $(I_1 \neq 0, I_2 = 0)$ and $(I_1 = 0, I_2 \neq 0)$. The East pole characterises the in-phase dynamics, $(I_1 = I_2, \phi = 0)$, whereas the West pole describes the anti-phase dynamics ($I_1 = I_2, \phi = \pi$). The dynamics associated with these points is sketched in[1] and it is reviewed in the next paragraph.

4.3. *Dynamics of relative variables*

First consider the following change of variables $(I_1, \phi_1, I_2, \phi_2) \rightarrow (A, \Phi, I, \phi)$:

$$\begin{aligned} A &= I_1 + I_2 \\ \Phi &= \frac{\phi_1 + \phi_2}{2} \\ I &= \frac{I_1 - I_2}{2} \\ \phi &= \phi_1 - \phi_2 \end{aligned} \tag{31}$$

It is straightforward to write the Hamiltonian in new coordinates:

$$H = \tilde{\omega}_0 A + \chi\left(\frac{A^2}{4} + I^2\right) - \tilde{\gamma}\sqrt{\frac{A^2}{4} - I^2}\cos\phi \tag{32}$$

Hence, the dynamics of the relative variables is deduced as follows:

$$\begin{aligned} \dot{\phi} &= I\left(2\chi + \tilde{\gamma}\frac{\cos\phi}{\sqrt{\frac{A^2}{4} - I^2}}\right) \\ \dot{I} &= -\tilde{\gamma}\sqrt{\frac{A^2}{4} - I^2}\sin\phi \end{aligned} \tag{33}$$

The dynamics of the global variables could also be written, leading to $A = const$, $\Phi = \Omega t + F(I(t), \phi(t))$, where $\Omega = \tilde{\omega}_0 + \chi A/2$ and F is some functional which is not written explicitly. Therefore the equilibria of the relative dynamics, ($\dot{\phi} = 0, \dot{I} = 0$) correspond to periodic dynamics of the coupled oscillators.

In the linear case, $\chi = 0$, the only stationary states are ($I = 0, \phi = 0$ or π) , corresponding to the well-known linear modes, i.e. respectively the in-phase and the anti-phase periodic oscillations. Since there are no other equilibria, the remaining dynamics consist of closed circles going around the W and the E poles, as represented on Fig. 4(a). The circle passing through the N and S poles deserves some attention. In fact the parametrisation of this motion is singular, as seen in eqs. (33) with $\phi = \pi/2$ or $3\pi/2$. Nevertheless one shows that this trajectory on the sphere describes a motion where one oscillator transfers all of its energy to the other, back and forth in time. For weak coupling, this is nothing but the *beating* phenomenon which is better represented in normal mode coordinates.

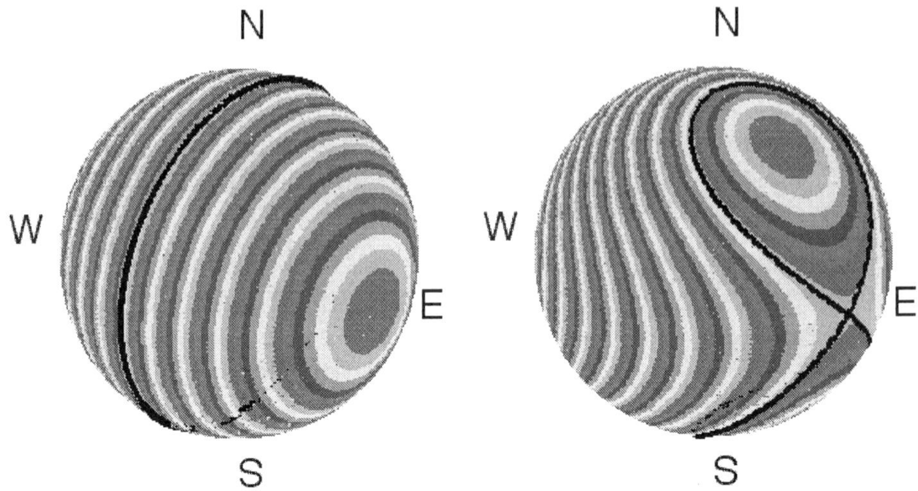

(a) (b)

Figure 4. Energy levels given by eq. (32) are represented on the 2-oscillator sphere in grey levels, for $A = 2$, $\tilde{\omega}_0 = 0.5$ and for different values of $\tilde{\gamma}$ and χ (See also Fig. 5). (a) $\tilde{\gamma} = 1$ and $\chi = 0$. This is the linear dynamics. The "beating" trajectory is represented in black. (b) $\tilde{\gamma} = 1$ and $\chi = -1$. Nonlinear dynamics with two local modes. The separatrix is the black line.

When the nonlinearity χ is turned on, the stationary points at E and W poles continue to exist (for any χ indeed). However, in view of eqs. (33) with $\phi = 0$ or π , new equilibria can appear. It is not difficult to work out that it occurs when the following condition is fulfilled:

$$\frac{\tilde{\gamma}}{A|\chi|} < 1 \tag{34}$$

In this case, two new stationary (i.e. periodic) solutions are created, near $\phi = 0$ if $\chi < 0$ (soft oscillators) , or near $\phi = \pi$, if $\chi > 0$ (hard oscillators). Stability analysis can be performed, but we give only the results which should be clear when looking at the 2-oscillator sphere. In the first case $\chi < 0$ (e.g. for two pendula), the in-phase oscillations becomes unstable after the bifurcation, giving rise to two stable solutions (see Fig. 4(b)). The case $\chi > 0$ is analoguous, but the anti-phase oscillations become unstable. The physical interpretation of this pitchfork bifurcation is very interesting: when condition (34) is fulfilled, e.g. when the coupling v.s. amplitude is small, two new periodic solutions emerge, which are characterised by the fact that the amplitude of one oscillator is larger than the other one. This is called a *local mode.* This local mode bifurcation has applications in various contexts, e.g. in chemistry[31], or more recently in the study of Bose-Einstein condensates[15]. Interestingly the latter work analyses clearly the onset of chaotic dynamics in this system when a little dissipation is added.

In fact the concept of local mode is nothing but a precursor of that of *discrete breathers* (spatially localised and time-periodic oscillations in network of oscillators) which is amply discussed in this volume. From this point of view, proving existence of discrete breathers (DB) amounts to show that the concept of local mode holds uniformly with respect to the size of an extended networks of oscillators. Moreover the formation of DB can be induced in a similar way as for local modes. Indeed, the in-phase (or anti-phase) instability of two oscillators which is ruled by (34) generalises for a chain of oscillators. Then it is called "modulational instability". For instance the in-phase oscillations of a chain of pendula become unstable by decreasing the coupling v.s. amplitude ratio. Beyond this bifurcation the dynamics exhibits self-localisation of the amplitude (named sometimes "auto-focalisation" in the context of the nonlinear Schrodinger equation). This mechanism can be seen as a first step towards the creation of DB [17,18,24].

Consider now the limit of zero coupling, $\gamma = 0$. In this limit, the local modes coincide with the N and S poles of the 2-oscillator sphere (see Fig. 5(c)). By contrast to the linear case (Fig. 4(a)), here the phase portrait is completed by horizontal circles. Moreover the equator ($I = 0$) corresponds to a continuous family of periodic solutions, where the amplitude of the two oscillators are the same, but the relative phase is a constant ϕ . As soon as the coupling γ is turned on, this degenerate situation disappears, as explained e.g. in[1], and at least two periodic solutions subsist for $\gamma \neq 0$. These are the in-phase and the anti-phase modes, one being stable and the other being unstable depending on the sign of χ. Therefore we recover a picture similar to the pendulum phase portrait, namely two stationary points of different stability connected by a separatrix, but now on a sphere (see Fig. 5(d)).

This picture can be submitted to the interpretation of energy barrier. To fix the ideas, suppose $\chi < 0$ as for the pendulum (the case $\chi > 0$ is similar). Figure 5(d) shows that the anti-phase state behaves like a trapped mode, with a relatively small island of stability. Increasing the relative momentum I allows one to cross the separatrix so that the new trajectory alternates

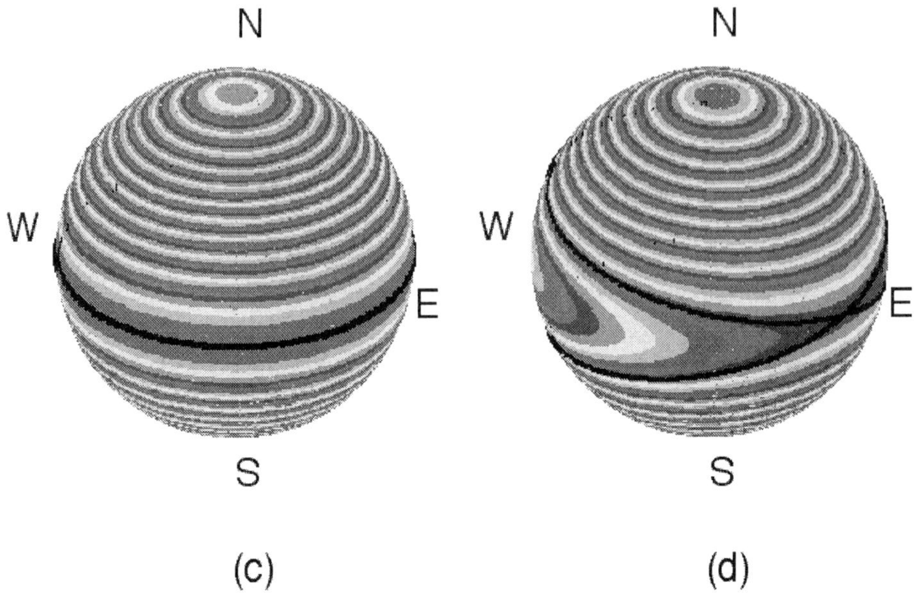

Figure 5. (Continuation of Fig. 4). (c) $\tilde{\gamma} = 0$ and $\chi = -1$. This is the uncoupled dynamics. The equator in black line represents degenerate periodic solutions. (d) $\tilde{\gamma} = 0.1$ and $\chi = -1$. Weak coupling. The separatrix is the black line.

between states resembling the two normal modes (i.e. the trajectory passes near the E and W poles on the 2-oscillator sphere). This can be considered as a precursor of a moving discrete breather. In this case it is natural to define the Peierls-Nabarro barrier as the energy difference between the in-phase and the anti-phase periodic solutions on this sphere. Notice also that in this situation there is an energy exchange *between normal modes*, by contrast with the linear case where energy transfer between local modes exist, as recalled above. On the other hand, by increasing the coupling from this situation (Fig. 5(d)), one retrieves the phase portrait of Fig. 5(b). Here one could argue, however, that the pinned modes are the local modes, with a relatively small region of stability delimited by the separatrix of the unstable in-phase mode. In this case the PN barrier should be defined as the energy variation between the in-phase and one of the local mode. Then crossing this separatrix gives a motion which alternates between both local modes. This might be called "nonlinear beating", which can also be considered as another precursor of moving DB.

In conclusion, we see that projecting the phase space of the two identical coupled pendula (oscillators) on the 2-oscillator sphere enables one to discuss various physical phenomena, at least in the weakly nonlinear regime and for weak coupling. In fact, the 2-oscillator sphere applies to more cases. First, it is straightforward, and quite interesting, to extend this study to two coupled non-identical oscillators. Indeed, in this case the Birkhoff-truncated Hamiltonian (27) generalises to:

$$H = \omega_1|\psi_1|^2 + \omega_2|\psi_2|^2 + \frac{\chi_1}{2}|\psi_1|^4 + \frac{\chi_2}{2}|\psi_2|^4 - \tilde{\gamma}\mathrm{Re}\,(\psi_1\overline{\psi_2}) \qquad (35)$$

The basic property that $|\psi_1|^2 + |\psi_2|^2$ is constant is still valid and thus the symmery reduction works equally well. Analysis of this system has been performed in[9]. The main result of this paper is that, contrary to the previous case, if the oscillators are different, then the motions for which complete energy transfer between the two oscillators is possible (i.e. a trajectory passing through the N and S poles of the 2-oscillator sphere) are restricted to selective values of A (and other conditions which are not reported here). In particular, it is found in ref.[9] that a necessary condition for it to happen is

$$A = -2\frac{\omega_1 - \omega_2}{\chi_1 - \chi_2} \qquad (36)$$

This fine-tuning phenomenon has been named Targeted Energy Transfer (TET) by these authors. This concept is interesting as it allows one to think of new mechanisms to transfer selectively energy in inhomogeneous systems (which is the typical case in natural systems). It can also be used as a theoretical model for selective electron transfer[8]. Further studies about this topic are exposed in this volume and in[26].

These results extend also for other reasons which are explained in next section.

4.4. *Parametrising periodic solutions with their area*

In the previous section energy barriers between different dynamical states of two coupled oscillators were described in the framework of Hamiltonian (27), namely the truncated Birkhoff normal form of two coupled oscillators. In this section we discuss to which extent these results still hold when this framework is abandoned.

Firstly, it should be clear that the concept of local modes still holds for any pair of weakly coupled oscillators (e.g. two pendula with arbitrary amplitude) for the following reasons. The simplest physical argument to understand existence of local modes is anharmonicity: if the amplitudes of the two pendula are quite different, for instance one large and one small, then the detuning in frequency prevents energy exchanges since the system is out of resonance. The energy of the large amplitude pendulum is self-trapped and a local mode is created. This intuitive reason can be supplied by another argument, which as a matter of fact is one of the key ingredients which lead to the first proof of existence of DB[27]. Consider, in the zero coupling limit, one pendulum at rest and one pendulum oscillating with a large amplitude. This configuration defines actually a family of non-degenerate periodic solutions, parametrised for example by their frequencies. This family can be continued to non-zero coupling by using the implicit function theorem in a space of periodic functions. By continuity of the amplitude, the continued family is a set of local modes, i.e. the amplitude of the oscillator which was initially at rest remains small compared with the other oscillator. Another set of local modes is obtained by permuting the two oscillators.

Therefore, for arbitrary nonlinearity and weak coupling there exist two families of local modes. The latter can be parametrised by their frequencies, or by their amplitudes, or by their energies, etc... In view of what has been discussed in the previous section, there should be also a family of unstable periodic solutions whose separatrices determine boundaries for the stability regions of the two local modes. Therefore, once again the picture of the energy barrier sets in. Local modes can be interpreted like pinned modes; increasing their "momentum" allows one to cross the separatrix and to set them into motion, obtaining a dynamical state with energy exchange between the two local modes. The energy barrier should be calculated as the energy difference between the unstable mode and one of the (stable) local mode. However, now a difficulty arises. This energy barrier seems arbitrary since it depends on the parametrisation chosen for the family of local modes. The energy barrier even seems to drop to zero when the parametrisation is the energy itself ! Should it mean that the energy needed to "depin" a local mode is not well defined ? Can it be reduced arbitrarily, for example by an appropriate choice of the initial phase of the periodic solution which is trapped ? Such considerations have been proposed in the litterature. This difficulty is wiped out, however, by considering MacKay's idea to use *area* of the periodic solutions as a good parametrisation, in order to define energy barriers, and more generally to design some effective dynamics[1,26,28]. (The notion of "area" of a periodic solution is recalled in the Appendix. For a single DOF Hamiltonian, it is just $\int pdq$ over one period, i.e. 2π times the action.)

Considering sets of loops with constant area is quite a general idea, and its application goes much beyond the problems introduced in this paper. Let us give a general argument for it, and then come back to the application of two weakly coupled oscillators.

The key idea is twofold. A first point is that comparing periodic solutions with same area amounts to compare, in a sense, equilibria of some Hamiltonian. And for equilibria, e.g. in the case of kinks reviewed in section 3, the comparison of energies follows a clear procedure. Stationary states are found by solving $\nabla_z H = 0$. If this equation has one solution, say z_1, then the corresponding energy is $E_1 = H(z_1)$. So, proceeding likewise for other solutions, energy differences are defined univoquely. Moreover, as it was briefly explained in the case of kinks and in footnote (d), an effective potential (or even an effective Hamiltonian dynamics) can be derived, in principle, given a family of stationary states which form a quasi-invariant manifold. Now, to convert to equilibria the case of periodic solutions of Hamiltonian systems, the trick is to think of the variational formulation of the dynamics (cf. appendix). A periodic solution $z = (p, q)$ of H has the property to extremalize the functional:

$$W = \frac{1}{T}\int_0^T (H(p,q)dt - pdq) \tag{37}$$

defined on a space of loops (periodic functions). So, if the area is kept fixed,

$$\int_0^T pdq = a \tag{38}$$

then periodic solutions are critical points of the following Hamiltonian:

$$\mathsf{H}(z) = \frac{1}{T}\int_0^T H(z(t))dt \tag{39}$$

which can be defined on the same space of loops. Given a loop, this Hamiltonian is interpreted as its averaged energy. (One readily checks that in this formulation the frequency $1/T$ comes out as the Lagrange multiplier linked to the constraint (38)). In this way, the periodic solutions naturally becomes "stationary states" of some Hamiltonian, and thereby energy barriers are easily defined.

The second point of the idea is that if area is used as a "collective coordinate" to describe the evolution of a set of loops (e.g. approximate DB) then its time evolution is trivial, at least in first approximation: it is constant in time. This can be shown simply, but this fact is also related to a well-known result in Hamiltonian dynamics, that the area is an adiabatic invariant. So its motion is much slower than the evolution of other variables.

Now, how this idea helps to extend results discussed in the previous section ? Consider two weakly coupled nonlinear oscillators, and describe each one by mean of its (non-truncated) Birkhoff normal form. The Hamiltonian of the coupled system is thus written as:

$$H = h(|\psi_1|^2) + h(|\psi_2|^2) + \epsilon K(\psi_1, \psi_2, \overline{\psi}_1, \overline{\psi}_2) \tag{40}$$

The function h has been introduced in eq. (7). The coupling function K is multiplied by ϵ to recall that it is weak.

Consider first $\epsilon = 0$, and assume a periodic motion in (40), so $\psi_n(t) = \sqrt{I_n} e^{-i(\omega t + \phi_n)}$ $(n = 1, 2\ )$. Next compute the area:

$$\begin{aligned} a &= \int_0^{\frac{2\pi}{\omega}} (i\overline{\psi}_1 \dot{\psi}_1 + i\overline{\psi}_2 \dot{\psi}_2) dt \\ &= \omega \int_0^{\frac{2\pi}{\omega}} (|\psi_1|^2 + |\psi_2|^2) dt \\ &= 2\pi(|\psi_1|^2 + |\psi_2|^2) \end{aligned} \tag{41}$$

Therefore in this case there is a perfect agreement between the approach based on invariance of $|\psi_1|^2 + |\psi_2|^2$ (section 4), and the method of parametrising periodic solutions with area.

When $\epsilon \neq 0$, it can be shown that $a = 2\pi < |\psi_1|^2 + |\psi_2|^2 > +O(\epsilon^2)$, where $< . >$ denotes the average value along one period. In fact, related to the adiabatic invariance of a mentioned above, it can be also shown that $|\psi_1|^2 + |\psi_2|^2$ varies slowly $O(\epsilon^2)$ over one period. Therefore projecting the dynamics of two weakly coupled oscillators on the 2-oscillator sphere gives a good approximation, even for strong nonlinearity. The closer one starts to a periodic motion, the better the approximation. Then this approximation is valid over a time interval which is quite long compared with the period of the single oscillator. It would be interesting to work out an accurate estimate of this time interval, but to our knowledge it is an open problem.

5. Peierls-Nabarro potential for discrete breathers

The concepts and the methods reviewed in the previous sections can be applied to a network of oscillators. The case of two weakly coupled oscillators can be generalised by considering the dynamics of the relative phases of N excited oscillators sitting in a large lattice of oscillators (the non-excited oscillators having small amplitudes and being non-resonant with the excited ones). This problem is dealt with, at least partially, in[1], by constructing an effective Hamiltonian. This method can be used also to build a collective coordinate approach of approximate traveling DB, and to define the concept of PN barrier for DB. This approach has been developed in[28] and it will be presented quite succintly in this section.

Let us rewrite the Hamiltonian of a chain of pendula in a slightly more general form than (12):

$$H = \sum_{n \in \mathbb{Z}} \left[\frac{p_n^2}{2} + V(x_n) + \frac{\gamma}{2}(x_{n+1} - x_n)^2 \right] \tag{42}$$

As for the case of kinks discussed in section 3 , cf. eq. (18), the starting point is to assume a family of approximate DB of the form

$$x_n^{(A,Q)} = u(A, n - Q, \omega t) \tag{43}$$

interpolating between exact DB which are known to exist, for example for $Q = m$ and for $Q = m + \frac{1}{2}$ (describing respectively a stable site-centred DB, and an unstable bond-centred DB). Here parameter A indexes the amplitude of DB. (Note that ω is not a free parameter, but it typically depends on the other parameters (A, Q)) Next a momentum coordinate k is associated to Q. A possibility which works in some cases is to consider[28,35]:

$$x_n^{(A,Q,k)} = u(A, n - Q, \omega t - kn) \tag{44}$$

Now comes the requirement which has been motivated in the preceding section: The area of this family of DB should be fixed to a constant value. The latter is obtained by calculating the following expression:

$$a = \int_0^T (\sum_n p_n^{(A,Q,k)} \dot{x}_n^{(A,Q,k)}) dt \tag{45}$$

with $T = 2\pi/\omega$. In the ideal case one should obtain from (45) that a is proportional to, or related to A by a simple functional relation. If this is not the case, then the parametrisation of (44) has to be redefined in terms of (a, Q, k).

Then an effective Hamiltonian can be computed as the averaged energy for each members of the family of DB:

$$H_a^{eff}(k, Q) = \frac{1}{T} \int_0^T H(\{x_n^{(a,Q,k)}, p_n^{(a,Q,k)}\}) dt \tag{46}$$

Therefore if the effective Hamiltonian can be decomposed as

$$H_a^{eff}(k, Q) = E_{cin}(k) + V_{\text{PN}}(Q) \tag{47}$$

then $V_{\text{PN}}(Q)$ defines the PN potential for this family of DB[g]. Typically the situation is analogous to what is obtained for kinks: $V_{\text{PN}}(Q)$ is integer periodic, and the difference between its extrema defines the PN barrier.

This scheme has been applied to some definite examples[28,35] which are not reproduced here, except for Fig. 6 which illustrates some results. On the other hand the following application shows that the above theory can be helpful in analysing moving D B -in particular using the concept of PN barrier- even in a situation where the knowledge of the approximate family (43) is absent.

We consider the Fermi-Pasta-Ulam chain (FPU), with the following Hamiltonian:

$$H = \sum_n \left(\frac{p_n^2}{2} + V(x_{n+1} - x_n)\right) \tag{48}$$

with $V(x) = \alpha\frac{x^2}{2} + \beta\frac{x^4}{4}$. Discrete breathers have shown up in numerical simulations of the FPU system more than a decade ago[37,34], but proof of their existence (as exact periodic solutions) was performed only recently[20,7]. Two types of DB were found in numerics and they were approximated by caricaturing their shapes as $u_n^P = A(\cdots, 0, -\frac{\sqrt{3}}{2}, \frac{\sqrt{3}}{2}, 0, \cdots)$ and $u_n^{ST} = A(\cdots, 0, -\frac{1}{2}, 1, \frac{1}{2}, 0, \cdots)$, known respectively as the Page mode (P), which is stable, and the Sievers-Takeno mode (ST), which is unstable. This approximation is shown to be good for large $A^2\alpha/\beta$.

Traveling DB were also observed in numerics of FPU, e.g. in[34], most easily by perturbing the unstable P-mode. On the other hand, it would be interesting to test our study of the PN barrier on this case and create a moving DB by depinning the stable ST-mode.

Let us estimate the PN barrier between the ST-mode and the P-mode. Each of these DB modes can be represented by the function

$$x_n = A\, u_n f(\omega t) \tag{49}$$

where u_n is the spatial motif, A is the amplitude (whose ω depends) and $f(\omega t)$ accounts for the time evolution. A good approximation is a Jacobi elliptic function but this will not be important. The area of the DB is then estimated as follows:

$$a = 2\pi\omega A^2 < (f')^2 > (\sum_n u_n^2) \tag{50}$$

[g]To deduce an effective dynamics from this effective Hamiltonian may require a bit more work, since k and Q may not be canonically conjugated. This problem is treated in[28].

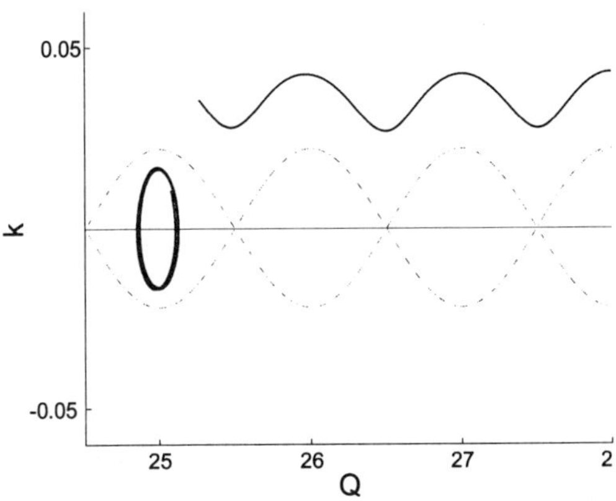

Figure 6. Illustration of the PN barrier for discrete breathers. In this example there is a good agreement between the full dynamics (solid lines) and the effective dynamics (characterised only by its separatrix, in dashed lines) of the Salerno model[32]. The latter is in fact a 1-parameter family of models interpolating between the Ablowitz-Ladik model and the Discrete Nonlinear Schrodinger system (which is a variant of Hamiltonian (27) for a chain of oscillators.). This figure is reproduced from[28].

In the present approximation, $\sum_n u_n^2 = \frac{3}{2}$ for both the P and the ST modes. Therefore they have same area if

$$\omega_P(A_P)A_P^2 = \omega_{ST}(A_{ST})A_{ST}^2 \tag{51}$$

Assuming that the frequency functions are the same, $\omega_P(A) = \omega_{ST}(A)$ implies that $A_P = A_{ST}$ is needed to get the same area. Thus the frequencies are also equal. Let us denote the period by T. Then the PN barrier can be computed as the energy difference between

$$\frac{1}{T}\int_0^T H(x^{(ST)}) = \frac{a}{2T} + A^2(\frac{5}{2}\alpha + A^2\frac{82}{32}\beta) \tag{52}$$

and

$$\frac{1}{T}\int_0^T H(x^{(P)}) = \frac{a}{2T} + A^2(\frac{9}{4}\alpha + A^2\frac{81}{32}\beta) \tag{53}$$

which gives

$$\Delta E_{\text{PN}} = \frac{A^2}{8}(\alpha + A^2\frac{\beta}{8}) \tag{54}$$

Now let us perturb the stable P mode by giving it a kick of momentum $p(\cdots, 0, 1, -2, 1, 0, \cdots)$. This induces a variation in kinetic energy of $\Delta E_{cin} = 3p^2$. Matching ΔE_{PN} and ΔE_{cin} gives the minimum p to launch a DB: $p_m = \sqrt{\frac{\Delta E_{\text{PN}}}{3}}$.

This estimate can be tested with the parameter values of ref.[34], namely $\alpha = 0.25$, $\beta = 0.405$ and $A = 1$. It gives $p_m \approx 0.15$. This seems to be in good agreement with the numerics. We find numerically that $p_m = 0.14$ is enough to depin the ST mode, whereas $p_m = 0.13$ is not. Therefore our estimate makes sense, although it is an upper bound and a finer comparison between theory and numerics could be worth. Also, in this perspective, the PN barrier can be computed with more accuracy by improving the approximations of the spatial patterns u_n of the P and ST modes.

Finally we remark that this application to the FPU system, which consists in looking only at the energy difference between the stable and the unstable DB, cannot be applied blindly. For example the estimate (54) indicates that the PN barrier should decrease with α, which intuitively should promote the depinning. Paradoxally we observe (numerically) a failure of DB propagation in the limit $\alpha \to 0$. So other criteria than energy barrier are necessary to understand DB mobility. This should be reflected by studying the full effective dynamics of these DB.

Acknowledgments

I am grateful to R. MacKay for his fruitful suggestions and for helpful discussions about these topics. Stimulating discussions with S. Aubry are also acknowledged. Work supported in part by the European Community's Human Potential Programme under contract HPRN-CT-1999-00163, LOCNET.

Appendix

Many physical systems cannot be modeled by Hamiltonian systems. However in many cases, especially in mechanics, hydrodynamics, optics, etc.,

Hamiltonian formulation is a nice idealisation of physical systems, which can be considered as an ideal starting point. It gives strong constraints on its mathematical structure and this proves very usefull to solve problem.

Hamiltonian systems can be described at different levels of abstraction. The symplectic formulation is very usefull, but will not be explicitly used here. A good introduction to it can be found in the standard reference of Arnold[2]. Another way to describe Hamiltonian dynamics, which is quite general, is to start with a variational principle. (Again, ref.[2] is a perfect introduction). It says that $(p(t), q(t))$ is a solution of the Hamiltonian $H(p, q, t)$ if it stationarises the action defined by:

$$W(z) = \int_{t_0}^{t_f} (H(p, q, t)dt - pdq)$$

in the space of C^1 trajectories with fixed ends $\delta q(t_0) = \delta q(t_f) = 0$. Then, from the equation $\delta W = 0$ one deduces:

$$\begin{aligned} \dot{p} &= -\partial_q H \\ \dot{q} &= \partial_p H \end{aligned} \tag{55}$$

These are the well-known canonical equations. In fact, one of the advantages of the variational formulation is that it opens easily the way to non-canonical formulations of the Hamiltonian dynamics. This can be quite usefull, even for effective dynamics of discrete breathers[28]. The interested reader will find a nice and extensive development of the non-canonical formalism in[13]. In this paper we stick mainly with canonical coordinates. The advantage of a canonical change of coordinates is that the new Hamiltonian is obtained by simple substitution of the new variables in the old ones. Moreover the form of the equations of motion (55) are unchanged with the new variables. This property is used several times in this paper. To check that the transformation $(p, q) \to (u(p, q), v(p, q))$ preserves the canonical form, it suffices to show that $\sum_n du_n \wedge dv_n = \sum_n dp_n \wedge dq_n$. For instance, this is easily checked in the case of the change of variables given by (31), or in the one given by (4).

In Newtonian mechanics it can be easier to write the Lagrangian than the Hamiltonian: $L(x, \dot{x}) = E_{cin} - E_{pot}$, where E_{cin} is written in terms of the time-derivatives of the positions. Then the Hamiltonian is obtained by the function $H(p, x) = pv(x, p) - L(x, v(x, p))$ where $v(x, p)$ is obtained by inverting $p = \frac{\partial L}{\partial \dot{\theta}}(x, v)$. The latter is called a Legendre transform. For example for the pendulum, the Lagrangian is readily written by considering

$E_{cin} = m/2\,(ld\theta/dt)^2$, $E_{pot} = mgl(1-\cos\theta)$. Then the Legendre transform gives eq. (1).

References

1. T. Ahn, R.S. MacKay and J-A. Sepulchre, Dynamics of relative phases: generalised multibreathers, *Nonlinear Dynamics* **25** (2001) 157–182.
2. V.I. Arnold, *Mathematical Methods of Classical Mechanics*, Springer-Verlag, New-York-Heidelberg-Berlin, 1978.
3. V.I. Arnold, V.V. Kozlov and A.I. Neishtadt, *Mathematical Aspects of Classical and Celestial Mechanics*, Springer-Verlag, Berlin-Heidelberg, 1997.
4. S. Aubry, in *Solitons and Condensed Matter*, vol. 8 of *Solid State Sciences*, edited by A. Bishop and T. Schneider, (Springer, Berlin, 1978) p. 264.
5. S. Aubry, G. Abramovici, Chaotic trajectories in the standard map: the concept of anti-integrability, Physica D 43 (1990) 199–219.
6. S. Aubry, T. Cretegny, Mobility and reactivity of discrete breathers, *Physica D* **119** (1998) 34–46
7. S Aubry , G Kopidakis, V Kadelburg, Variational Proof for Hard Discrete Breathers in some classes of Hamiltonian Dynamical Systems, *Discrete and Continuous Dynamical Systems* B **1** (2001) 271-298
8. S. Aubry, G. Kopidakis, A nonlinear dynamical model for ultrafast catalytic transfer of electrons at zero temperature, submitted to *Int. J. Mod. Phys. B* (2002).
9. S. Aubry, G. Kopidakis, A.M. Morgante, G.P. Tsironis, Analytic conditions for Targeted Energy Transfer between nonlinear oscillators or discrete breathers, *Physica B* **296** (2001) 222–236.
10. O. Bang and M. Peyrard, Higher order breather solutions to a discrete Klein-Gordon model, *Physica D* **81** (1995) 9–22.
11. O.M. Braun and Y.S. Kivshar, Nonlinear dynamics of the Frenkel-Kontorova model, *Phys. Reports* **306** (1998) 1–108.
12. D. Cai, A. R. Bishop, and N. GronbechJensen, Localized states in discrete nonlinear Schrodinger equations, *Phys. Rev. Lett.* **72** (1994) 591-595.
13. J.R. Cary and R.G. Littlejohn, Noncanonical Hamiltonian Mechanics and its application to magnetic field line flow, *Annals of Physics*, **151** (1983) 1-34.
14. Ch. Claude, Yu.S. Kivshar, O. Kluth and K.H. Spatschek, Moving localized modes in nonlinear lattices, *Phys. Rev. B* **47** (1993) 14228–14232.
15. P. Coullet and N. Vandenberghe, Chaotic dynamics of a Bose-Einstein condensate in a double-well trap, *J. Phys. B: At. Mol. Opt. Phys.* **35** (2002) 1593-1612.
16. D. Chen, S. Aubry and G.P. Tsironis, Breather Mobility in Discrete ϕ^4 nonlinear lattices, *Phys. Rev. Lett.* **77** (1996) 4776–4779.
17. T. Cretegny, T. Dauxois, S. Ruffo and A. Torcini, Localization and equipartition of energy in the β-FPU chain: Chaotic breathers, *Physica D* **121** (1998) 109-126.
18. I. Daumont, T. Dauxois and M. Peyrard, Modulational instability: first step

towards energy localization in nonlinear lattices, *Nonlinearity* **10** (1997) 617-630.

19. S. Flach, C.R. Willis, Movability of localized excitations in nonlinear discrete systems, *Phys Rev Lett* **72** (1994) 1777.
20. G. James, Existence of breathers on FPU lattices, Comptes Rendus Acad Sci Paris, 332 (2001) 581-586.
21. P.G. Kevrekidis and M.I. Weinstein, Dynamics of lattice kinks, *Physica D* **142** (2000) 113–152.
22. Y.S. Kivshar and D.K. Campbell, Peierls-Nabarro potential barrier for highly localized nonlinear modes, *Phys. Rev. E* **48** (1993) 3077-3082.
23. T.A. Kontorova and Ya.I. Frenkel, *Zh. Eksp. Teor. Fiz.* **8** p.89, and **8** p.1340 (1938) (in russian).
24. Y.A. Kosevich and S. Lepri, Modulational instability and energy localization in anharmonic lattices at finite energy density, *Phys. Rev. B* **61** (2000) 299-307.
25. A.J. Lichtenberg and M.A. Lieberman, *Regular and Stochastic Motion*, AMS 38, Springer, NY-Heidelberg-Berlin, 1983.
26. R.S MacKay, Lecture Notes on Slow Manifolds, in preparation.
27. MacKay R S and Aubry S, Proof of existence of breathers for time-reversible or Hamiltonian networks of weakly coupled oscillators, *Nonlinearity* **7** (1994) 1623–1643.
28. R.S. MacKay and J-A Sepulchre, Effective Hamiltonian for traveling discrete breathers, *J.Phys. A: Math. and Gen.* **35** (2002) 3985–4002.
29. J.E. Marsden and T.R. Ratiu, *Introduction to Mechanics and Symmetry*, Springer-Verlag, NY, 1998.
30. M. Peyrard and M.D. Kruskal, kink dynamics in the highly discrete Sine-Gordon system, *Physica D* **14** (1984) 88–102.
31. M. Quack, Spectra and dynamics of coupled vibrations in polyatomic molecules, *Ann. Rev. Phys. Chem.* **41** (1990) 839–874.
32. M. Salerno, Quantum deformations of the discrete nonlinear Schrodinger equation, *Phys. Rev. A* 46 (1992) 6856-6859.
33. Sanders J A and Verhulst F, *Averaging Methods in Nonlinear Dynamical Systems*, (Springer, New York, 1985).
34. K.W. Sandusky, J.B. Page and K.E. Schmidt, Stability and motion of intrinsic localized modes in nonlinear periodic lattices, *Phys. Rev. B* **46** (1992) 6161–6168.
35. J-A Sepulchre, Theoretical analysis of traveling discrete breathers in the FPU chains, in preparation.
36. A. Scott, *Nonlinear Science. Emergence and Dynamics of Coherent Structures*, (Oxford University Press, Oxford, 1999).
37. Sievers A J and Takeno S, Intrinsic localized modes in anharmonic crystals, *Phys. Rev. Lett.* **61** (1988) 970–973.
38. C. Willis, M. El-Batanouny and P. Stancioff, Sine-Gordon kinks on a discrete lattice. I. Hamiltonian formalism, *Phys. Rev. B* **33** (1986) 1904–1911.

PHASE TRANSITIONS IN HOMOGENEOUS BIOPOLYMERS: BASIC CONCEPTS AND METHODS. *

N. THEODORAKOPOULOS †

Theoretical and Physical Chemistry Institute,
National Hellenic Research Foundation,
Vasileos Constantinou 48, 116 35 Athens, Greece

The basic models of helix-coil transitions in biomolecules are introduced. These include phenomenological, zipper (Bragg-Zimm) models of polypeptides, loop-entropy (Poland-Scheraga) and Hamiltonian (Peyrard-Bishop) models of homogeneous DNA denaturation. The transfer integral approach to one-dimensional thermodynamics is presented in some detail, including the necessary extensions to deal with the singular integral equations arising in the case of on-site potentials with a flat top. The (non-)applicability of the theorems which prohibit phase transitions in one-dimensional systems is discussed.

1. Introduction

The purpose of these notes is to provide a brief introduction to concepts and methods employed in the description of thermodynamic phase transitions in model biomolecular systems. This is one of the areas where the LOCNET network has been active, with an emphasis on modelling the fundamental interactions which provide a basis for understanding both the cooperative behavior and the nonlinear dynamics (what I will call "Hamiltonian" models). My primary aim is therefore to present a working introduction to this area, including some necessary details on methods and tools; I hope that young researchers who enter the field will find this material useful.

On the other hand there is a long and distinguished tradition in the field, which has achieved remarkable progress, based on a phenomenological description of the statistical properties of helix formation and growth and the entropies associated with them. I have therefore chosen to include a section on fundamentals of these "zipper" and "loop"-based models[1]. Again, the

*Proceedings of the conference on " *Localization and energy transfer in nonlinear systems*", June 17-21, 2002, San Lorenzo de El Escorial, Madrid, Spain. To be published by World Scientific.
†Work partially supported by EU contract HPRN-CT-1999-00163 (LOCNET network).

hope is that researchers who are active in developing microscopic Hamiltonian models will be guided by the succesful features of phenomenological, "zipper" and "loop" models. In addition, there is a simple, utilitarian reason for doing this: most of the experimental data presently available has been analyzed, directly or indirectly, in terms of such models. Therefore, anyone seriously interested in comparing theory and experiment should be familiar with them.

The plan of this review is as follows: Section 2 deals with zipper and loop models of polypeptides and DNA, respectively. Section 3 describes the Hamiltonian approach to DNA denaturation, including a somewhat detailed introduction to the transfer integral method. An appendix discusses the (non-)applicability of theorems which prohibit phase transitions in one dimension to the models presented.

2. Zippers and Loops

2.1. *Helix-Coil transitions in polypeptides*

2.1.1. *Background*

Synthetic polypeptides, i.e. macromolecules consisting of identical amino-acid residues, are ideal for studying the transition from the alpha-helical to coil-like structure. Understanding of this transition is central to controlling the stability of secondary protein structure[2]. Residues in helical regions give rise to distinct experimental signatures (e.g. viscocity, optical rotation). At a given macromolecular size N - which can be controlled in synthetic polypeptides - one can measure the helix fraction as a function of temperature. Typically[3], that fraction completes the transition from 1 to 0 over a fairly narrow temperature range - a few degrees K in the case of long chains. Chemists describe the process $A(helix) \longleftrightarrow B(coil)$ as an equilibrium between the two species,

$$K \equiv \frac{c_B}{c_A} \equiv e^{-\Delta G/T} \tag{1}$$

where the helix fraction is given by

$$\Theta \equiv \frac{c_A}{c_A + c_B} = \frac{1}{1+K} \,. \tag{2}$$

The sign convention is as follows: I am looking at the conversion of helix (A) to coil (B); therefore $\Delta G = G_B - G_A = \Delta H - T\Delta S$, and $\Delta H > 0$, i.e. the helix is energetically favored.

The value $\Theta = 0.5$ defines the midpoint of the transition, T_m. Assuming (although this is not exact, and sometimes not even a good approximation) that the enthalpy and entropy differences do not depend very much on temperature, leads to

$$\left(\frac{d\Theta}{dT}\right)_{\Theta=0.5} = -\frac{1}{4}\frac{\Delta H}{T^2} . \tag{3}$$

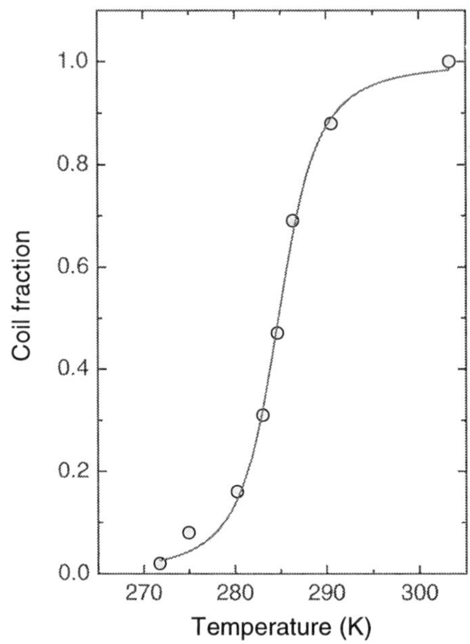

Figure 1. Coil fraction vs. temperature for a polypeptide (poly-γ-benzyl-L-glutamate) of controlled ($N = 1500$) length; The curve has been calculated in the framework of the generalized zipper model - cf. Eq. 16 below - (redrawn after Ref. [3]).

The inverse of Eq. (3) measures the width of the transition (in degrees K). A sharp transition (of a few degrees K) has a high [van't Hoff] ΔH (of the order of 100 Kcal/mol), indicating that perhaps as many as 100 hydrogen bonds are cooperatively broken during the transition.

2.1.2. *"Zipper" model family: underlying concepts*

Helix *initiation* and helix *growth* are viewed[4] as distinct processes:

- Growth: an existing helix may grow further at the *nth* site, or shrink. This is viewed as a forward and reverse reaction, with a rate ratio $s = \exp(-\Delta G^*/T)$, which reflects the difference in local free energies between the helix and coil states. If the ratio is greater than unity, the helix has a tendency to grow ("zip"). If it is less than unity, the helix will shrink ("unzip"). At temperatures near the transition, $s \approx 1$. The enthalpy difference $\Delta H^* < 0$ corresponds to the energy of a single hydrogen bond formed in the process of helix growth.

- Nucleation: in order to initiate a helix, 3 residues have to organize themselves. Again, viewing nucleation as a forward / reverse reaction, introduces a dimensionless $\sigma = \exp(-\Delta G_{init}/T)$. The large difference in the free energy comes mostly from the entropy loss associated with the organization of the 3-4 residues involved in the first turn of the helix.

I now present an outline of theoretical models[4], in order of increasing complexity:

2.1.3. *"0-th order" - The "all or nothing" (AON) model:*

Only two states are significant within this model. The pure coil, with relative statistical weight equal to unity; and the helix with N residues, with a relative weight σs^N. Intermediate states are suppressed, presumably due to high rate barriers. This gives a helix fraction

$$\Theta = \frac{1}{N}\frac{N\sigma s^N}{1+\sigma s^N} \tag{4}$$

and a slope at midpoint

$$\left(\frac{d\Theta}{dT}\right)_{\Theta=0.5} = \frac{N}{4}\frac{\Delta H^*}{T_m^2} \, . \tag{5}$$

There is strong cooperativity.

2.1.4. *Further considerations: the zipper model*

The model allows a single connected helical region of any length $n \leq N$. The statistical weight (Boltzmann factor) is -according to the general considerations, cf. above-, σs^n, and the helix can commence at any of the first $A_n = N - n + 1$ positions. This gives a partition function

$$Z = 1 + \sum_{n=1}^{N} A_n \sigma s^n \tag{6}$$

and a helical fraction

$$\Theta = \frac{1}{Z}\sum_{n=1}^{N} n A_n \sigma s^n = \frac{s}{Z}\frac{\partial Z}{\partial s} \quad , \tag{7}$$

where the partition sum can be evaluated to give

$$Z(N) = 1 + \sigma s^{N+2} - (N+1)s + \frac{N}{(s-1)^2} \quad . \tag{8}$$

2.1.5. *The generalized zipper model*

The only difference is topological; then macromolecule consists of any number of helical and coil regions which may alternate freely. One associates the following weights:

- 1 if coil comes after helix or coil;
- s if helix comes after helix;
- σs if helix comes after coil (nucleation).

The model thus implements the ideas presented in section 2.1.2 without imposing any further constraints. The state of the residue at site i can be described by a 2-vector ν_i, and the partition function is given by

$$\begin{aligned} Z_N &= \sum_{\{\nu_1\}\ldots\{\nu_N\}} < \nu_1|T|\nu_2 >< \nu_2|T|\nu_3 > \ldots < \nu_{N-1}|T|\nu_N > \\ &= \sum_{\{\nu_1\},\{\nu_N\}} < \nu_1|T^N|\nu_N > \end{aligned} \tag{9}$$

where the matrix elements of $\mathbf{T}$ express the Boltzmann factors specified above, i.e.

$$\mathbf{T} = \begin{pmatrix} s & 1 \\ \sigma s & 1 \end{pmatrix} \ . \tag{10}$$

To evaluate the partition sum, I apply periodic boundary conditions (convenient, not a must, and certainly wrong for small chains) and obtain

$$Z_N = Tr\mathbf{T}^N = \lambda_0^N + \lambda_1^N \tag{11}$$

where the eigenvalues are given by

$$\begin{aligned} \lambda_{0,1} &= \frac{1}{2}\left[1 + s \pm \Delta\right] \\ \Delta &= \sqrt{(1-s)^2 + 4\sigma s} \end{aligned} \tag{12}$$

and, in the large N limit, Z is dominated by the largest eigenvalue, λ_0.

Note that the partition function (not the $\mathbf{T}$ -matrix!) can be mapped onto the one of the ferromagnetic Ising model with exchange interaction J and magnetic field h, with the identifications

$$s \Leftrightarrow e^{-2\beta h} \tag{13}$$

$$\sigma \Leftrightarrow e^{-2\beta J} \tag{14}$$

$$\lambda_{magnetic} = e^{\beta(J+h)}\lambda_{helix-coil} \; . \tag{15}$$

To obtain the helix fraction, note that if the probability of obtaining a helical segment of length k is given by $\phi_k(\sigma)ks^k$, where ϕ is the coefficient of s^k in the partition sum. This gives

$$\Theta = \frac{1}{N}\frac{1}{Z}\sum_{k=1}^{N} \phi_k(\sigma)ks^k = \frac{1}{N}\frac{s}{Z}\frac{\partial Z}{\partial s} \tag{16}$$

One can now verify that as $s \to 1$, $\Delta \to 2\sqrt{\sigma}, \Theta \to 1/2$; for $\sigma << 1$ (cf. below),

$$\left(\frac{d\Theta}{dT}\right)_{\Theta=0.5} = \frac{1}{4\sqrt{\sigma}}\frac{\Delta H^*}{T_m^2} \; . \tag{17}$$

The experimental situation[3] for long ($N = 1500$) chains is summarized in Fig. 1. Fits can be obtained with $\Delta H^* = -3.8kJ/mole$ (cf. calorimetric measurements $\Delta H = -3.97kJ/mole$) and $\sigma = 1.6 \times 10^{-4}$. One usually interprets $1/\sqrt{\sigma}$ as number of residues cooperatively involved in the transition (cf. AON theory, Eq. 5). This interpretation also follows from the Ising model, where the inverse correlation length is given (in lattice constants) by

$$1/\xi = \lambda_1 - \lambda_0 = 2\sqrt{\sigma} \;\; (\text{at } s = 1) \; . \tag{18}$$

The following simple conformational argument provides an independent estimate for $\sigma = e^{-\Delta G_{init}/T} \approx e^{\Delta S_{init}}$. Initiation of the helix involves organization of J (=3 or 4) residues, each one by 2 dihedral angles. Typically a dihedral angle can take 3 independent orientations in space. This gives a total of 3^{2J} states, or an entropy loss $\Delta S_{init} = -2J\ln 3$. (=-6.6 for $J = 3$, or -8.8 for $J = 4$). This compares favorably with $\ln(2 \times 10^{-4}) = -8.5$.

Similarly, one can relate the entropy loss involved in helix *growth*, to the energy of the H-bond. At the transition, $\Delta S^* = \Delta H^*/T_m = -1.85$. This is roughly comparable to the estimate $-2\ln 3 \approx -2.20$, obtained by considering the 2 dihedral angles which must be organized to admit a residue into the helix.

2.1.6. *A useful shortcut*

It is possible to obtain the thermodynamics of the generalized zipper model without recourse to the transfer matrix formalism. I present this "hand-waving" [5], because it will be useful for DNA loops (cf. below).

The fundamental "entity" of the macromolecule is a helical region of length n, followed by a coil region of length m. This "helix-coil" entity is characterized by a free energy

$$g(n,m) = -T\ln\sigma - nT\ln s \tag{19}$$

where the two terms correspond to the contributions of helix nucleation and growth, respectively; it occurs with a probability

$$P_{n,m} = \exp\{-[g(n,m) - g_0]/T\} \tag{20}$$

where $g_0 \equiv -T\ln z$ is the equilibrium free energy per site of the full macromolecule, to be determined by the normalization condition

$$\sum_{n,m=1}^{\infty} P_{n,m} = 1 \quad . \tag{21}$$

Both the n and the m- summations can be done trivially as long as $s < z$ and $z > 1$. The condition (21) can then be written as

$$\frac{s}{z-s}\frac{1}{z-s} = \frac{1}{\sigma} \quad , \tag{22}$$

whose roots are identical to those of (13) obtained via the transfer matrix; the largest root is the one which satisfies the condition $z > \max(1,s)$ (cf. above).

2.2. *Loop entropies and DNA denaturation*

2.2.1. *Background*

Thermal DNA denaturation occurs when the two strands of the double helix separate upon heating. In real DNA the phenomenon of multistep melting is ubiquitous, reflecting the inhomogeneity of the molecule. "Homogeneous", synthetic DNA, which consists of a few thousand identical base pairs has been studied experimentally[6] and shown to exhibit a very sharp transition; the qualitative shape of the coil fraction curve, as obtained by optical density or viscocity measurements, is similar to that of Fig. 1; however, the observed temperature width is of the order of one degree. It is reasonable to speculate that in the thermodynamic limit the transition would be of the first order. Poland and Scheraga[7](PS) proposed a simple model of the thermodynamics involved, based on the ideas discussed in Section 2.1.2, and the concept of loop entropy (cf. below). There is however a difference in the physical origin of the parameters involved; the helix growth probability $s = \exp\{-\epsilon/T\}$ now reflects the combined effect of both interactions which are significant in the bonded DNA state: the hydrogen bonding responsible for binding base pairs and the stacking interaction between adjacent bases.

2.2.2. *Loop entropies*

A denaturation loop of length m, i.e. a region of m sites where the double helix has locally melted, is characterized by the extra entropy it contributes[7]. This extra entropy has been calculated for lattice polymers and is of the form

$$S_L(m) = am + b - c\ln m \quad , \tag{23}$$

where a and b are constants and c depends on the dimensionality. In the case of Gaussian polymer chains, or random lattice walks, which ignore the effects of excluded volume, $c = d/2$. Taking account of excluded volume tends to increase the value of c. It will be seen below that this can have a decisive influence on the nature of the transition.

2.2.3. *The phase transition*

In an infinitely long DNA molecule, the fundamental entity is again a double-helical region of n sites (base pairs), followed by a denaturation loop of length m ($2m$ bases); putting together the contributions from helical and loop part (cf. Sections 2.2.1 and 2.2.2), I obtain the free energy of this entity as

$$g(n,m) = -T\ln\sigma + n\epsilon - mT\ln u^2 + Tc\ln m \tag{24}$$

where $u = \exp(a/2)$ is another constant, and I have dropped the irrelevant constant b. It is now possible to derive the the thermodynamics exactly as in section 2.1.6. Inserting (24) in the normalization condition (21) gives

$$\frac{z}{s} - 1 = \sigma U\left(\frac{z}{u^2}\right) \quad , \tag{25}$$

where $s = \exp(-\epsilon/T)$ is the only temperature dependent parameter, and

$$U(x) = \sum_{m=1}^{\infty} \frac{1}{m^c}\left(\frac{1}{x}\right)^m \quad . \tag{26}$$

Near $x = 1$ it is possible to approximate $U(x)$ by the expression[8]

$$U(x) \approx U_0 - U_1\left[1 - \frac{1}{x}\right]^{c-1} \tag{27}$$

where, for $c > 1$, $U_0 \approx U_1 \approx \zeta(c)$ [1].

I now follow the thermodynamic behavior near the putative singularity by defining[8] an $s_c \equiv \exp(-\epsilon/T_c)$ via

$$\frac{u^2}{s_c} - 1 = \sigma U(1) \tag{28}$$

and subtracting (28) from (25) to obtain

$$s_c - s + \Delta + \frac{\sigma s_c}{u^2} U_1 \Delta^{c-1} = 0 \quad , \tag{29}$$

where $z = u^2(1+\Delta)$ and I have only kept lowest order terms in the small quantities Δ and $s - s_c$. It is now straightforward to use (29) and obtain $\Delta(s)$; the helix fraction is then given by

$$\frac{\partial \ln z}{\partial \ln s} \propto \frac{\partial \Delta}{\partial s} \quad . \tag{30}$$

Two cases can be distinguished:

- $1 < c < 2$. The linear term in Δ can be neglected in (29). The helix fraction is proportional to $(T_c - T)^{(2-c)/(c-1)}$, i.e. it approaches zero continuously near the transition. In particular, if $c = 3/2$ (the value which corresponds to $d = 3$ and neglecting excluded volume effects), one obtains a second order transition.
- $c > 2$. The linear term in Δ dominates, and the transition becomes first order, i.e. the helix fraction drops abruptly to zero at the transition.

The above analysis shows how crucial the value of c is in determining the nature of the transition. It has been long known[8] that excluded volume effects, as calculated within the framework of self-avoiding walks, can increase the value of c to 1.75 for loops embedded in three-dimensional space. Recent research[9] suggests that c may be as high as 2.1.

3. Hamiltonian approach to DNA denaturation

3.1. *The model*

A "minimal" Hamiltonian model of homogeneous DNA denaturation has been proposed by Peyrard and Bishop[10](PB). The model assumes two parallel, harmonic chains, with lattice constant l, joined in the form of a ladder by anharmonic springs; the particular model proposes a Morse potential because of its analytical tractability, although any form with a repulsive core, a stable minimum and a flat top (e.g. Lennard-Jones) would be physically suitable. The emphasis is on modelling the unbinding of the two chains, not the helical aspect of the ordered state; this is done in the general spirit of the theory of critical phenomena, which has demonstrated that the "essentials" of the interactions completely determine the critical behavior. The Hamiltonian

$$H_{tot} = \frac{m}{2} \sum_n \left[\dot{u}_n^2 + \dot{v}_n^2 + \omega_0^2 \left(u_n - u_{n-1}\right)^2 + \omega_0^2 \left(v_n - v_{n-1}\right)^2 \right]$$

$$+\sum_n V(u_n - v_n) \tag{31}$$

describes the motion of the two bound chains with coordinates $\{u_n\}, \{u_n\}$; dots denote time derivatives; only the motion transverse to the chains is considered; bases have equal masses m and are connected by harmonic springs of equal strength, determined by the frequency ω_0; the energy scale of the Morse potential

$$V(x) = D(e^{-ax} - 1)^2 \tag{32}$$

is given by D and its spatial range by $1/a$.

Transformation to center-of-mass and relative coordinates, $Y_n = (u_n + v_n)/2, y_n = u_n - v_n, M = 2m, 1/\mu = 2/m$ decouples center-of-mass from relative motion, i.e.

$$H_{tot} = H_0(Y) + H(y) \quad , \tag{33}$$

where

$$H_0(Y) = \sum_n \left[\frac{P_n^2}{2M} + \frac{1}{2}M\omega_0^2(Y_n - Y_{n-1})^2\right] \quad , \tag{34}$$

$P_n = M\dot{Y}_n$ is the canonical momentum conjugate to Y_n, and

$$H(y) = \sum_n \left[\frac{p_n^2}{2\mu} + \frac{1}{2}\mu\omega_0^2(y_n - y_{n-1})^2 + V(y_n)\right] \quad , \tag{35}$$

where $p_n = \mu\dot{y}_n$ is the canonical momentum conjugate to y_n.

3.2. *Statistical Mechanics*

H_0 is just the Hamiltonian of a harmonic chain with the total base pair mass $2m$ per site. It gives an additive nonsingular contribution to all thermal properties. It will be neglected in what follows. The classical thermodynamics of H is described by the canonical partition function

$$Z = \int \prod_{n=1}^{N} dp_n dy_n e^{-\beta H} \quad . \tag{36}$$

One can immediately perform the Gaussian integrals over momentum space and obtain

$$Z = Z_K Z_P, \tag{37}$$

where each integration in the kinetic part contributes a $(2\pi\mu/\beta)^{1/2}$ factor to the partition function, i.e.

$$Z_K = (2\pi\mu/\beta)^{N/2} \quad . \tag{38}$$

The nontrivial part is

$$Z_P = \int \left(\prod_{n=1}^{N} dy_n \right) T(y_1, y_2) \cdots T(y_{N-1}, y_N) T(y_N, y_{N+1}) \quad , \tag{39}$$

where

$$T(x, y) = e^{-\beta \left[\frac{\mu \omega_0^2}{2} (y-x)^2 + V(x) \right]} \quad . \tag{40}$$

3.3. *Transfer integral: the formalism*

3.3.1. *Definitions and Notation*

Consider the eigenvalue problem defined by the asymmetric kernel T (the kernel can be easily symmetrized but need not be so; in fact, working with the asymmetric kernel is technically advantageous in examining the validity of some approximations, cf. below):

$$\int_{-\infty}^{\infty} dy \, T(x, y) \, \Phi_\nu^R(y) = \Lambda_\nu \Phi_\nu^R(x) \tag{41}$$

$$\int_{-\infty}^{\infty} dy \, T(y, x) \, \Phi_\nu^L(y) = \Lambda_\nu \Phi_\nu^L(x) \, , \tag{42}$$

where left and right eigenstates have been assumed to be normalized; note that the normalization integral is $\int dx \Phi_\nu^L(x) \Phi_\nu^R(x)$. Orthogonality

$$\int_{-\infty}^{\infty} dx \, \Phi_\nu^L(x) \, \Phi_{\nu'}^R(x) = \delta_{\nu\nu'} \tag{43}$$

and completeness

$$\sum_\nu \Phi_\nu^L(x) \, \Phi_\nu^R(y) = \delta(x - y) \tag{44}$$

relationships are assumed to hold. Note that this is not obvious for the class of potentials of interest here. This is a point which will be further taken in section 3.6. I will further use the notation

$$\Lambda_\nu = e^{-\beta \epsilon_\nu} \tag{45}$$

(sensible as long as the eigenvalues are nonnegative).

3.3.2. *The partition function*

The integrand of (39), as written down has a problem: it includes a reference to the displacement y_{N+1} of the $N + 1st$ particle, which has not yet been defined. For a large system, this is best remedied by means of periodic boundary conditions (PBC), i.e. by demanding that $y_{N+1} = y_1$. Alternatively, the integration may be extended to one more variable, dy_{N+1}, with the simultaneous introduction of a factor $\delta(y_{N+1} - y_1)$ to take care of PBC. This however is the same as the sum in the left-hand-side of (44). I then obtain

$$Z_P = \sum_\nu \int dy_1 \cdots \underbrace{dy_{N+1} \Phi^L_\nu(y_1)}\, T(y_1, y_2) \cdots \underbrace{T(y_N, y_{N+1}) \Phi^R_\nu(y_{N+1})}\,. \tag{46}$$

The braces make clear that I can perform the integral over dy_{N+1} and obtain a factor $\Lambda_\nu \Phi^R_\nu(y_{N+1})$, using the defining property of right-hand eigenfunctions. The process can be repeated N times, each time giving a further factor Λ_ν and a right eigenfunction with an argument whose index is smaller by one. At the end, I am left with

$$Z_P = \sum_\nu \int dy_1 \Phi^L_\nu(y_1) \Lambda^N_\nu \Phi^R_\nu(y_1) = \sum_\nu \Lambda^N_\nu\,. \tag{47}$$

In the thermodynamic limit, Z_P is dominated by the largest eigenvalue Λ_0 or, equivalently, the lowest ϵ_0:

$$\lim_{N\to\infty} \frac{1}{N} \ln Z_P = \ln \Lambda_0 = -\beta\epsilon_0 \tag{48}$$

3.3.3. *The order parameter*

$$\begin{aligned}
< y_i > &= \frac{1}{Z_P} \int dy_1 \cdots dy_N T(y_1, y_2) \cdots T(y_{i-1}, y_i) y_i \\
&\quad T(y_i, y_{i+1}) \cdots T(y_N, y_{N+1}) \\
&\equiv \frac{1}{Z_P} \sum_\nu \int dy_1 \cdots dy_{N+1} \Phi^L_\nu(y_1) \underbrace{T(y_1, y_2) \cdots T(y_{i-1}, y_i)}_{i-1} y_i \\
&\quad \underbrace{T(y_i, y_{i+1}) \cdots T(y_N, y_{N+1})}_{N-i+1} \Phi^R_\nu(y_{N+1})\,,
\end{aligned} \tag{49}$$

after insertion of a complete set of states (cf. above); the braces denote the number of times I can perform an integration and obtain, respectively, a right eigenfunction with an argument smaller by one, or a left eigenfunction with an argument larger by one, as well as a factor Λ_ν. The remaining integral must be performed explicitly:

$$\begin{aligned}
< y_i > &= \frac{1}{Z_P} \sum_\nu \Lambda^N_\nu M_{\nu\nu} \\
&\approx M_{00}
\end{aligned} \tag{50}$$

where the second line is exact in the thermodynamic limit, and I have used the abbreviation

$$M_{\nu\mu} = \int_{-\infty}^{\infty} dy \Phi_\nu^L(y) y \Phi_\mu^R(y) \quad . \tag{51}$$

3.3.4. *Correlations*

With $i < j$,

$$\begin{aligned} < y_i y_j > \equiv & \frac{1}{Z_P} \int dy_1 \cdots dy_N T(y_1, y_2) \cdots T(y_{i-1}, y_i) y_i T(y_i, y_{i+1}) \\ & \cdots T(y_{j-1}, y_j) y_j T(y_j, y_{j+1}) \cdots T(y_N, y_{N+1}) \\ = & \frac{1}{Z_P} \sum_\nu \int dy_i \cdots dy_j \Lambda_\nu^{i-1} \Phi_\nu^L(y_i) y_i T(y_i, y_{i+1}) \\ & \cdots T(y_{j-1}, y_j) y_j \Lambda_\nu^{N-j+1} \Phi_\nu^R(y_j) \end{aligned} \tag{52}$$

where the straightforward integrations, i.e the first $i-1$ and the last $N-j+1$ have already been performed (cf. above). In order to perform the remaining integrations, I insert two more factors of 1, after y_i and before y_j, i.e. integrals $\int \delta(y_i - \bar{y}_i)$ and $\int \delta(y_j - \bar{y}_j)$, respectively; exploiting the presence of the δ functions, I may substitute the variables y_i and y_j by $\bar{y}_i$ and $\bar{y}_j$ respectively. This translates to two more sums over complete sets of states and another $j - i$ integrals which can now be performed:

$$\begin{aligned} < y_i y_j > = & \frac{1}{Z_P} \sum_{\nu,\mu,\rho} \int d\bar{y}_i d\bar{y}_j dy_i \cdots dy_j \Lambda_\nu^{i-1} \Phi_\nu^L(\bar{y}_i) \bar{y}_i \Phi_\mu^R(\bar{y}_i) \Phi_\mu^L(y_i) \\ & T(y_i, y_{i+1}) \cdots T(y_{j-1}, y_j) \Phi_\rho^R(y_j) \Phi_\rho^L(\bar{y}_j) \bar{y}_j \Lambda_\nu^{N-j+1} \Phi_\nu^R(\bar{y}_j) \\ = & \frac{1}{Z_P} \sum_{\nu,\mu,\rho} \Lambda_\nu^{N+i-j} \Lambda_\rho^{j-i} \int d\bar{y}_i d\bar{y}_j \Phi_\nu^L(\bar{y}_i) \bar{y}_i \Phi_\mu^R(\bar{y}_i) \\ & \delta_{\mu,\rho} \Phi_\rho^L(\bar{y}_j) \bar{y}_j \Phi_\nu^R(\bar{y}_j) \\ = & \frac{1}{Z_P} \sum_{\nu,\mu} \Lambda_\nu^{N+i-j} \Lambda_\mu^{j-i} \left| M_{\nu\mu} \right|^2 . \end{aligned} \tag{53}$$

In the thermodynamic limit the $\nu = 0$ term dominates; the resulting factor cancels against the denominator and leaves

$$< y_i y_{i+r} >= \sum_\mu \left| M_{0\mu} \right|^2 e^{-\beta(\epsilon_\mu - \epsilon_0) r} \tag{54}$$

where I have used Dirac shorthand for the matrix element and set $j = i + r$. The first term ($\mu = 0$) in the above sum corresponds to $< y >^2$ and should properly be subtracted from both sides; This leaves

$$< \delta y_i \delta y_{i+r} > \equiv < y_i y_{i+r} > - < y_i >< y_{i+r} > \tag{55}$$
$$= \sum_\nu {}' \, |M_{0\nu}|^2 \, e^{-\beta(\epsilon_\nu - \epsilon_0)r}$$

where now the ground state is excluded from the summation. The above result identifies the correlation length ξ, i.e the typical length over which the decay of correlations takes place, as

$$\frac{\xi}{l} = \frac{1}{\beta(\epsilon_1 - \epsilon_0)} \tag{56}$$

where the subscript 1 stands for the first excited state (dominant exponential in the limit of large r).

3.4. *TI results: Gradient-expansion approximation*

Suppose that the displacement field does not change appreciably over a lattice constant. This is certainly reasonable

at low temperatures. Note that this does not exclude large displacements per se. Nonlinearity is explicitly allowed, but the displacement field must be smooth. The assumption is certainly reasonable at low temperatures.

I set $y = x + z, \Phi^R \to \phi$ and rewrite (41) as

$$e^{-\beta[\epsilon_\nu - V(x)]}\phi_\nu(x) = \int_{-\infty}^{+\infty} dz \, e^{-\frac{1}{2}\beta\mu\omega_0^2 z^2} \left\{ \phi_\nu(x) + z\phi'_\nu(x) + \frac{1}{2}z^2\phi''_\nu(x) \right\}$$
$$= \left[\frac{2\pi}{\beta\mu\omega_0^2}\right]^{1/2} \left\{ \phi_\nu(x) + \frac{1}{2\beta\mu\omega_0^2}\phi''_\nu(x) \right\} \tag{57}$$

where higher terms in the gradient expansion have been neglected and the Gaussian integrals have been performed; this is meaningful as long as the width of the Gaussians is smaller than the range of the Morse potential, i.e.

$$\beta\mu\omega_0^2/a^2 > 1 \quad . \tag{58}$$

The factor in front of the r.h.s. of (57) can be absorbed in the eigenvalue by defining $\tilde{\epsilon}_\nu = \epsilon_\nu + 1/(2\beta)] \ln[2\pi/(\beta\mu\omega_0^2)]$. Now, for many practical purposes, when it comes to calculating matrix elements, the relevant magnitude of $\epsilon - V(x)$ is D, the depth of the Morse well (or some other characteristic energy in the case of another potential). The key to this statement is that one does not need to consider large negative values of x, where $V(x)$ is huge, because at such x, both the exact eigenfunction Φ and its approximation

ϕ can be expected to be negligible. If then $\beta D \leq 1$[a] it is reasonable to expand the exponential and keep only the first term. Dividing both sides by β, I obtain a Schrödinger - like equation,

$$-\frac{1}{2\mu(\beta\omega_0)^2}\phi_\nu''(x) + [V(x) - \tilde{\epsilon}_\nu]\,\phi_\nu(x) = 0 \quad . \tag{59}$$

Before continuing the discussion of (59) and its properties, I pick up the bits and pieces (cf (37), (38), (48)) of the thermodynamic free energy (per site)

$$f = -\frac{1}{\beta N}\ln(Z_K Z_P) \equiv -\frac{1}{\beta}\ln\left(\frac{2\pi}{\beta\omega_0}\right) + \tilde{f} \quad , \tag{60}$$

where $\tilde{f} = \tilde{\epsilon}_0$. The first term in (60) is the free energy of the small oscillations (transverse phonons in this context). It is a term smooth in temperature (constant specific heat!) and therefore irrelevant to any phase transition. Any nontrivial physics is hidden in the second term, which is identical with the the smallest eigenvalue of (59).

A couple of comments are in order. First, (59) would be a literal (i.e. quantum-mechanical) Schrödinger equation, if I substituted $1/(\beta\omega_0)$ by $\hbar$. I will come back to that point. Second, I can get a dimensionless potential (and eigenvalue) by dividing both sides of (59) by D. In other words, the relevant dimensionless parameter is

$$\delta^2 = \begin{cases} \frac{2\mu}{a^2\hbar^2}\cdot D \ \text{(quantum mechanics)} \\ \\ \frac{2\mu\beta^2\omega_0^2}{a^2}\cdot D \ \text{(statistical mechanics).} \end{cases} \tag{61}$$

In terms of δ, the bound state spectrum of (59) is given [11] by

$$\frac{\tilde{\epsilon}_n}{D} = 1 - \left[1 - \frac{n + 1/2}{\delta}\right]^2$$
$$n = 0, 1, ..., int(\delta - 1/2)\,. \tag{62}$$

There is at least one bound state if $\delta > 1/2$. For $1 \geq \delta > 1/2$ there is *exactly* one bound state. And if δ becomes equal to, or smaller than $1/2$, there is no bound state at all. *The value* $\delta_c = 1/2$ *is "critical"*. In quantum mechanical language, if a particle has a mass which is lighter than a critical mass $\mu_c = \hbar^2 a^2/(8D)$, it cannot be confined in the Morse well. Quantum fluctuations will drive it out[b]. In the context of statistical mechanics, δ_c

---

[a]Note that, in connection with (58), this defines a temperature window $D < k_B T < \mu\omega_0^2/a^2$ for the validity of the overall approximation scheme.

[b]This is a general property of asymmetric one-dimensional wells; symmetric wells will support a particle in a bound state, no matter how low its mass.

corresponds, via (61), to a critical temperature $T_c = 2(\omega_0/a)\sqrt{2\mu D}$. The free energy is given by

$$\frac{\tilde{f}}{D} = \begin{cases} 1 & T > T_c \\ 1 - \left(1 - \frac{T}{T_c}\right)^2 & T < T_c \ , \end{cases} \tag{63}$$

where in the upper line I have made use of the fact that the bottom of the continuum part of the spectrum is at $\epsilon = D$. The free energy f is non-analytic at $T = T_c$, where its second derivative is discontinuous (i.e. there is a jump in the specific heat). This corresponds to a second order transition, according to the Ehrenfest classification scheme[c].

In order to gain some further insight into the physics involved[d] it is useful to examine the average displacement (50), determined by the ground-state (GS) eigenfunction

$$\phi_0(x) = e^{-\zeta/2}\,\zeta^{\delta-1/2} \tag{64}$$

where $\zeta = 2\delta e^{-ax}$. It is straightforward to see that, as T approaches T_c from below, the eigenfunction extends towards larger and larger positive values of x:

$$\phi_0(x) \propto e^{-\lambda x} \tag{65}$$

where

$$\lambda = \frac{1}{\delta - \delta_c} \tag{66}$$

is a (transverse) characteristic length which measures the spatial extent of the GS eigenfunction. As a consequence, we can estimate that $< y >$, which is dominated by the large values of the argument, will also behave as

$$< y > \sim (\delta - \delta_c)^{-1} \sim \left(1 - \frac{T}{T_c}\right)^{-1} . \tag{67}$$

---

[c]Note that the term "second order" is meant literally in this case, not just as a metaphor for the absence of a latent heat (for which the term "continuous transition" would be appropriate).

[d]The mathematical analogy between the behavior of the spectral gap which occurs in a point ($d = 0$) system and the singularity in the free energy of a classical chain ($d = 1$) is an example of a deeper analogy which relates quantum to thermal fluctuations; the formal correspondence $\hbar \leftrightarrow 1/(\beta\omega_0)$ manifests a far-reaching analogy between d-dimensional quantum mechanics and $(d+1)$-dimensional classical statistical mechanics. The analogy is most fruitful at $d = 1$, because of the interplay and the richness of exact available results which based either in the transfer-matrix approach of 2-dimensional classical statistics or on the Bethe-Ansatz developed for 1-d quantum spin systems.

As the critical temperature is approached from below, particles cease to be confined to the minimum of the Morse well. They perform larger and larger excursions to the flatter part of the potential. At T_c the transition is complete; the average transverse displacement is infinite. Particles move, on the average, on the flat top of the Morse potential. Unwinding ("melting") of the DNA has occurred.

In the language of critical phenomena $< y >$ is the order parameter. In the "usual" phase transitions, one goes from an ordered to a disordered phase. The order parameter m vanishes at the transition point, i.e $m \propto (T_c - T)^\beta$ with a positive critical exponent β (not to be confused with the inverse temperature: standard notation of critical phenomena!). DNA melting is really an instability[12] - rather than an "order-disorder" transition. It is therefore not surprising that the corresponding critical exponent β extracted from (67) is negative (-1).

Experimental data on DNA denaturation do not deliver $< y >$ directly. The "experimental order parameter" is the helical fraction, i.e. the probability that a given base pair is still bound; technically one uses an (instrumentation-dependent) cutoff y_0 and measures $P(y > y_0, T)$. For the model presented here, this function approaches zero smoothly (linearly) as $T \to T_c$, independently of the choice of y_0.

Eq. (56) states that the correlation length is also contolled by the gap in the eigenvalue spectrum; as the transition is approached,

$$\frac{\xi}{l} = \frac{1}{\beta D}\left(1 - \frac{T}{T_c}\right)^{-2} \tag{68}$$

which identifies a critical exponent $\nu = 2$ for the divergence of the correlation length. The picture of thermal denaturation which emerges is one of ordered regions, where helical structure persists; these regions are interrupted by droplets of the high-temperature phase, i.e. "denaturation bubbles" of typical size ξ.

3.5. *A first order transition?*

It is possible to generalize the theory in order to take account of the fact that the stacking energy is a property of successive base pairs, rather than individual bases. A practical way of doing this is to substitute the second term in the Hamiltonian (35) by

$$\frac{1}{2}\mu\omega_0^2 \left[1 + g(y_n + y_{n-1})\right] (y_n - y_{n-1})^2 \quad , \tag{69}$$

where [13]

$$g(x) = e^{-\alpha x} \quad . \tag{70}$$

The effect of Eqs. (69)-(70) is to interpolate between the original value of the elastic coupling if *either* (or both) of the two base pairs n, $n-1$ is unbound (in which case $y_n \to \infty$ or $y_{n-1} \to \infty$), and twice that value if both are bound; in the latter case, typically, $y_n \approx 0$; the much higher values of the stacking energy, which (70) in principle allows, are statistically irrelevant due to the repulsive core of the Morse potential. Within the gradient expansion approximation, it can be shown[15] that the main effect of the nonlinear stacking energy on the thermodynamics is to generate an effective, on-site, "thermally activated" barrier

$$U(y) = \frac{T}{2D} \ln\left(1 + e^{-2\alpha y}\right) \quad . \tag{71}$$

which appears in Eq. 59 and acts in addition to the Morse potential.

It has been shown[14,15,17] that the character of the transition changes dramatically as the value of the stacking parameter ratio α/a decreases (corresponding to a longer range in the effective potential). Although the transition remains asymptotically second order, the limiting asymptotic behavior becomes relevant only within an exponentially small range of the temperature difference $T_c - T$. For all practical purposes, the transition is first order, with a finite melting entropy $\Delta S = A_0 D/T_c$, where A_0 is a numerical constant of order unity[15].

It should be noted that the interpolation (70) is not unique; an interpolation function of the type $g(x) = 1$, $x < x_0$, $g(x) = (x_0/x)^2$, $x > x_0$, leads - depending on the other parameters - to a rich variety of critical behavior, ranging from a first-order transition to continuously varying critical exponents.[16]

With the above modification (69), it has become possible[17] to describe, at least in principle, the series of multistep melting observed in real, heterogeneous DNA.

3.6. *TI beyond the gradient expansion*

It was stated in Section 3.3.1 that the TI formalism rests on the assumption that the integral equations (41) and (42) - have a complete, orthonormal set of eigenfunctions. Within the gradient approximation approach this was demonstrated by construction - since the integral equation was reduced to a Schrödinger-like equation. In many cases however, the gradient expansion is not valid at all. It is therefore necessary to develop alternative, mostly numerical methods for computing TI thermodynamics. For such applications it is expedient to consider the symmetrized, dimensionless version of the kernel (40), i.e.

$$T_s(x,y) = e^{-\left[(y-x)^2/R+V(x)+V(y)\right]/(2T)} \tag{72}$$

and the associated integral equation

$$\int_{-\infty}^{\infty} dy T_s(x,y)\phi(y) = \Lambda\phi(x) \quad , \tag{73}$$

where $R = Da^2/(\mu\omega_0^2)$, $T = 1/(\beta D)$ and the Morse potential is now dimensionless, $V(x) = (1-e^{-x})^2$, as are the displacement variables x, y.

Due to the flat top of the Morse potential, the kernel (72) is not of the Hilbert-Schmidt type[18]; therefore the integral equation (73) is singular and it can not be a priori stated that it possesses a complete orthonormal set of eigenstates; in other words, the prerequisites for directly applying the TI method are not strictly met. In the rest of this section I will outline a mathematically consistent procedure of examining the spectral gap of (73), based on finite-size scaling concepts[19].

Due to the presence of the Gaussian factors in the kernel, it is possible to approximate the integral in the left-hand-side of (73) by using a Gauss-Hermite grid of size N, i.e.

$$\int_{-\infty}^{\infty} d\bar{y}\, e^{-\bar{y}^2} f(\bar{y}) \approx \sum_{m=1}^{N} w_m f(\bar{y}_m) \tag{74}$$

where the positions $\{\bar{y}_m\}$ and weights $\{w_m\}$ are given by the appropriate Gauss-Hermite quadratures routine. The largest $\bar{y}_N \approx (2N+1)^{1/2} \equiv L$ can be used as estimate of the *transverse* "size of the system" employed at any given discretization. I emphasize transverse because the length of the chain is infinite, i.e. the thermodynamic limit has already been taken.

I use "rescaled" variables, i.e. $y = \rho\bar{y}$, $\rho = (2RT)^{1/2}$, divide both sides of (73) by $\rho\sqrt{\pi}$, and use the approximation (74). The result is an approximation of (73) by the matrix eigenvalue equation

$$\sum_{j=1}^{N} D_{ij} A_j^\nu = \tilde{\Lambda}_\nu A_i^\nu \tag{75}$$

where

$$D_{ij} = \left(\frac{w_i w_j}{\pi}\right)^{1/2} e^{\bar{y}^i \bar{y}^j} e^{-(\bar{y}^i-\bar{y}^j)^2/2} e^{-\left[V(\rho\bar{y}^i)+V(\rho\bar{y}^j)\right]/(2T)} \tag{76}$$

and $\tilde{\Lambda}_\nu \equiv \Lambda_\nu/(2\pi RT)^{1/2} \equiv e^{-\epsilon_\nu/T}$.

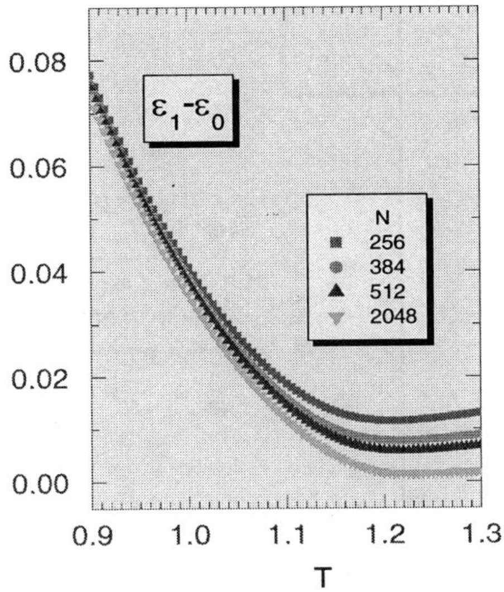

Figure 2. The gap between the two lowest eigenvalues of the matrix eigenvalue problem (75), for a variety of N values. For a given N, the gap has a minimum at a certain temperature T_m.

It is now possible to solve numerically the real, symmetric matrix eigenvalue problem (76) for a range of temperatures and a sequence of increasingly fine grids. Results for the difference between the two lowest eigenvalues are shown in Fig. 2 for $R = 10.1$. For any given size L, the gap has a minimum $\Delta\epsilon_m(L)$ at a certain temperature $T_m(L)$. Fig. 3 demonstrates that (i) the value of the gap approaches zero quadratically as $L \to \infty$ (with an accuracy of 10^{-5}), and (ii) the sequence of $T_m(L)$'s also approaches a limiting value $T_c = 1.2275$ quadratically.

It is natural to identify the limiting temperature T_c, where the spectral gap of the limiting, infinite-dimensional matrix eigenvalue equation (75) vanishes, as the transition temperature of the original TI equation (73). Further application of finite-size scaling methods demonstrates[19] that the various critical exponents coincide with those obtained within the gradient expansion method.

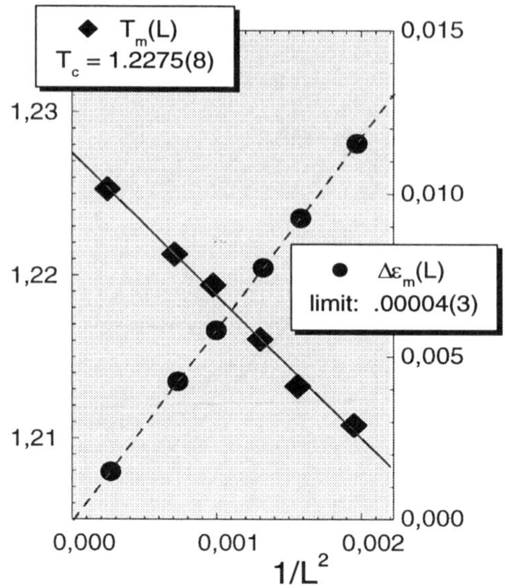

Figure 3. The magnitude of the gap minimum (circles, right y-axis scale) approaches zero as the system size goes to infinity. The sequence of the temperatures corresponding to the gap minima, $T_m(L)$ (diamonds, left y-axis scale), can be used to provide an estimate of the critical point T_c.

Acknowledgments

I thank M. Peyrard and T. Dauxois for many helpful discussions and comments.

Appendix: Phase transitions in one-dimensional systems

I briefly discuss why the general prohibitions on phase transitions in one dimension are inapplicable to both the PS and the PB models of DNA denaturation.

Van Hove's theorem[21] states that no phase transitions occur in 1-d particle systems with short-range pair interactions. The PB model has on-site potential - i.e. the theorem is not applicable. It is however worth noting that similar mathematical proofs, have been given for systems with periodic on-site potentials[22]. Such proofs however seek to prove analyticity of the eigenvalue spectrum and hence absence of a phase transition; as such, they tend to exclude potentials which give rise to singular TI equations.

The PS model is not a Hamiltonian model and therefore van Hove's theorem is again not applicable.

Landau's theorem[23] is significantly stronger. It states that "macroscopic phase coexistence cannot occur at finite temperatures in one dimensional systems". It is less obvious why it should not apply. I therefore outline the proof. Consider a system with N sites, which may exist in either phase A or phase B. Let θ be the fraction of phase A; furthermore, let there be $m << N$ contacts between the phases, each of energy ϵ. These can be steplike (Ising) or continuous domain walls. The free energy of the configuration is given by

$$F = N\theta f_A + N(1-\theta) f_B + F_{DW} \qquad \text{(A.77)}$$

where

$$F_{DW} = m\epsilon - k_B T S_{DW}(m, N) \qquad \text{(A.78)}$$

and the (dimensionless) entropy is given by

$$S_{DW}(m, N) = \ln\left[\frac{N!}{m!(N-m)!}\right] \approx m \ln\left[\frac{Ne}{m}\right] \quad . \qquad \text{(A.79)}$$

Minimization of the total free energy with respect to m yields a macroscopic average number (ie. a *finite density*) of domain walls

$$\bar{m} = N e^{-\epsilon/(k_B T)} \, . \qquad \text{(A.80)}$$

The system breaks up into m regions of finite size $e^{\epsilon/(k_B T)}$. Macroscopic phase separation can only occur at zero temperature (as the domain size goes to infinity).

Landau's argument covers a wide range of systems, e.g. double-well on-site potentials (Ising universality class), or periodic on-site potentials. It does not cover the PB case, because the DW has infinite energy[20]. It does not apply to the PS case because the loop entropy is not proportional to the size of the loop and therefore (A.77) does not hold. On the contrary, the theorem is applicable to the generalized zipper model, as its authors had correctly noted[4].

References

1. A definitive review of the subject has been given by D. Poland and H. A. Scheraga, *Theory of helix-coil transitions in biopolymers*, Academic (1970).
2. *e.g.* T.E. Creighton, *Proteins*, W.H. Freeman (1992).
3. B.H. Zimm, P. Doty and K. Iso, *Proc. Nat. Acad. Sc. (USA)*, **45**, 1601 (1959); V.A. Bloomfield, *Am. J. Phys.* **67**, 1212 (1999).
4. B.H. Zimm and J.R. Bragg, *J. Chem. Phys.* **31**, 526 (1959).

5. M. Ya. Azbel, *Phys. Rev.* **A10**, 1671 (1979).
6. (a) A. Wada, S. Yabuki and Y. Husimi, *CRC Crit. Rev. Biochem.* **9**, 87 (1980); (b) J. de Ley, *J. Theor. Biol.* **22**, 89 (1969); (c) R.B. Inman and R.L. Baldwin, *J. Mol. Biol.* **8**, 452 (1964).
7. D. Poland and H. A. Scheraga, *J. Chem. Phys.* **45**, 1464 (1966).
8. M.E. Fisher, *J. Chem. Phys.* **45**, 1469 (1966).
9. Y. Kafri, D. Mukamel and L. Peliti, *Phys. Rev. Lett.* **85**, 4988 (2000).
10. M. Peyrard and A.R. Bishop, *Phys. Rev. Lett.* **62**, 2755 (1989).
11. L. D. Landau and E. M. Lifshitz, *Quantum Mechanics*, Pergamon (1977).
12. A related instability is the wetting of interfaces, where many of the ideas discussed here have been developed, cf. D. M. Kroll and R. Lipowski, *Phys. Rev. B* **28**, 5273 (1983), R. Lipowski, *Phys. Rev. B* **32**, 1731 (1985).
13. T. Dauxois, M. Peyrard and A. R. Bishop, *Phys. Rev. E* **47**, R4 (1993).
14. T. Dauxois and M. Peyrard, *Phys. Rev. E* **51**, 4027 (1995).
15. N. Theodorakopoulos, T. Dauxois and M. Peyrard, *Phys. Rev. Lett.* **85**, 6 (2000).
16. R. K. P. Zia, R. Lipowski and D. M. Kroll *Am. J. Phys.* **56**, 160 (1988); N.Theodorakopoulos (unpublished).
17. D. Cule and T. Hwa, *Phys. Rev. Lett.* **79**, 2375 (1997).
18. Y-L Zhang, W-M Zheng, J-X Liu, Y. Z. Chen, *Phys. Rev. E* **56**, 7100 (1997).
19. N. Theodorakopoulos (to be published).
20. T. Dauxois, N. Theodorakopoulos and M. Peyrard, *J. Stat. Phys.* **107**, 869 (2002).
21. L. van Hove, Physica **16**, 137 (1950).
22. J.A. Cuesta and A. Sanchez, *J. Phys. A: Math. Gen.* **35**, 2373 (2002).
23. L. D. Landau and E. M. Lifshitz, *Statistical Physics*, Pergamon (1980).

CHARGE TRANSPORT IN A NONLINEAR, THREE-DIMENSIONAL DNA MODEL WITH DISORDER

JFR ARCHILLA

Nonlinear Physics Group of the University of Sevilla, Departamento de Física Aplicada I, ETSI Informática, Avda Reina Mercedes s/n, 41012- Sevilla, Spain
Email: archilla@us.es

D HENNIG AND J AGARWAL

Freie Universität Berlin, Fachbereich Physik, Institut für Theoretische Physik, Arnimallee 14, 14195-Berlin, Germany

We study the transport of charge due to polarons in a model of DNA which takes in account its 3D structure and the coupling of the electron wave function with the H-bond distortions and the twist motions of the base pairs. Perturbations of the ground states lead to moving polarons which travel long distances. The influence of parametric and structural disorder, due to the impact of the ambient, is considered, showing that the moving polarons survive to a certain degree of disorder. Comparison of the linear and tail analysis and the numerical results makes possible to obtain further information on the moving polaron properties.

1. Introduction

Charge transport along DNA is a subject of particular interest for two main reasons: on the one hand, it plays a fundamental role in biological functions as repair and biosynthesis; on the other hand, because of possible applications in molecular electronics and as molecular wires [1]. Results on experimental DNA conductivity are controversial. It has been reported that it is good conductor [2], insulator [3] and semiconductor [4]. A possible explanation, among others, for these striking differences can be the different long-range correlations of the DNA sequences [5].

In this work we propose a nonlinear mechanism for charge transport along DNA in the framework of the base pair picture [6,7], taking into account its spatial structure and the coupling of the spatial and electron variables. It turns out to be an efficient mechanism for charge transport which survives to a certain degree of diagonal and structural disorder.

2. Model

We consider a semi–classical three–dimensional, tight–binding model for the DNA molecule, the sketch of the system being shown in Fig. 1. The lattice

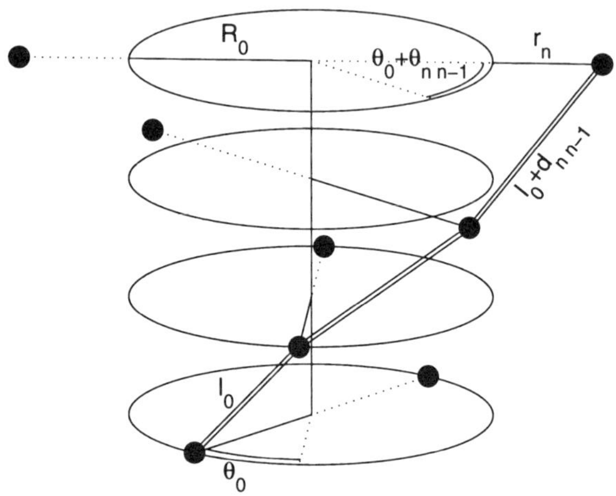

Figure 1. Sketch of the helical structure of the DNA model, the bases being represented by bullets. Geometrical parameters R_0, θ_0, l_0 and the radial and angular variables r_n and $\theta_{n,n-1}$, respectively are indicated.

oscillators are treated classically while the charge is described by a quantum system. The justification is that the nucleotides are large molecules with mass numbers of about 300. This also implies that molecular motions are small and slow compared to the one of the charge particle. The Hamiltonian of the whole system is given by $\hat{H} = \hat{H}_{\rm el} + \hat{H}_{\rm rad} + \hat{H}_{\rm twist}$, where $\hat{H}_{\rm rad}$ and $\hat{H}_{\rm twist}$ are, *de facto*, classical Hamiltonians and we can omit the hat on them. The Hamiltonians corresponding to the distances between nucleotides in each base pair $H_{\rm r}$, and the twist motion $H_{\rm twist}$ are given by

$$H_{\rm rad} = \sum_n \frac{(p_n^{\rm r})^2}{2\,M} + \frac{M\,\Omega_r^2 r_n^2}{2}; \quad H_{\rm twist} = \sum_n \frac{\left(p_{n,n-1}^{\theta}\right)^2}{2\,J} + \frac{J\,\Omega_\theta^2 \theta_{n,n-1}^2}{2} \qquad (1)$$

where $\{r_n\}$ represent the stretchings from the equilibrium distance, M the reduced mass or each base pair, Ω_r the linear radial frequency, $\{\theta_{n,n-1}\}$ are the angles between two consecutive base pairs with respect to their equilibrium value $\theta_0 = 36°$, J is the inertia moment, and Ω_θ, the twist linear frequency.

The electronic part is given by a tight–binding system of the form

$$\hat{H}_{\text{el}} = \sum_n E_n |n\rangle\langle n| - V_{n-1,n}|n-1\rangle\langle n| - V_{n+1,n}|n+1\rangle\langle n| \,. \qquad (2)$$

In the state $|n\rangle$ the charge carrier is localized at the nth base pair. The quantities $V_{n-1,n}$ represent the nearest–neighbour transfer integrals along the base pairs and E_n are the on–site matrix–elements. We write the electronic state as $|\Psi\rangle = \sum_n c_n(t)|n\rangle$, where $c_n(t)$ is the probability amplitude of finding the electron at the state $|n\rangle$. The interaction between the electronic variables and the structure variables r_n and $\theta_{n,n-1}$ arises from the dependence of the electronic parameters E_n and $V_{n,n-1}$ on the spatial coordinates, given by $E_n = E_n^0 + k\, r_n$ and $V_{n,n-1} = V_0\,(1 - \alpha\, d_{n,n-1})$, where $d_{n,n-1}$ is the first order Taylor expansion of the distances between consecutive nucleotides, with respect to their equilibrium value. Is is given by

$$d_{n,n-1} \simeq \frac{R_0}{l_0}\,[(1-\cos(\theta_0))\,(r_n + r_{n-1}) + \sin(\theta_0)\,R_0\,\theta_{n,n-1}] \,. \qquad (3)$$

Realistic parameters for DNA molecules are given in Refs. [8,9,10]. We scale the time according to $t \to \Omega_r\, t$, which allow us to write the Hamiltonian in terms of dymensionless quantities [10]. The classical Hamiltonian is defined as $H_{\text{class}} = \langle\Psi|\hat{H}|\Psi\rangle$, the dynamical equations for r_n and $\theta_{n,n-1}$ are $\dot{p}_n^{\text{r}} = M\,\ddot{r}_n = -\partial H_{\text{class}}/\partial r_n$ and $\dot{p}_{n,n-1}^{\theta} = J\,\ddot{\theta}_{n,n-1} = -\partial H_{\text{class}}/\partial\theta_{n,n-1}$, while the evolution equations for the electron variables c_n are obtained from the Schrödinger equation $\mathrm{i}\hbar\,(\partial\Psi/\partial t) = \hat{H}_{\text{el}}|\Psi\rangle$, which is equivalent to $\dot{c}_n = -(\mathrm{i}/\hbar)(\partial H_{\text{class}}/\partial c_n^*)$. In this way we can obtain the scaled dynamical equations, which are

$$\begin{aligned}
\mathrm{i}\,\tau\dot{c}_n &= (E_n^0 + k\,r_n)\,c_n \\
&\quad - (1-\alpha\,d_{n+1,n})\,c_{n+1} - (1-\alpha\,d_{n,n-1})\,c_{n-1} && (4)\\
\ddot{r}_n &= -r_n - k\,|c_n|^2 \;-\; \alpha\,\frac{R_0}{l_0}\,(1-\cos\theta_0) \\
&\quad \times \left\{ [c_{n+1}^* c_n + c_{n+1} c_n^*] + [c_n^* c_{n-1} + c_n c_{n-1}^*] \right\} && (5)\\
\ddot{\theta}_{n,n-1} &= -\Omega^2\,\theta_{n,n-1} \;-\; \alpha\,V\,\frac{R_0^2}{l_0}\,\sin\theta_0\,[c_n^*\,c_{n-1} \;+\; c_n\,c_{n-1}^*]\,, && (6)
\end{aligned}$$

where $\tau = \hbar\,\Omega_r/V_0$ represents the time scale separation between the fast electron motion and the slow bond vibrations.

The values of the scaled parameters are [10] $\tau = 0.2589$, $\Omega^2 = [0.709 - 1.417]\times 10^{-2}$, $V = 0.0823$, $R_0 = 34.862$ and $l_0 = 24.590$, the time unit being $\sim 1.6\,\text{ps}$. Note that the time scales for the different variables differ in an order of magnitude: the fastest variables are the $\{c_n\}$, with a characteristic frequency of order $1/\tau \sim 4$, followed by $\{r_n\}$ with the unity, and $\{\theta_{n,n-1}\}$ with $\Omega \sim 0.08$

There are no reliable values for the electron–radial and electron–twist coupling parameters, k and α. The criterion we have taken for their values is the consistency of the numerical simulations with the hypothesis, i.e., the deformations of the helix are small, and the linear approximations of the distance $d_{n,n-1}$ and the trigonometric functions remain valid. Typical values are $k \sim 1$ and $\alpha \sim 0.002$.

3. Stationary states

We suppose initially that r_n and $\theta_{n,n-1}$ are constant, i.e., the Born-Oppenheimer approximation. This allows the obtention of expressions for them from Eqs. (5)–(6) which are inserted in Eq. (4), leading to a nonlinear Schrödinger equation for the electronic amplitudes:

$$\begin{aligned} \mathrm{i}\,\tau\,\dot{c}_n = & \Big[E_n^0 - k^2\,|c_n|^2 - k\,\alpha\,\frac{R_0}{l_0}\,(1-\cos\theta_0) \\ & \times \big\{ [c_{n+1}^* c_n + c_{n+1} c_n^*] + [c_n^* c_{n-1} + c_n c_{n-1}^*] \big\} \Big]\, c_n \\ & - (1-\alpha\, d_{n+1\,n})\, c_{n+1} - (1-\alpha\, d_{n,n-1})\, c_{n-1}\,, \end{aligned} \tag{7}$$

whith $\{d_{n,n-1}\}$ given by Eq. (3), depending algebraically on the $\{c_n\}$ through $\{r_n\}$ and $\{\theta_{n,n-1}\}$.

To obtain stationary localized solutions we use a numerical method [11,12]. We substitute $c_n = \Phi_n \exp(-\mathrm{i}\,E\,t/\tau)$ in Eq. (7), with Φ_n constants, and obtain a nonlinear difference system $E\,\Phi = \hat{A}\,\phi$, with $\Phi = (\Phi_1, \ldots, \Phi_N)$. Its solutions are attractors of the map:

$$\Phi \to \Phi' = \hat{A}\,\Phi / ||\hat{A}\,\Phi||\,, \tag{8}$$

$||\cdot||$ being the quadratic norm. We start with a completely localized state at a site n_0, i.e., $\Phi_n = \delta_{n,n_0}$ and apply the map above until $\Phi' = \Phi$. In this way we obtain stationary localized solutions and their energy E.

In the ordered case, with $E_n^0 = E_0, \forall n$, E_0 can be made zero with the ansatz $c_n \to c_n \exp(-\mathrm{i}E_0 t/\tau)$. We obtain symmetric polarons with width of about 20 sites, combined with a local compression of order ~ 0.15 and a local unwinding of $\sim 1°$ and energies $E \lesssim -2 \sim -0.2\,\mathrm{eV}$. Considering static diagonal disorder with random E_n^0 the localization is enhanced, due to Anderson localization, the polaron being asymmetric, with a specific shape that depends on each particular disorder implementation. Similar results are obtained with structural disorder, i.e., random equilibrium values R_n^0 and θ_n^0. The Floquet analysis shows that these polarons are linearly stable [10].

4. Charge transport

To activate the polaron motion we need to perturb the zero velocities of the ground state with a localized, spatially antisymmetric mode. This is obtained either with the time–consuming pinning–mode method [13], either with the simpler discrete gradient method, i.e., perturbing the variables r_n with velocities parallel to $\{r_{n+1} - r_{n-1}\}$ [14]. The main difference is that the first guarantees the mobility, while the second does not, but in practice works well very often. Values of the modulus of the *kick* velocity $\lambda_{\mathrm{r}} = (\sum \dot{r}_n^2(0))^{1/2} \sim 0.02$, equivalent to a kinetic energy of $\sim 200\,\mathrm{meV}$ are appropriate to obtain good mobility with low radiation.

We obtain propagation of the localized electron amplitudes with constant velocity. A complementary compression and unwinding of the helix travels with it, while a localized oscillation of the angular variables remains at the initial positions. If we increase the diagonal disorder with $E_n \in [-\Delta E, \Delta E]$, the charge transport still takes place, until values of $\Delta E \sim 0.05$. Thereafter mobility becomes impossible and the polaron is pinned to the lattice. With disorder, however, the velocity of the first momentum of the electronic occupation amplitude $n_c(t) = \sum_n n\,|c_n(t)|^2$ is not uniform, as shown in Fig. 2.

We also consider structural disorder with random distributions of the base pair spacings and angles around their ordered case values, with different mean standard deviations Δ. The conclusion is that, up to values of $\Delta = 10\%$, it does not significantly affect polaron mobility, although there is a slight reduction of its velocity.

It is interesting to remark that for the values of α coherent with our small amplitude hypotheses it is not possible to move the polaron by *kicking* the angular variables. These results are coherent with the fact that the Floquet analysis shows that the pinning mode appears only in the radial variables and its eigenvalue separates from the optimal value around 1 when the disorder is increased [10].

5. Linear and tail analysis

Although our system is studied in the nonlinear regime, and the results obtained throughout this work are essentially nonlinear, there is a number of linear techniques that can be applied as a reference and as a tool for obtaining useful information. They are the study of the linear system itself and the tail analysis.

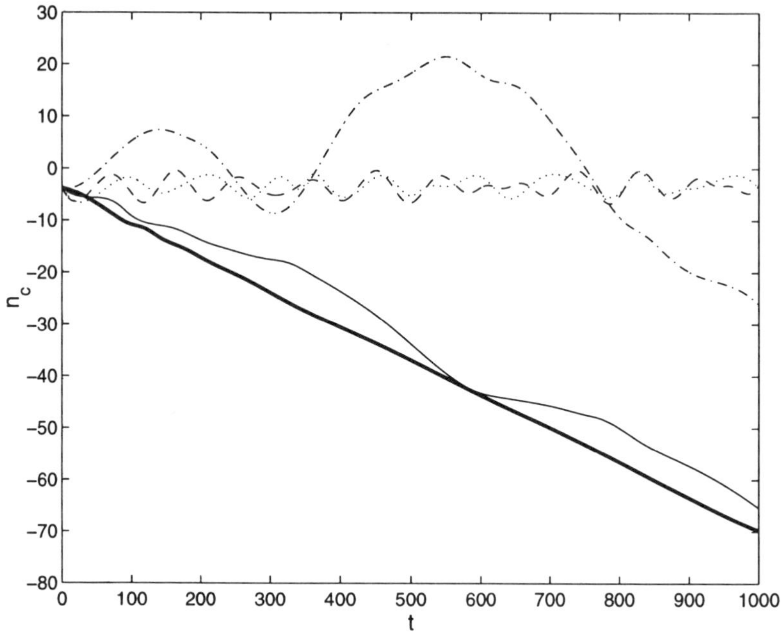

Figure 2. Diagonal disorder. The position of the center of the electron breather as a function of time and for different amounts of disorder. Full line (Ordered case), short dashed line ($\Delta E = 0.025$), dashed-dotted line ($\Delta E = 0.050$), dotted line ($\Delta E = 0.100$) and long dashed line ($\Delta E = 0.500$).

5.1. *The linear system*

If in Eqs. (4)–(6) we cancel out the nonlinear terms, we obtain:

$$\mathrm{i}\,\tau\,\dot{c}_n = -c_{n+1} - c_{n-1}\,, \quad \ddot{r}_n = -r_n\,, \quad \ddot{\theta}_{n\,n-1} = -\Omega^2\,\theta_{n\,n-1}\,. \tag{9}$$

In the first equation the terms E_n, which are all equal to some value E_0 in an homogeneous system, have been eliminated through the ansatz $c_n \to c_n \exp(-\mathrm{i}\,E_0\,t/\tau)$. The system becomes decoupled and the variables r_n and $\theta_{n\,n-1}$ correspond to independent linear oscillators with frequencies $w_r = 1$ and $\Omega = 0.0842$, respectively, in the scaled units. To obtain the ground state we substitute in Eq. (9) $c_n(t) = \phi_n \exp(-\mathrm{i}\,E\,t/\tau)$, ϕ_n being time independent. We obtain the stationary discrete Schrödinger equation: $E\,\phi_n = -\phi_{n+1} - \phi_{n-1}$. Substitution of the linear modes $\phi_n = \exp(\mathrm{i}\,q\,n)$ leads to: $E = -2\,\cos q$. Therefore, the linear energy spectrum runs from -2 to 2. The minimum energy, $E = -2$, corresponds to the wave vector $q = 0$, i.e., all the oscillators vibrating in phase. The nonlinear ground states described above have $E \lesssim -2$ and derive from this mode.

5.2. *Tail analysis*

For sites far enough from the polaron center, which can be only a few sites, we can still apply the linear analysis. Now, the substitution of the tail mode $c_n(t) = \phi_n \exp(-\xi\, n - \mathrm{i}\, q\, n)$, with $n > 0$ and $\xi > 0$, leads to: $E = -2\cos(q)\cosh(\xi)$ and $2\sin(q)\sinh(\xi) = 0$. The second equation, implies that only two wave vectors are possible, $q = 0$, with negative energy, and $q = \pi$, with positive one. The energy of the first, which is the one that appears throughout this paper because it is movable, is given by: $E = -2\cos(\xi) < -2$. Therefore, the nonlinearity, produces localized states with larger values of the frequency $w_p = |E|/\tau$ outside the linear spectrum. The distance of the energy calculated in the full system from $E_{q=0} = -2$, is a measure of the degree of nonlinearity. It is interesting to relate the values of the energies with the polaron breadth. Here, we define it loosely as three times the number of sites necessary for $|\phi_n|^2$ to decrease to a 5%. The factor 3 is there to allow for both and for the nonlinear center of the polaron. This leads to $\Delta n \sim 4.5/\xi$. An approximate table is:

ξ	2	1	0.5	0.15	0.075
Δn	2.25	4.5	9	30	60
E	-7	-3	-2.25	-2.02	-2.006

where the dimensionless E units are equivalent to 0.1 eV. This makeshift method gives a good estimate of the breadth of the excitation and produces a good fitting a few units far from its center. The numerically calculated polarons found in this work correspond to $\Delta n \sim 20$.

The tail analysis can also be applied to the moving polaron, far enough from its center. Let us propose a localized traveling wave of the form: $c_n = \exp\{-\xi\,(n+v\,t) - \mathrm{i}\,(q\,n + w_p\,t)\}$, which represents a traveling localized wave moving to the left behind the polaron, with positive ξ, q, and velocity v. Substitution in Eq. (9) leads to the following equations: $E \equiv w_p\tau = -2\cos(q)\cosh(\xi)$ and $\tau\,\xi\,v = 2\sin(q)\sinh(\xi)$. They are undetermined, as there are four unknowns: E, q, ξ and v, and do not allow us to determine the velocity. This is little surprise, as the polaron velocity depends on the energy given with the perturbation to the static one, and this is done to the spatial variables. However, these equations can be used to obtain an estimate of the moving polaron energy and wave number, using the decay length ξ corresponding to the static one, as there is no appreciable change of shape, and the velocity observed within the simulations. The conclusion is that there is only a negligible increase of the energy, i.e., the energy given by the *kick* is stored in the spatial variables. The fact that there is a imaginary part of the energy $-i\,\xi\,v\,\tau$ is, as it is known, consequence of the fact that we are dealing with only a part of the system [15], in which the energy is decreasing as the polaron moves to the left.

6. Conclusions

We have considered a model for charge transport along DNA, taking into account the 3D structure. We obtain linearly stable, localized stationary states. We are able to move them using the pinning mode and the discrete gradient methods. For the ordered case, the translational velocity is constant, while for the disordered case the electron motion is irregular and if the disorder is high enough it is impossible. We have performed a linear and tail analysis of the system, which makes possible to obtain the phonon spectrum, measure the degree of nonlinearity, obtain estimates of the polaron breath and energy. It shows that the charge transports very little energy, being most of it associate with the spatial coordinates. The main conclusion of the whole work can be that our proposed mechanism is an efficient means for charge transport along ordered and disordered DNA.

Acknowledgments

The authors are grateful to partial support under the LOCNET EU network HPRN-CT-1999-00163. JFRA acknowledges DH and the Institut für Theoretische Physik for their warm hospitality

References

1. M Ratner. *Nature*, 397:480, 1999.
2. B Alavi P Tran and G Gruner. *Phys. Rev. Lett.*, 85:1564, 2000.
3. E Braun, Y Eichen, U Sivan, and G Ben-Yoseph. *Nature*, 391:775, 1998.
4. D Porath, A Bezryadin, S de Vries, and C Dekker. *Nature*, 403:635, 2000.
5. P Carpena, P Bernaola-Galvan, PC Ivanov, and HE Stanley. *Nature*, 418:955, 2002.
6. S Cocco and R Monasson. *J. Chem. Phys.*, 112:10017, 2000.
7. M Barbi, S Cocco, and M Peyrard. *Phys. Lett. A*, 253:358, 1999.
8. M Peyrard and MD Kruskal. *Physica D*, 14:88, 1984.
9. L Stryer. *Biochemistry*. Freeman, New York, 1995.
10. D Hennig, JFR Archilla, and J Agarwal. Physica D, in press, 2003.
11. G Kalosakas, S Aubry, and GP Tsironis. *Phys. Rev. B*, 58:3094, 1998.
12. NK Voulgarakis and GP Tsironis. *Phys. Rev. B*, 63:14302, 2001.
13. D Chen, S Aubry, and GP Tsironis. *Phys. Rev. Lett.*, 112:139, 1996.
14. M Ibañes, JM Sancho, and GP Tsironis. *Phys. Rev. E*, 65:041902, 2002.
15. C Cohen-Tannoudji, B Du, and F Laloé. *Quantum Mechanics*, volume 1. John Wiley & Sons Inc, 1977.

BREATHERS AND CONFORMATIONAL TRANSITIONS IN MOLECULAR CRYSTALS

MARIETTE BARTHES
GDPC - Université Montpellier-2 -
34095 Montpellier - FRANCE

The interplay of Amide vibrational modes with lattice phonons induces strongly anharmonic properties in crystals with chains of H-bonded molecules, like Acetanilide, N-methylacetamide, L-Alanine and some polyAlanines. Theoretical models predicting nonlinear excitations are able to take into account these anomalous spectra with "unconventional" sidebands, temperature dependent intensity and intense overtones.
However, structural instabilities have been recently observed in some of these crystals. The unusual spectroscopic properties of L-Alanine are presented and discussed in the light of the new structural data.

It has been shown that the properties of non-harmonic vibrational modes implying the >-N-H...O=C< intermolecular hydrogen bond observed in some molecular crystals are well described within the theory of the Davydov soliton [1], or vibrational polaron [2], intrinsic localized mode (ilm) or breather [3].

The first observed molecular crystal having such anomalous Amide-1 vibration is Acetanilide, widely investigated [4] and discussed [5]. Related spectral anomalies have been also discovered in N-Methylacetamide [6], L-Alanine [7], and poly-Alanines [8]. However, the interpretation of the observed anomalies as experimental evidence for polarons has been in every case controversed. In our recent measurements of crystallized L-Alanine ($+H_3N$-C_2H_4-CO_2-) by different experimental techniques (Infrared, Raman scattering, Birefringence, X-ray crystallography, DSC) we observe together the anomalous hindered rotation of the NH3+ group whose properties are well described by a breather model [9], and effects resulting from some subtle symmetry breaking [9][10], which slightly change the molecular structure while preserving the crystal space group symmetry [11]. So the question of the part of breathers in the driving mechanism for this kind of conformational transition is open.

1. Experimental evidence for a strongly nonharmonic libration of NH_3^+

In crystallized L-Alanine, the infrared absorption band related to the NH_3^+ torsional mode at 490-500cm-1 is shown to have a splitting below about ≈ 220K (Fig.1), and both components have an increasing integrated intensity when decreasing the temperature[9]. This torsional mode has an intense overtone at ≈ 950-972 cm-1, (Fig.2), whose intensity also strongly increases on cooling. The Log of the normalized integrated intensity of the NH_3^+ torsion and its overtone are plotted vs T^2. A temperature dependence proportional to ≈ $\exp [- T^2 / \theta^2]$, predicted by the vibrational polaron model is obeyed below about 250K (Fig.3).

A polaron can result from the coupling of the NH3 group libration to one or more of the three H-bonds stretches . The sideband on the low energy side of the main band may be tentatively explained as the experimental signature of the vibrational polaron.

The gap in energy between the "exciton" and the polaron is expected to obey the law : $\Delta E = \nu_e - \nu_p = (\chi^2 / 2w) - 2J$, where χ is the exciton -phonon coupling, w is the H-bond elastic constant and J is the dipole coupling. Moreover, using the χ value so deduced, it is possible to calculate the frequency of the Nth overtone $\nu (N)_p$ using the theoretical model developed by A.Scott et al [12] :

$\nu (N)_p = \nu_0 . N - 1/2 (\gamma . N^2)$, with $\gamma = \chi^2 /w$ and ν_0 the internal mode frequency without any coupling with phonons. Finally, this value may be compared to the experimentally measured frequency of the overtone .

Using the present experimental results, we assume that the polaron frequency is $\nu_p \approx 495$ cm-1 , $\nu_e \approx 500$ cm-1 and $\Delta E = 5$ cm-1 .

Using the values J ≈ 1.5 to 4 cm-1(as in ACN) we estimate the anharmonic factor : $\gamma = x2/w \approx 16$ to 26 cm-1, and the frequency of the first overtone 964 cm-1 < $\nu (2)_p$ < 974 cm-1, which is very close to the experimental value of ≈ 972 cm-1 at 10K. Assuming $\chi = 6 \times 10^{-11}$ N, the H bond average force constant is estimated to be about 8N/m, slightly higher than in ACN and consistent with the shorter H-bond distance in Alanine.

From the fit of the intensity curve with the law $I \approx \exp (- T^2 / \theta^2)$ it is possible also to deduce $\theta \approx 80$K , which is a reasonable value for the Debye temperature in this type of crystal. So the application of the polaron theory to the spectroscopic properties of the $NH3^+$ group in L-Alanine is quite convincing: the $NH3^+$ libration is highly anharmonic and behaves like a breather, with a specific temperature dependence and an intense overtone.

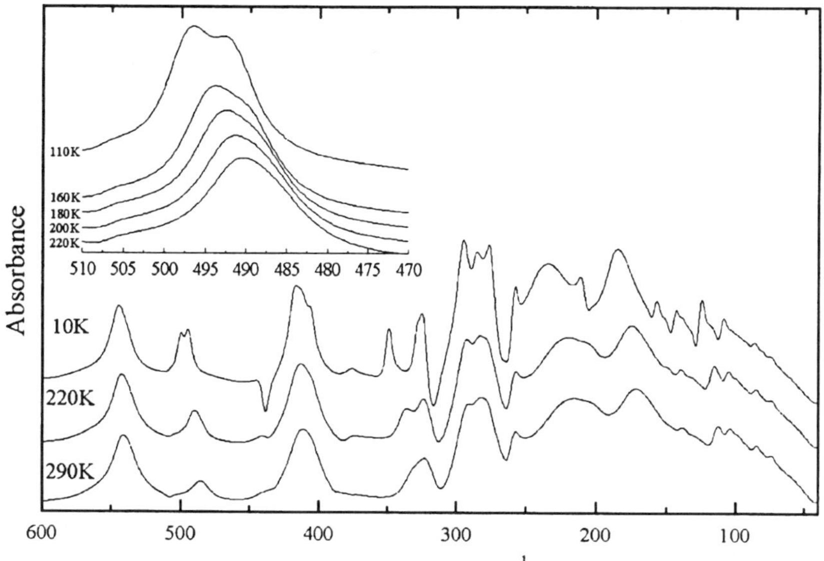

Fig.1 : IR spectra of polycrystalline L-Alanine at various temperature. The inset shows the evolution of the $NH3^{+}$ - torsional mode.

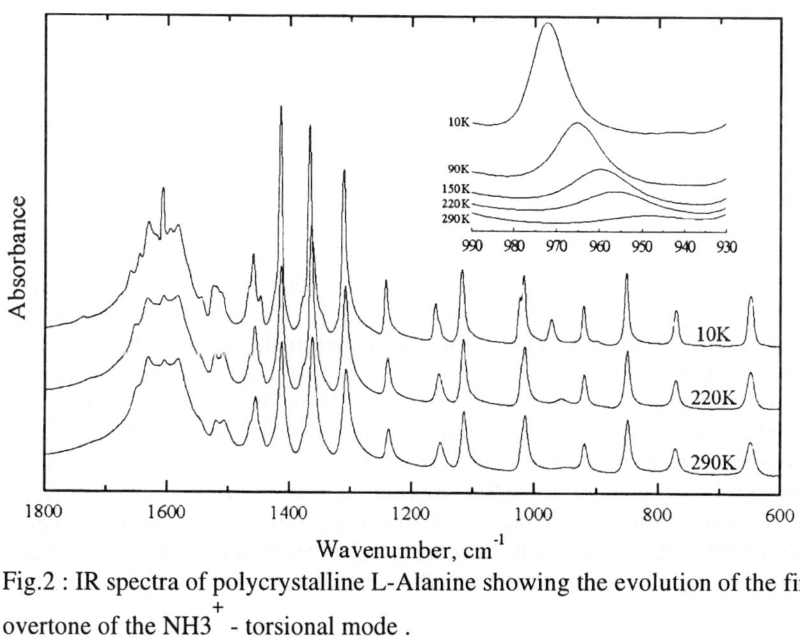

Fig.2 : IR spectra of polycrystalline L-Alanine showing the evolution of the first overtone of the $NH3^{+}$ - torsional mode .

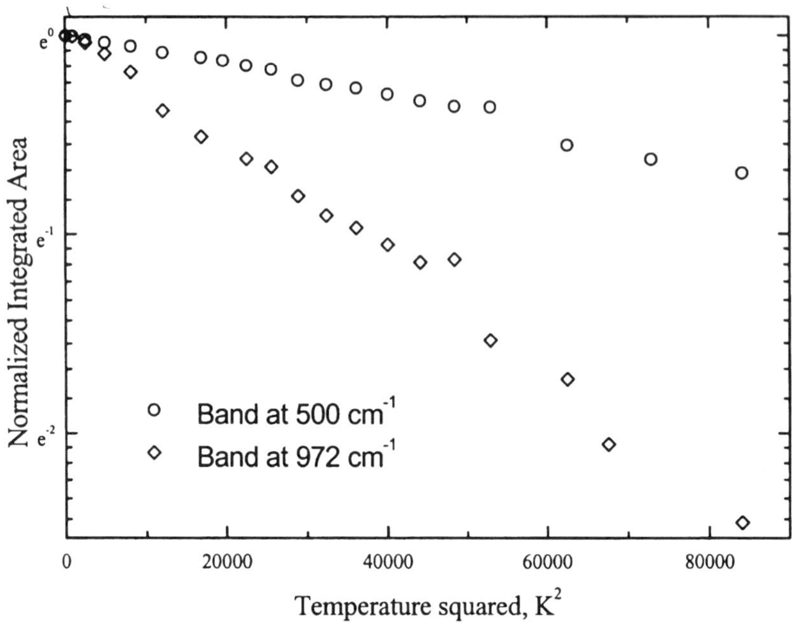

Fig. 3 : Log of the normalized integrated intensity of the NH_3^+ torsion and its overtone vs T^2. A temperature dependence proportional to $\exp[-T^2/\theta^2]$, predicted by the vibrational polaron model is obeyed below about 250K.

2 - Experimental evidence for a structural instability

The crystallographic structure of L-Alanine has been determined with X-ray and neutron scattering, at room temperature and 23K .The same orthorhombic space group has been deduced ($P2_12_12_1$) in each experiment, and no structural change was detected[13].Formerly, we could not observe any indication of structural change on high resolution birefringence curve or on DSC signal either.

However, studying the transmitted light by a single crystal of L-Alanine between crossed polarizers (zero at room temperature), a subit increase of intensity below about 220 - 240K appears, indicating that a slight distorsion in the crystallographic axes occurs at this temperature[9]. A new preliminary calorimetric study, done with a more precise calorimeter (TA Instruments) shows a very small step of the specific heat (less than 0.01W/g) around 240K (Fig.4).

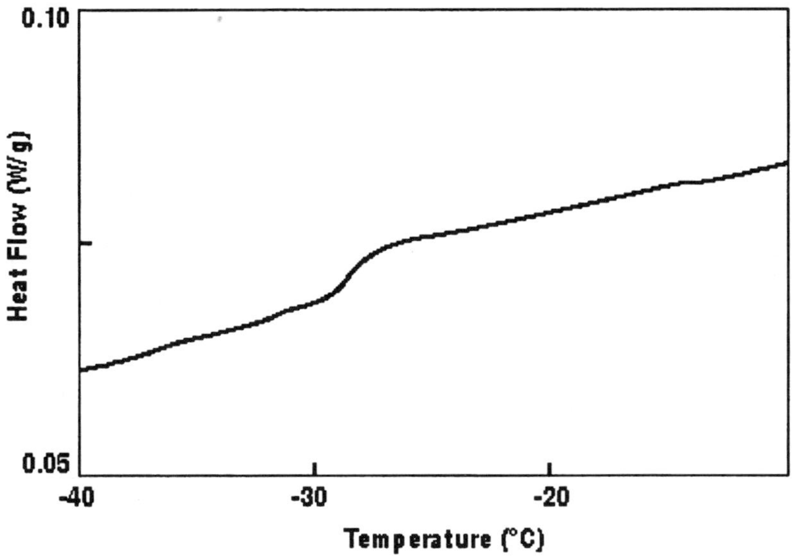

Fig. 4 - Differential scanning calorimetry of L-Alanine, showing a small discontinuity, without any latent heat.This step on the specific heat is an indication of a second order phase transition

A new polarized Raman scattering study of L-Alanine single crystals, with b(cc)b,b(ca)b and then c(bb)c scattering geometry has been performed [10], and confirms the anomalous behavior of the $NH3^+$ torsional mode (Fig.5), and its splitting (observable only with b(ca)b geometry. Moreover, most of the frequency of the external modes show a singularity around $\approx$ 220K (Fig.6) . The interplay between the two lowest-energy librons at 42 and 48 cm-1 is also confirmed.

The existence of a symmetry breaking at $\approx$ 220K in L-Alanine is now established by the three above experimental proofs (depolarization of the transmitted light, DSC, singularity on all the low frequency Raman modes).

Our very recent structural X-Ray investigations at ESRF (with ID-10A beam line) and with a four circles diffractometer operating at different temperatures confirm that a change of conformation of the L-Alanine molecule occurs below $\approx$ 220K, with an anomalous increase of the **c** lattice parameter, but the macroscopic crystal symmetry is preserved, and the best refinement of the structure is obtained assuming the same orthorhombic space group [11].

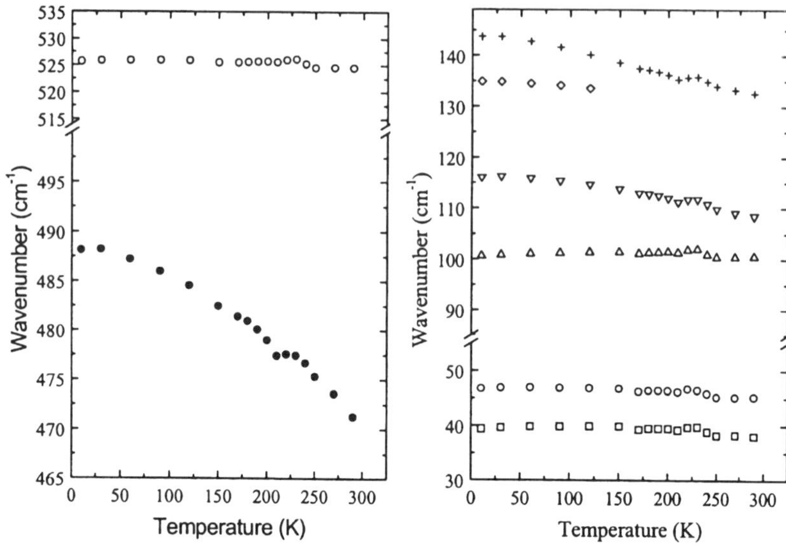

Figure 5: Left : Peak frequency of the NH_3^+ torsional mode (470-490 cm-1) and of the CO_2^- rock (≈ 525 cm-1). Right : Peak frequencies of the low-frequency b(cc)b Raman lines plotted versus the temperature

3 - Interplay of breathers and pretransitional effects

As soon as the existence of a structural instability is established in L-Alanine, the presence below the critical temperature of a new IR active component of the NH_3^+ hindered rotation could be understood without the necessity of invoking the breather theory. Moreover, the structural change consists in reversibly distorting the NH_3^+ pyramid and implies a modification of the electric dipole moment of each molecule[11].

As a consequence, the anomalous IR mode related to the NH_3^+ torsion could reflect the increase of the new order parameter of this low temperature phase, and its intensity could obey a law ~ $(T_c - T)^{1/2}$, instead of the exponential law predicted by the polaron theory.

From an experimental point of view and within the instrumental accuracy, it is difficult to determine which of both laws is obeyed in the studied temperature interval.

However, this increasing intensity on cooling should be compensated by a loss of intensity of another component of the torsional mode.This loss could be observed neither in IR nor in Raman scattering. A lowering of the symmetry with two different types of molecules could explain the splitting, but is contradicted by the crystallographic former and new data[11].

It seems that the polaron theory better takes into account together the splitting, the change of intensity, the properties of the strong overtone of the NH_3^+ torsion, but does not explain the discontinuity observed at $\approx 220K$.

At this point it is useful to note the convergence of different theories : nonlinear Hamiltonians are incorporated into Schrödinger equations driving to breathers and solitons solutions; coupling terms into the phenomenological Landau developments near structural and magnetic phase transitions (coupling between primary and secondary order parameters , Lifschitz invariants...) lead to specific excitations (amplitudons, phasons) and kinks (discommensurations, Bloch walls, ...).

So the concept of nonlinear excitations is common to both types of theoretical models. Structural phase transition usually results in macroscopic symmetry breaking , and is more familiar at three dimensions. A linear chain undergoing critical fluctuations or a regular array of breathers which modify its structure, undergoes a "dimerization" instead of a "phase change" .

In the particular case of L-Alanine we are confronted to a structural subtle change of the shape of each molecule, while the crystal unit cell symmetry is preserved. This conformational transition is accompanied by a very unusual behaviour of the NH_3^+ libration, unequally coupled to lattice phonons by the three H-bond networks which unequally contract on cooling. The split NH_3^+ libration has the properties of a breather and it is necessary to determine their part in the pretransitional regime of such a conformational transition which preserves the space group symmetry.

The specific character of such a gradual conformational change [11] should be studied in more details because L-Alanine, besides its importance for biology and nonlinear physics, is also a possible candidate for the Abdus Salam transition [14], theory which aims to explain the homochirality of biomolecules.

References

[1] A.S.Davydov , Solitons in Molecular Systems - D.Reidel Pub. Comp. - Dordrecht - 1985
[2]D.M. Alexander and J.A.Krumhansl, Phys. Rev. B **33**, 7172 (1986)
[3]Scott A.C. , Nonlinear Science - Oxford University Press - 1999
[4] G. Careri et al. , Phys. Rev. B **30**, 4689 (1984)
[5]see for example :
- B.Swanson and C. Johnston, Chem. Phys. Lett. **114**, 547 (1985)
- W.Fann et al Phys Rev Lett **64**, 607 (1990)
- S. Johnson et al., Phys. Rev. Lett.**74**, 2844 (1995)
- A. Spire et al , Physica D **137**, 392 (2000)

[6] G. Araki et al., Phys. Rev. B **43** 12662 (1991)
[7] Migliori A. et al., Phys.Rev. B **38**, 13464 (1988)
[8] R. Lohikoski et al ., in Time Resolved Vibrational spectroscopy VI . Ed. A. Lau, Springer Verlag Berlin 1994.
[9] M. Barthes et al., J. Phys. Chem. A, **106**, 5230 (2002)
[10]A. F. Vik et al , J. Phys. IV (France) accepted.
[11] M. Barthes et al. , The European Physical Journal B, to be published.
[12] A.C.Scott et al., Phys.Rev.B **32**, 555 (1985).
[13] R.Destro et al., J.Phys.Chem.**92**, 966 (1988).
[14] A. Salam, Phys. Lett. B **288**, 153 (1992)

WHAT DRIVES PROTEIN FOLDING AND PROTEIN FUNCTION?

L. CRUZEIRO-HANSSON
CCMAR and FCT, University of Algarve,
Campus de Gambelas, 8000 Faro,
PORTUGAL
E-mail: lhansson@ualg.pt

The difficulties with the current views on how proteins work are pointed out and vibrational energy transfer is proposed as an essential driver for protein folding and protein function.

1. The protein folding problem.

Proteins act as the machines of life, they drive essentially all the physical and chemical processes that go on in living cells: they catalyse reactions, pass signals and provide basic structure. Physically, they are polymers whose units are the aminoacids. In cells, the specificity of proteins is determined by the sequence of aminoacids. An average protein in the human genome has 240 aminoacids of usually 20 different types. The question of how proteins happen to choose the right conformation from the enormous number of potentially wrong ones is known as the *protein folding problem.* The current answer to this problem is to assume that the (free) energy landscape of a protein is funnel shaped [1] so that no matter what extended structure proteins start off in, thermal fluctuations will always drive them to the structure at the bottom of the funnel, which is assumed to correspond to the native structure.

While proteins need to have a well defined structure to work properly, their function is associated with conformational changes which are also well-defined transformations from one structure to another. These specific conformational changes are triggered by the binding of ions or small molecules, or by chemical reactions, such as the hydrolysis of Adenosinetriphosphate (ATP). A problem very similar to the protein folding problem can also be raised in the case of conformational changes, namely, how do proteins arrive at the same final conformation when they can potentially assume so many others? The funnel hypothesis can easily be extended to accommodate the conformational question as well by assuming that the action of the triggers leads to a new funnel, the bottom of which corresponds to the conformation the protein has in the presence of the triggers. Thus, if the funnel hypothesis is correct, protein folding and protein conformational changes can take place just by thermal fluctuations, i.e. a stochastic relaxation will always drive proteins to the structure at the bottom of the funnel.

Although models based on the funnel hypothesis have been applied to non-Arrhenius rate laws, chevron plots, mutational effects and the relation between equilibrium and kinetic measurements and are especially successful in estimating the rates of folding of proteins[2], the funnel hypothesis has yet to be proven. Molecular dynamics simulations with potentials that take into account the full atomic constitution of proteins lead to the possibility that proteins get trapped in structures that are far from the native structure and yet have similar energies (see [3], especially note 22). These results can be rationalized by assuming that the energy landscape of proteins is not funnel-shaped, but multi-funnel-shaped (see figure 1). Although a systematic probing of the energy landscape of most proteins is still precluded by the shear size of the problem, such a study of an all-atom model of polyalanine AC(ala)$_8$NHMe lead to a double funnel[4], even for this comparatively small model protein. A multi-funnel energy landscape with at least two funnels has also been found experimentally for prions, which are known to possess two or more conformations, one in which they are able to perform their functions and other(s) in which they aggregate and cause the death of cells[5].

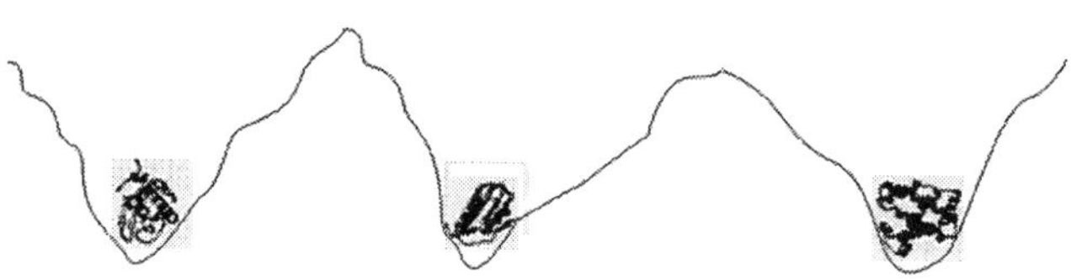

Figure 1. Artistic view of a multi-funnel energy landscape. According to this hypothesis, proteins can have very different structures with similar energies, in the same thermodynamic equilibrium conditions. Prions follow the multi-funnel picture.

In the multi-funnel hypothesis, each funnel corresponds to a structure of the protein that is sufficiently different from the native to be virtually inaccessible from it via thermal fluctuations at normal temperatures. A multi-funnel-shaped energy landscape leaves open the question of how a particular funnel, that which corresponds to the native structure, is chosen among all the other funnels. This choice cannot be the result of a random process, i.e. cannot be driven by thermal fluctuations and implies the existence of transient, deterministic, forces that are not generated by all-atom potentials. These forces, although they may be very short-lived, are fundamental for the proper function of proteins since they are responsible for the specificity of protein structure and the specificity of protein conformational changes. In the next section, the events that precede protein folding and function are analyzed and a possible source for the deterministic forces is proposed.

2. The Davydov model.

The conformational changes associated with protein work are concerted motions of hundreds or even thousands of atoms and can be considered classical events. However, the triggers of protein work are quantum events. This is particularly clear when the trigger is a chemical reaction like the hydrolysis of ATP: a chemical reaction is a quantum process and the immediate outcome of a quantum process is a quantum state. Thus, even if, at first sight, the possibility that proteins have a quantum stage may seem very speculative, in fact, at least in some cases, it is trivial. Although perhaps not so intuitive, quantum excitations can also be created by ligand binding[6]. Therefore, the question is not *whether* proteins have a quantum stage but what form this stage assumes, how long it lasts and what role it has. In the Davydov/Scott model studied here[7,8], it is assumed the energy released in the hydrolysis of ATP (or by the binding of a ligand) is deposited in a well-known vibrational mode of the peptide groups called amide I that consists essentially of the stretching of the C=O bond. In the Davydov/Scott model this vibrational excitation propagates from one group to the next because of the dipole-dipole interaction between the groups. But it also interacts with the neighbouring hydrogen bonds, leading to a deformation of the lattice and a lower energy state. The state constituted by an amide I vibration and its associated hydrogen bond distortion is known as the Davydov soliton because, in the absence of a thermal bath, it propagates coherently along a protein α-helix, i.e. without changing shape and with a constant velocity.

For a one amide I quantum the equations of motion are:

$$i\hbar\frac{d\varphi_n}{dt} = -V\left(\varphi_{n-1} + \varphi_{n+1}\right) + \chi\left(u_{n+1} - u_{n-1}\right)\varphi_n \tag{1}$$

$$M\frac{d^2u_n}{dt} = \chi\,(|\,\varphi_{n+1}\,|^2 - |\,\varphi_{n-1}\,|^2) + \kappa\,(u_{n+1} + u_{n-1} - 2\,u_n) \qquad (2)$$

where φ_n is the probability amplitude for an amide I excitation at peptide group n, $-V$ is the dipole-dipole interaction between neighbouring peptide groups, χ is an anharmonic parameter arising from the coupling between the vibrational excitation and the lattice displacements, u_n is the displacement from equilibrium positions of peptide group n, M is the mass of the peptide group and κ is the elasticity constant of the peptide lattice.

The numerical simulation of these equations (and of extensions of these equations to the three interacting chains that are present in an α-helix[8]) confirmed Davydov's analytical studies in the continuum limit[7] by showing that solitons can form and travel along the chains. It was also found that in the discrete chains there is a threshold in the nonlinearity parameter χ above which localized excitations can form. This thresholds tends to zero as the length of the system increases.

3. Behaviour at finite temperature.

The results mentioned so far are valid in the absence of thermal noise. Coupling to a thermal bath was first modelled by extending the equations of motion (1-2) into Langevin equations[9] by adding stochastic forces $F(n)$ and damping terms $-\Gamma\frac{du_n}{dt}$ to the right hand side of eq. (2) so that these terms obey the fluctuation-dissipation relations: $< F_n(t)\,F_m(t') >= 2M\Gamma k_B T\delta_{nm}\delta(t-t')$, k_B being the Boltzmann constant and T the temperature. The result is that soliton solutions disperse in a few picoseconds at biological temperatures[9]. On the other hand, exact quantum Monte Carlo (QMC) simulations[10] showed that the distortion induced by the excitation increases with temperature, while the reverse is obtained by averages performed with snapshots from the Langevin equations[11]. This conflict was resolved by showing that the coupling of a classical bath to a mixed quantum-classical system leads to a classical behaviour of the quantum part[12]. Monte Carlo methods, which can be used to study the equilibrium regime of quantum systems, mixed quantum/classical systems and classical systems, showed that while the excitation states of a classical amide I excitation at finite temperature are delocalized, the states of a quantum amide I excitation are predominantly localized[13]. A set of dynamic equations that leads to the same equilibrium averages as the QMC simulations[14] is as follows:

$$E\varphi_n = -V\,(\varphi_{n-1} + \varphi_{n+1}) + \chi\,(u_{n+1} - u_{n-1})\,\varphi_n \qquad (3)$$

$$M\frac{d^2u_n}{dt} = \chi\,(|\,\varphi_{n+1}\,|^2 - |\,\varphi_{n-1}\,|^2) + \kappa\,(u_{n+1} + u_{n-1} - 2\,u_n) + \qquad (4)$$
$$+F(n) - \Gamma\frac{du_n}{dt}$$

It has been shown that the Born-Oppenheimer-like approximation implicit in Eqs.(3-4) is valid if the quantum quasiparticle is much faster than the lattice [15]. The dynamics at finite temperature, predicted by Eqs.(3-4) is as shown in figure 2. There are two causes for the hopping motion of the

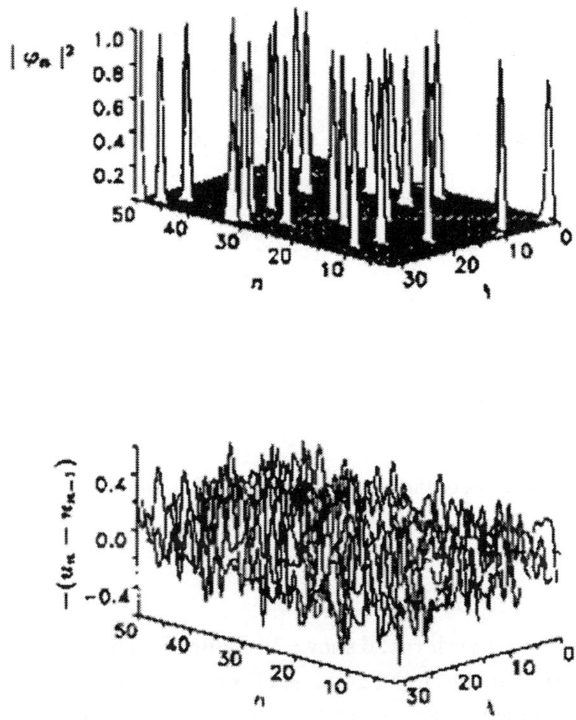

Figure 2. Dynamics of the amide I excitation at finite temperature: localized amide I states jump stochastically on top of a dynamically disordered lattice.

amide I seen in figure 2: one is the dynamics of the lattice and the other is transitions between quantum levels. Figure 3 shows four eigenstates of the amide I excitation for a typical lattice conformation at finite temperature. All states are essentially localized at a single site and the "broadest" state

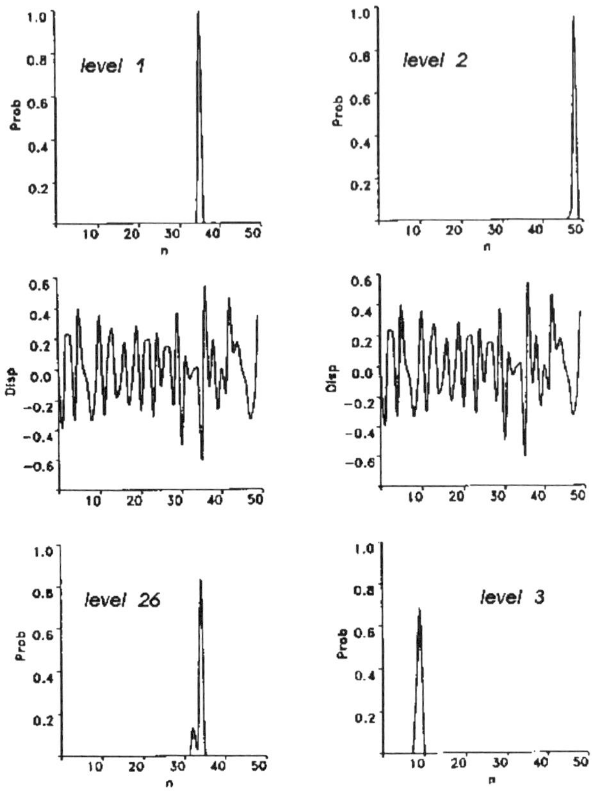

Figure 3. Four eigenstates of the amide I excitation when the lattice distortion shown.

in this case was found for level 26 shown in Figure 3. The lower energy state tends to be centred at the site of greatest effective compression of the lattice. As the lattice changes, because of thermal noise, the compression jumps from one site to another in a non-continuous way, and the amide I excitation follows it around. This is one reason for the stochastic hopping. A second reason is that, for one given lattice configuration, there are N possible energy levels for the amide I excitation, each level located at different sites. Therefore, a quantum transition also generates a transfer of the amide I vibration from one site to another, making a second contribution to the stochastic hopping seen in figure 2.

4. Discussion.

The Davydov/Scott model can explain how amide I excitations, generated at a protein active site, either by chemical reactions such as the hydrolysis of ATP or the binding of ligands, can travel, without dissipation and in a few picoseconds, to any other region of the protein where their energy is needed to do work. This work is associated with conformational changes, something the Davydov/Scott model does not describe. The proposal here is that conformational changes take place when the amide I excitations are annihilated and the energy stored in them is transferred to the more classical degrees of freedom of the protein. This step can generate the transient, deterministic, forces necessary to choose one specific funnel out of all the potential ones. The idea is that protein folding and protein function are driven by two types of forces: first, deterministic forces arising from vibrational excited states and second, random forces generated by the interactions described by all-atom potentials. The deterministic forces are created by the binding of ions and small molecules to specific locations of the unfolded protein and, although short-lived, are responsible for the selection of a particular funnel. Once a protein is at the mouth of a funnel, stochastic relaxation driven by protein atom-atom interactions drives the protein to the bottom of the funnel selected and to its native structure. This latter process is the rate-limiting step, but protein structure is defined in the first step.

This view implies the existence of a few sites in the protein that are very important for its proper function and this is indeed what is found from mutational experiments which show that most mutations do not change the structure and do not affect the function of proteins, while a few, as yet unpredictable ones, do. Also, a 30% sequence identity between two proteins usually means that their structure is very similar and, as the size of proteins *increases*, the percentage sequence identity necessary for structural similarity *decreases*[16]. Both of these facts indicate that a few aminoacids control protein structure. The suggestion is that the controlling aminoacids are the sites for generation of vibrational excited states and/or for transfer of the excitation to the classical degrees of freedom.

Protein misfolding is associated with many diseases[17]. It was first thought that there was something special about the aminoacid sequence of prions (the proteins responsible for BSE and vCJD) and of other proteins associated with misfolding diseases that explained why they aggregate. But it has been shown that proteins not associated with these diseases, such as myoglobin, can be made to aggregate and form amyloid fibrils, just as prions do[18], indicating that these states are possible for most proteins. Theories are tested by the predictions they lead to. *One prediction* of the multi-funnel energy landscape theory is that just as most proteins can form aggregates, also most proteins can acquire different structures, in the same thermodynamic conditions, as prions do. *A second prediction* is that many random sequences, if they can be synthesized using the cell machinery, will have well defined structures, i.e. the selection of aminoacid sequences in cells is not driven by the fact that only those selected will have a well-defined structure. Instead, it is predicted that the selection of aminoacid sequences is only driven by function, that is, only structures that can perform a function that the cell needs are selected. Finally, if what the triggers of protein function do is create vibrational excited states, then *a third prediction* is that it should be possible to create these states directly via interaction with a radiation field and have proteins perform their functions in the absence of their usual chemical triggers.

References

1. P.G. Wolynes *et al.*, *Science* **267**, 1619 (1995).
2. K.A. Dill and H.S. Chan, *Nature Struct. Biol.* **4**, 10 (1997).
3. T. Lazaridis and M. Karplus, *Science* **278**, 1928 (1997).
4. P.N. Mortenson and D.J. Wales, *J. Chem. Phys.* **114**, 6443 (2001).
5. P.M. Harrison, P. Bamborough, V. Daggett, S. B. Prusiner and F. E. Cohen, *Curr. Opin. Struct. Biol.* **7**, 53 (1997).
6. G. Careri and J. Wyman, *Proc. Natl. Acad. Sci. U.S.A.* **81**, 4386 (1984).
7. A.S. Davydov, *Biology and Quantum Mechanics* (Pergamon, New York, 1982).
8. A. Scott, *Phys. Rep.* **217**, 1 (1992).
9. P.S. Lomdahl and W.C. Kerr, *Phys. Rev. Lett.* **55**, 1235 (1985).
10. X. Wang, D.W. Brown and K. Lindenberg, *Phys. Rev. Lett.* **62**, 1796 (1989).
11. L. Cruzeiro-Hansson, *Physica D* **68**, 65 (1993).
12. L. Cruzeiro-Hansson, *Phys. Rev. Lett.* **73**, 2927 (1994).
13. L. Cruzeiro-Hansson and V.M. Kenkre, *Phys. Lett. A* **203**, 362 (1995).
14. L. Cruzeiro-Hansson, *Europhys. Lett.* **33**, 655 (1996).
15. L. Cruzeiro-Hansson and S. Takeno, Phys. Rev. E **56**, 894 (1997).
16. A. Fersht, *Structure and Mechanism in Protein Science*, (W.H. Freeman & Co, N.Y., 1999).
17. C.M. Dobson, *Nature* **418**, 729 (2002).
18. M. Fändrich, M. A. Fletcher and C. M. Dobson, *Nature* **410**, 165 (2001).

SOME EXACT RESULTS FOR QUANTUM LATTICE PROBLEMS

J. CHRIS EILBECK

Department of Mathematics,
Heriot-Watt University,
Edinburgh EH14 4AS, UK
E-mail: J.C.Eilbeck@hw.ac.uk

We study the exact eigenvalue spectrum for a variety of quantum lattice models, concentrating on the Quantum Discrete Nonlinear Schrödinger (QDNLS) model. In particular we discuss eigenstates which are the quantum equivalent of the classical breather. These results extend previous work, by considering larger number of quanta and bigger lattices.

1. Introduction

Physics is increasingly moving towards nano-scale technology, and with this comes the need to understand small lattices or quasi-lattices supporting a small number of quanta. Often this gives the quantum equivalent of a classical breather, though with some interesting differences. In this short paper we review some work on quantum lattice problems, concentrating mainly on the quantum discrete nonlinear Schrödinger (QDNLS) model. The corresponding classical DNLS lattice is discussed elsewhere in this book. Our eventual aim is to understand such systems as quantum dot arrays, Bose-Einstein Condensate lattices, and even models for Quantum Computers.

We adopt here a pedagogical approach with many details omitted–see[2,5,6] for a fuller description of the background material in our approach, and [3,4] for some other interesting developments by other authors.

Consider the classical DNLS model[1] in one spatial dimension with nearest-neighbour interactions

$$\mathrm{i}\frac{dA_j}{dt} + A_{j-1} + A_{j+1} + \gamma|A_j|^2 A_j = 0,$$

where $j = 1, 2, \ldots, f$ are the lattice points, the A_j's are the complex mode amplitudes at each site j, and γ is the anharmonic parameter. This equation can be derived from the following Hamiltonian

$$H = -\sum_{j=1}^{f} \left[\frac{1}{2}\gamma |A_j|^4 + A_j^*(A_{j+1} + A_{j-1}) \right],$$

where the canonical variables are A_j, A_j^*, and periodic boundary conditions are assumed $(A_{j+f} = A_j)$. As shown elsewhere in this volume, this model supports both stationary and mobile breathers, strongly localized solutions with an internal mode of oscillation. How do these properties carry over to the quantum case?

A corresponding Quantum DNLS Hamiltonian is

$$\hat{H} = -\sum_{j=1}^{f} \left[\frac{\gamma}{2} b_j^\dagger b_j^\dagger b_j b_j + b_j^\dagger (b_{j-1} + b_{j+1}) \right] \tag{1}$$

where the boson annihilation (b_j) and creation $(b_j^\dagger)$ operators destroy or create a boson at site j according to the following rules.

$$b_j|n_j\rangle = \sqrt{n_j}|n_j - 1\rangle, \quad b_j|0\rangle = 0, \quad b_j^\dagger|n_j\rangle = \sqrt{n_j + 1}|n_j + 1\rangle$$

where $|n_j\rangle$ is the number of bosons at lattice point j. The b_j satisfy the boson commutation relations $b_j b_k^\dagger - b_j^\dagger b_k = \delta_{j,k}$

An important feature is that the Hamiltonian (1) conserves the *number* of bosons in the system

$$\hat{N} = \sum_{j=1}^{f} b_j^\dagger b_j.$$

The methods we discuss can be extended to a range of other number-conserving models, for example the following boson models

- The Quantum Ablowitz-Ladik (QAL) model

  $$\hat{H} = -\sum_{j=1}^{f} \left[a_j^\dagger (a_{j+1} + a_{j-1}) \right],$$

  where $a_j^\dagger$ and a_j are operators satisfying the commutation relations $[a_j^\dagger, a_k^\dagger] = [a_j, a_k] = 0$, $[a_j, a_k^\dagger] = \left(1 + \frac{1}{2}\gamma a_j^\dagger a_k\right)\delta_{jk}$.
- The Salerno system. This is a q-deformation of the QDNLS system which interpolates between the QAL and QDNLS systems. Now $[a_j, a_k^\dagger] = \left(1 + \frac{1}{2}(\gamma - \epsilon) a_j^\dagger a_k\right)\delta_{jk}$, and the Hamiltonian is

  $$\hat{H} = -\sum_{j=1}^{f} \left[a_j^\dagger (a_{j+1} + a_{j-1}) - \right.$$

$$- \frac{2\epsilon}{(\gamma - \epsilon)} \left(\frac{\ln\left(1 + \frac{1}{2}(\gamma - \epsilon) a_j^\dagger a_j\right)}{\ln\left(1 + \frac{1}{2}(\gamma - \epsilon)\right)} + a_j^\dagger a_j \right) \Bigg] .$$

Note that the corresponding number operator will be different in all these cases, details can be found in [6,5]. The methods also extend to a number of fermion models such as a fermionic polaron model and the Hubbard model.

2. Eigenvalues of the QDNLS Hamiltonian

We now describe our computational method. Since the number is conserved, we can block-diagonalize the Hamiltonian matrix using states which are simultaneously eigenstates of $\hat{H}$ and $\hat{N}$

$$H = \left\langle \Psi \middle| \hat{H} \middle| \Psi \right\rangle = \begin{pmatrix} H_0 & 0 & & & \\ 0 & H_1 & 0 & & \\ & 0 & H_2 & 0 & \\ & & \ddots & \ddots & \ddots \end{pmatrix}$$

where H_n is the block describing states with a total of n bosons. Each eigenstate for a fixed value of n is formed as a linear combination of number states with a fixed n.

$$|\Psi_n\rangle = \sum_i c_i |\psi_n^{(i)}\rangle$$

The number states $|\psi_n^{(i)}\rangle$ are formed from the different ways we can distribute n bosons over the f sites on the lattice (we are assuming a finite lattice with periodic boundary conditions).

$$|\psi_n^{(i)}\rangle = |n_1^{(i)}\rangle |n_2^{(i)}\rangle \ldots |n_f^{(i)}\rangle = [n_1^{(i)}, n_2^{(i)}, \ldots, n_f^{(i)}], \text{ where } n = \sum_j n_j^{(i)}.$$

For example, $[2,2,0,0,0,1]$ means a state with 2 bosons on site 1, 2 bosons on site 2, and 1 boson on site 6. For a fixed value of n and f there are $(n+f-1)!/n!(f-1)!$ different number states, a quantity which expands rapidly with n and f.

We can further block-diagonalize the Hamiltonian by using the fact that it is translationally invariant, and hence we can simultaneously diagonalize with respect to the momentum operator[5,6]. As a simple example, consider a 1D periodic lattice of length $f = 3$ with $n = 2$ bosons. There are 6 possible number states $[2,0,0], [0,2,0], [0,0,2], [1,1,0], [0,1,1], [1,0,1]$, so H_2 in this case is 6×6. But we can block-diagonalize this into three 2×2

blocks $H_{2,k}$ using the translationally invariant states

$$|\Psi_2^{(a)}\rangle = [2,0,0] + t[0,2,0] + t^2[0,0,2]$$
$$|\Psi_2^{(b)}\rangle = [1,1,0] + t[0,1,1] + t^2[1,0,1]$$

with $t = 1, \exp 2\pi i/3, \exp -2\pi i/3$, so that $t^3 = 1$, with corresponding k (momentum) values $0, \pm 2\pi/3$ respectively. When f and n are large this can give a substantial saving in calculation time.

The problem of a non-translationally invariant $\hat{H}$, such as a lattice with a defect, is a more difficult problem and is currently under investigation.

2.1. *Quantum Mechanics in Maple*

We can further speed up our studies by using an algebraic manipulation package to manipulate the states and to calculate the Hamiltonian in algebraic form as a function of the parameters. In Maple, for example, we represent $[2,2,0,0,0,1]$ as `psi(2,2,0,0,0,1)`, where `psi()` is an "undefined" function. Then the operators $b_j^\dagger$ are defined something like

```
bd := proc(phi,j::nonnegint)
        nj:=op(j,phi);
        RETURN(sqrt(nj+1)*subsop(j=nj+1,phi))
end
```

with a corresponding definition for b, the QDNLS $\hat{H}$ is defined along the following lines

```
H := sum('gamma/2*bd(bd(b(b(phi,i),i),i),i)
      +bd(b(phi,cyc(i+1)),i)+bd(b(phi,cyc(i-1)),i)', 'i'=1..f)
```

where `cyc` deals with the periodic boundary conditions, i.e. `cyc(f+1)=1`, etc. This is only a brief sketch of the codes lying behind the calculations. It has recently been found that by careful optimization of the algorithms, a speedup of almost two orders of magnitude is possible. This, combined with the continuing improvement in micro-chip speeds, means that much bigger problems can be tackled than 10 years ago.

2.2. *The $n = 2$ case*

In this case each $H_{2,k}$ is tridiagonal. In the case of the QDNLS equation for large odd f, the value of $H_{2,k}$ is given by

$$H_{2,k} = \begin{pmatrix} \gamma & \sqrt{2}q^* & & & & \\ \sqrt{2}q & 0 & q^* & & & \\ & q & 0 & q^* & & \\ & & q & 0 & q^* & \\ & & & \ddots & \ddots & \ddots \end{pmatrix},$$

where $q = -(1 + \exp(ik))$. Eigenvalues and eigenvectors can be calculated numerically, or analytically, and some simple formula are known in the limit $f \to \infty$. An investigation of the 1D case for various models was made in 1992 by Eilbeck and Pego, but unfortunately this work has not yet been published. The results in the limit $f \to \infty$ for the QDNLS model are shown in Fig.1. In this figure, eigenvalues are plotted vertically for each k value

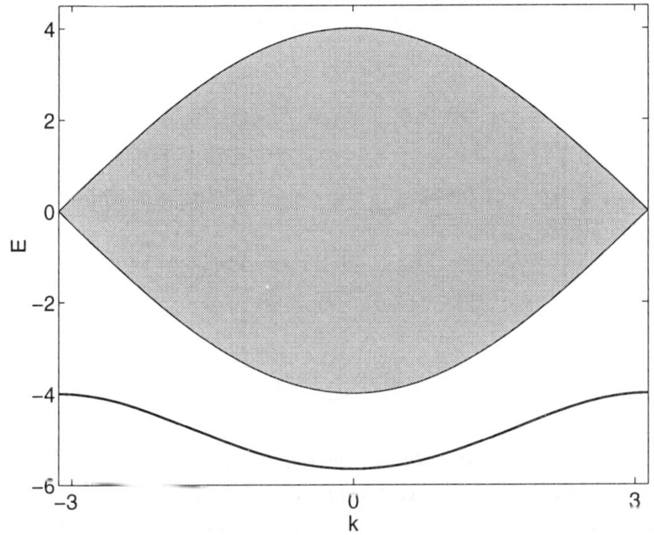

Figure 1. Eigenvalues $E(k)$ for QDNLS, $n = 2, \gamma = 4$. The lower band is the "breather" band.

on the horizontal axis. There is a continuum set and an isolated eigenvalue for each k. The equation for the lower band of isolated eigenvalues is, in the limit $f \to \infty$,

$$E = -\sqrt{\gamma^2 + 16\cos^2(k/2)}$$

The isolated band corresponds to an eigenvector of $H_{2,k}$ with the following form

$$v = \left(\frac{1}{\sqrt{2}}, \mu, \mu^2, \mu^3, \dots\right)',$$

where

$$\mu = \frac{-(\gamma + E)e^{ik/2}}{4\cos(k/2)}.$$

This eigenvalue and the corresponding eigenvector can be checked by direct calculation. It is also straightforward to check that $|\mu|^2 < 1$ for $\gamma > 0$. When $\gamma \gg \cos(k/2)$ we have

$$\mu \approx \frac{2}{\gamma}\cos(k/2)\exp(ik/2)$$

so that $\mu \to 0$ as $\gamma \to \infty$. Note also that $\mu = 0$ when $k = \pm\pi$, a result originally pointed out by S. Flach (private communication).

The ordering is such that the ith element of v multiplies the following translationary invariant states Ψ_i

$$\begin{aligned}
\Psi_1 &= [2,0,0,\dots] + e^{ik}[0,2,0,\dots] + e^{2ik}[0,0,2,\dots] + \dots \\
\Psi_2 &= [1,1,0,\dots] + e^{ik}[0,1,1,0,\dots] + e^{2ik}[0,0,1,1,0,\dots] + \dots \\
\Psi_3 &= [1,0,1,0,\dots] + e^{ik}[0,1,0,1,0,\dots] + e^{2ik}[0,0,1,0,1,0,\dots] + \dots \\
\dots &= \dots
\end{aligned}$$

so that the (unnormalized) eigenfunction is

$$\Psi = \sum_{i=1}^{\infty} v_i \Psi_i$$

This is a localized eigenfunction in the sense that there is a high probability of finding the two bosons on the *same* site, but with an equal probability of finding these two bosons at *any* site in the system. *We claim this is a quantum analogue of the classical localized breather.*

2.3. *Results for general* n, γ

There are some general results for general n, but only in the large γ limit[6,5]. In this case there is still a "breather" band with eigenfunction

$$\Psi \approx [n,0,0,\dots]+[0,n,0,\dots]+\dots+[0,0,\dots,n]+O\left(\gamma^{-1}\right)([n-1,1,0,\dots]+\dots),$$

but the continuum band in the $n = 2$ case now bifurcates into a number of separate bands. These can be understood by considering the γ-dependent term in the Hamiltonian

$$-\sum_{j=1}^{f} \left[\frac{\gamma}{2} b_j^\dagger b_j^\dagger b_j b_j \right].$$

Any site with $n_j > 1$ will contribute a value $\frac{1}{2}\gamma n_j(n_j - 1)$ to this sum. If all the bosons are on one site this gives $\frac{1}{2}\gamma n(n-1)$. If there are $n-1$ bosons at one site and 1 at another site, the contribution will be $\frac{1}{2}\gamma(n-1)(n-2)$. If there are $n-2$ bosons at one site and 2 at another site, the contribution will be $\frac{1}{2}\gamma[(n-2)(n-3)+2]$, and so on.

For example, consider the case $n = 4$. With $\gamma = 0$ we get the spectrum shown in Fig.2. Note that now we are working with a finite size lattice, the spectrum is for discrete k values, but it is clear that in the continuum limit we will get a single continuum band. If we now increase γ to 7 we get Fig.3.

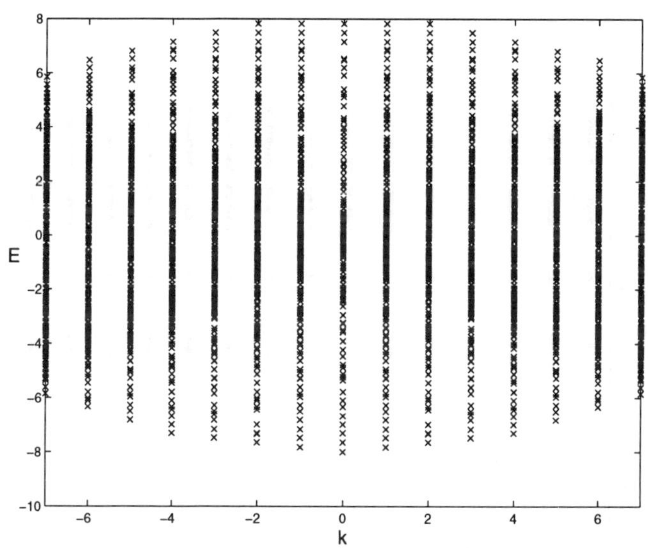

Figure 2. Example: 4 bosons, 15 sites, $\gamma = 0$

Now the single breather band has clearly split off below, and the main continuum band has started to split into two or more bands. If we increase γ to 14 we get Fig.4. Now there are 5 bands clearly visible. The lowest corresponds to the single 4-breather band, conveniently labelled (in the

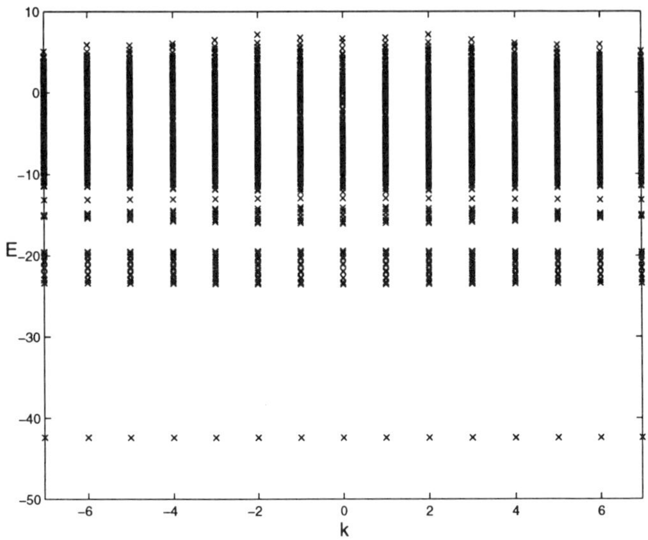

Figure 3. Example: 4 bosons, 15 sites, $\gamma = 7$

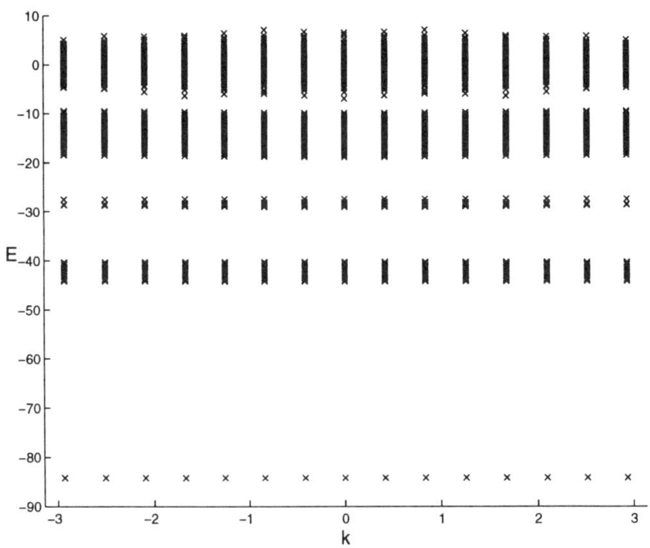

Figure 4. Example: 4 bosons, 15 sites, $\gamma = 14$

large γ limit) as $[4, 0, \ldots]$ (plus cyclic permutations). The next lowest is the "3-breather band plus single boson" band $[3, 1, 0, \ldots]$, plus permutations. The next narrow band is the "double 2-breather band" $[2, 2, 0, \ldots]$. Interestingly, this band shows some structure at higher magnification which will be reported elsewhere. Moving up, the penultimate band is the "2-breather band plus single bosons", $[2, 1, 1, 0, \ldots]$, and the top band consists of single bosons only, $[1, 1, 1, 1, 0, \ldots]$.

As a final example, we show the result of a calculation on a 2D lattice with $n = 2$. Fig.5. The "breather band" is now a 2-dimensional sheet, and

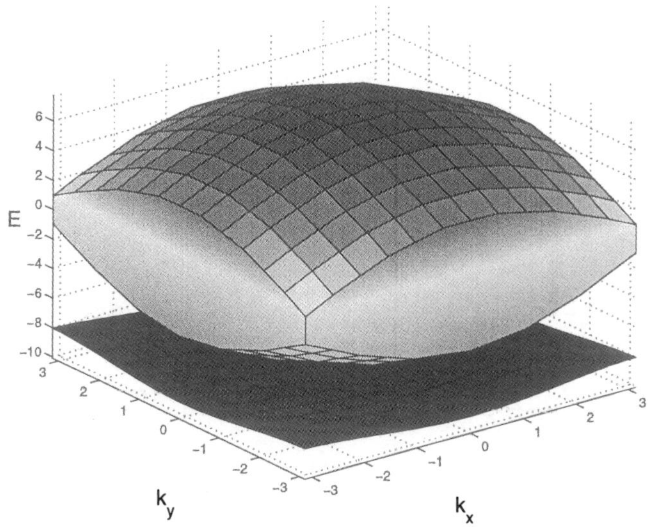

Figure 5. Example: 2 bosons, 13×13 lattice

the "continuum band" is a lens-shaped volume.

Further work is now concentrating on the fine structure of these bands in one and higher dimensions, both in QDNLS and in QAL and other models.

Acknowledgements

I am grateful for support from the EU under the LOCNET grant. I would also like to thank Oliver Penrose, Mario Salerno and Alwyn Scott for many helpful discussions.

References

1. J. C. Eilbeck, P. S. Lomdahl, and A. C. Scott. The discrete self-trapping equation. *Physica D*, 16:318–338, 1985.
2. J. C. Eilbeck and A. C. Scott. Quantum lattices. In K. H. Spatschek and F. G. Mertens, editors, *Nonlinear Coherent structures in Physics and Biology, NATO ASI Series B: Physics*, volume 329, pages 1–14. Plenum Press, 1994.
3. S. Flach and V. Fleurov. Tunnelling in the nonintegrable trimer - a step towards quantum breathers. *J. Phys.-Cond. Matt.*, 9:7039–7061, 1997.
4. R. S. MacKay. Discrete breathers: classical and quantum. *Physica A*, 288:174–198, 2000.
5. A. C. Scott. *Nonlinear Science*. OUP, Oxford, 1999.
6. A. C. Scott, J. C. Eilbeck, and H. Gilhøj. Quantum lattice solitons. *Physica D*, 78:194–213, 1994.

COLLECTIVE ROTATIONAL TUNNELLING AND QUANTUM SINE-GORDON SOLITONS

FRANÇOIS FILLAUX
LADIR-CNRS, UMR 7075 Université P. et M. Curie, 2 rue Henry Dunant, 94320 Thiais, France

The collective rotational dynamics in 1-D of methyl groups in the 4-methylpyridine crystal are analyzed. In the quantum sine-Gordon theory, heavy pseudo particles corresponding to coherent translation of large numbers of kinks or antikinks account for collective rotation. We speculate these excitations are also solutions of the fully periodical Hamiltonian for an isolated infinite chain of coupled methyl groups. In the quantum regime these spatially extended particles are not diffracted by the chain lattice and free translation takes place if the kinetic energy is within the tunnelling energy band. The periodicity of the eigen states imposes quantization of the kinetic momentum p of the particles to be an integer number of $\hbar/r$, where r is the radius of the tops. The calculated spectrum is in accordance with previously reported transitions for collective tunnelling.

1. INTRODUCTION

Nonlinear dynamics giving rise to spatially localized non-dissipative waves in an extended lattice are a source of new phenomena and technological principles in advanced materials research.[1] They are also supposed to be key elements in complex events on the molecular level of life functioning. Among nonlinear models, the sine-Gordon equation covers a vast area of applications: dislocations in solids, long Josephson junctions ferromagnets, charge-density waves, liquid crystals, methyl rotation, field theory, *etc.* Nonlinear waves such as kinks, antikinks and breathers are exact solutions of this integrable system in the continuous limit. In the classical regime, the stability and dynamics of these excitations are rather well documented but theoretical and experimental studies of the quantum regime are rather scarce. The 4-methylpyridine crystal (4MP or γ-picoline, C_6H_7N) is a unique example where the rotational dynamics of infinite chains of coupled methyl groups can be represented with the quantum sine-Gordon theory.[2,3,4,5,6] This theory accounts for a long list of experimental data. However, it was built with different pieces that could be mutually in conflict: the fully periodical Hamiltonian for an infinite chain of coupled rotors

–see below Eq. (1)– and the sine-Gordon Hamiltonian –see below Eq. (5)– that is an approximation, in the strong coupling limit, of the fully periodical case. This gives extended tunnelling states in an energy band structure, on the one hand, and discrete travelling states of the breather pseudo particle, on the other. However, previously reported measurements of the decay of the tunnelling band intensity on a rather long timescale of $\approx$ 70 h[4] is not consistent with tunnelling transitions arising from the ground state. In this talk, we consider an alternative representation of the tunnelling transitions with thermally activated pseudo particles composed of a large number of kinks or antikinks.

2. Theoretical model

The Hamiltonian for an isolated infinite chain of coupled rotors can be written as

$$H = \sum_j -\frac{\hbar^2}{2I_r}\frac{\partial^2}{\partial\theta_j^2} + \frac{V_0}{2}(1-\cos 3i\theta_j) + \frac{V_c}{2}\left[1-\cos 3i\,(\theta_{j+1}-\theta_j)\right], \quad (1)$$

where θ_j is the angular coordinate of the jth rotor in the one-dimensional chain with parameter L. V_0 is the on-site potential which does not depend on lattice position, and V_c is the coupling ("strain" energy) between neighboring rotors. The index i = 1, 2 ... determines the potential periodicity compatible with the C_{3v} symmetry of the methyl-groups.

In a previous work, rotational tunnelling was regarded as a one-dimensional band-structure problem.[2] The tunnel splitting varies continuously between two extremes (E_{ip} and E_{op}) located at the zone center, where methyl groups are tunnelling in-phase, and at the zone boundary, where tunnelling occurs out-of-phase:

$$\begin{aligned} H_{ip} &= -\tfrac{\hbar^2}{2I_r}\tfrac{\partial^2}{\partial\theta_j^2} + \tfrac{V_0}{2}(1-\cos 3i\theta) \\ H_{op} &= -\tfrac{\hbar^2}{2I_r}\tfrac{\partial^2}{\partial\theta_j^2} + \tfrac{V_0}{2}(1-\cos 3i\theta) + \tfrac{V_c}{2}(1-\cos 6i\theta)\,. \end{aligned} \quad (2)$$

These extremes correspond to the maxima of the density-of-states and transitions observed with the INS technique at (539 ± 4) and $(472 \pm 4)\mu$eV in pure 4-methylpyridine were attributed to in-phase and out-of-phase (E_{op}) tunnelling transitions of the chain, respectively. The potential terms V_0 and V_c were determined accordingly. Apart from a phase factor, the wave functions for the tunnelling states can be represented with the basis set for free rotors as

$$\varphi_{0A}(\theta) = (2\pi)^{-1/2} a_{0A0} + \pi^{-1/2} \sum_{n=1}^{\infty} a_{0A3n} \cos(3n\theta)$$
$$\varphi_{0E+}(\theta) = \pi^{-1/2} \sum_{n=1}^{\infty} a_{0E+n} \cos(n\theta) + a_{0E+2n} \cos(2n\theta) \quad (3)$$
$$\varphi_{0E-}(\theta) = \pi^{-1/2} \sum_{n=1}^{\infty} a_{0E-n} \sin(n\theta) - a_{0E-2n} \sin(2n\theta).$$

The tunnel splitting is $E_{0E\pm} - E_{0A}$. The wave function for the lowest state is φ_{0A}, whilst φ_{0E+} (symmetrical) and φ_{0E-} (anti-symmetrical) correspond to the degenerate tunnelling states. The inelastic neutron scattering function is

$$S_{0A0E\pm}(Q_r, \omega) = \left|\langle \varphi_{0E\pm}(\theta)| \exp(iQ_r \times r\theta) |\varphi_{0A}(\theta)\rangle\right|^2 \delta(E_{0A0E\pm} - \hbar\omega) .(4)$$

The momentum transfer vector $\mathbf{Q} = \mathbf{k_0} - \mathbf{k_f}$ with $|\mathbf{k_0}| = 2\pi/\lambda_0$ and $|\mathbf{k_f}| = 2\pi/\lambda_f$, where λ_0 and λ_f are the incident and scattered wavelengths, respectively. Q_r is the component of the momentum transfer vector perpendicular to the axis of rotation. $S_{0A0E\pm}(Q_r, \omega)$ is non-zero if $Q_r \times r$ is an integer number, say t.

In the strong coupling (or displacive) limit, $\theta_{j+1} - \theta_j$ is small and Eq. (1) can be expanded into the sine-Gordon equation

$$H \approx \sum_j -\frac{\hbar^2}{2I_r} \frac{\partial^2}{\partial\theta_j^2} + \frac{V_0}{2}(1 - \cos 3i\theta_j) + \frac{(3i)^2 V_c}{4}(\theta_{j+1} - \theta_j)^2. \quad (5)$$

In the continuous limit, when variations of θ from site to site are small, this Hamiltonian is integrable and all excitations are well-known.[7,8,9,10,11,12,13,14,15,16,17,18,19] Kinks, antikinks and breathers are elementary excitations. Rotons (phonons) are beyond the tunnelling frequency range. At a low temperature, the kink density vanishes at thermal equilibrium. Only breathers may survive. In the quantum regime there is only one mass state, owing to the threefold symmetry of methyl groups. The breather wave behaves as a dimensionless pseudo particle. It can travel freely along the chain if the associated de Broglie wavelength is an integer fraction of the lattice parameter. The kinetic energy spectrum is

$${}^{q}E_{B,l,n} = \sqrt{{}^{q}E_{B,l,0}^2 + n^2\hbar^2\omega_c^2}; \quad n = 0, \pm1, \pm2 \ldots \quad (6)$$

The INS band observed at 517 μeV was thus assigned to the $|0\rangle \to |1\rangle$ transition, in rather good agreement with the values for V_0 and V_c estimated independently from the tunnelling transitions. However, the theoretical framework is not totally satisfactory as Eqs (1) and (5) do not apply to the same dynamical regime. The tunnelling energy band is not included in the strong coupling limit that does not account for the periodicity of the coupling potential. In the next section we report experimental data that discard the energy band scheme.

3. SPECTRA

The first measurements ever reported of the tunnelling transition in 4-methylpyridine showed a single band at $\approx 520\ \mu$eV, with a limited resolution of $\approx 200\ \mu$eV.[20] With a better resolution of $\approx 15\ \mu$eV, this band was found to be split into several components.[21] As well as the main band at 510 μeV, weaker bands at 468 and 535 μeV were partially resolved. Further analysis of the main band at 510 μeV revealed that it could be decomposed into unresolved components at ≈ 515 and 500 μeV. However, with a better resolution of $\approx 9\ \mu$eV and a sample at the very low temperature of 0.5 K, the fourth component was not confirmed.[5] Finally, the spectra with the best resolution ever obtained of $\approx 1\ \mu$eV, shown in Figure 1, have revealed new features.[4]

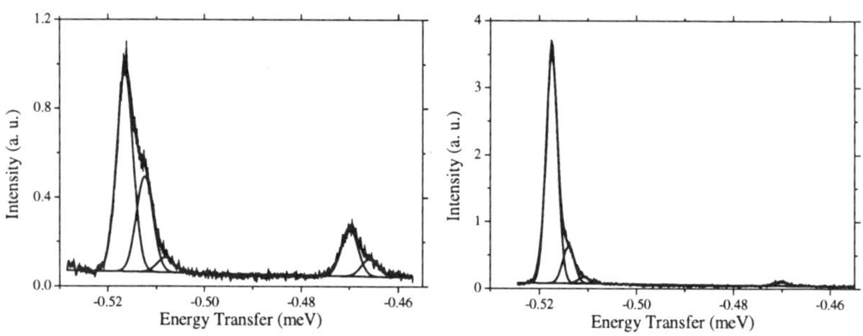

Figure 1. Inelastic neutron scattering spectra of 4-methylpyridine at 1.6 K, after ref. 4. Left: after ≈ 5 h. Right: after ≈ 70 h.

First, the main band previously assigned to the breather mode can be actually decomposed into 3 Gaussian profiles centered at 517.5, 513.8 and 510.6 μeV. The weaker band tentatively assigned to anti-phase tunnelling can be decomposed into two components at 470.5 and 467.4 μeV. Second, the weaker components of the breather mode at 513.8 and 510.6 μeV and the tunnelling bands disappear progressively with time, whilst the band intensity at 517.5 μeV increases. All observations confirm the assignment of the transition at 517.5μeV to the $|0\rangle \rightarrow |1\rangle$ transition of the breather travelling mode. The intensity decay suggests that other transitions arise from unstable species or states.

4. Multi-Kinks

Within the sine-Gordon theory, a kink or an antikink travelling along the chain can rotate the methyl groups by $\pm 2\pi/3$. This is analogous to a classical jump over the potential barrier. In contrast to the breather mode, these excitations are not parts of the ground state and must be created (for example thermally). The renormalized energy at rest of a kink is ${}^{q}E_{K0} \approx$ 11.5 meV,[2] and the energy required for the creation of a pair comprising a kink and an antikink from a breather is ≈ 5.32 meV. Transitions in the 0.5 meV are not appropriate for the creation of pairs.

In the quantum regime kinks and antikinks can be regarded as dimensionless particles and diffraction by the discrete lattice gives a quantization rule analogous to that for the breather travelling mode:

$$E_K(n) = \sqrt{{}^{q}E_{K0}^2 + n^2\hbar^2\omega_c^2}; \quad n = 0, \pm 1, \pm 2 \ldots \tag{7}$$

With this equation, the $|0\rangle \to |1\rangle$ transition is calculated at $\approx 720\ \mu$eV, quite far from the observed frequency. The $|0\rangle \to |2\rangle$ transition is at ≈ 2.7 meV, *etc.* Therefore, Eq. (7) is not appropriate for the transitions observed at 470 and 535 μeV.

In order to get rid of the translational quantization we consider spatially extended pseudo particles composed of several kinks or antikinks, which are solutions of the sine-Gordon equation.[23,24,25] We can thus build pseudo particles with well defined kinetic momentum values composed of any number of elementary excitations with the same velocity. It has been conjectured that kink positions should obey the Fermi statistics.[7] Consequently, the planar waves associated to the translation of each kink or antikink should have different spatial phases, presumably distributed at random. The amplitude of the planar wave associated to the composed particles should diminish as the number of kinks or antikinks increases and the quantization rule holding for dimensionless particles should be progressively relaxed. Then, free translation may occur at any energy and the discrete nature of the chain can be ignored.

The probability of creation, from the ground state, of pseudo particles composed of a large number of kinks or antikinks, all with the same velocity, is infinitely small at a low temperature. However, infrared and Raman spectra of 4-methylpyridine have shown that the sine-Gordon dynamics appears only below ≈ 100 K.[22] Above this temperature methyl groups are disordered and a large number of pseudo particles can be created as the sample is cooled down. In the classical sine-Gordon equation the number of kinks is strictly equal to the number of antikinks. This is no longer true for a disordered chain at high temperature and we speculate that after annihilation of all existing pairs, *via* collisions, the remaining unpaired kinks or antikinks can exchange kinetic momentum and thus give rise to non dispersive pseudo particles. In the sine-Gordon equation, there is no channel available for these excitation to decay. We suppose that in a real crystal, such excitations can have long enough life times to be observed.

If we ignore interactions between kinks or antikinks, the energy at rest of a N-soliton, made of N elementary excitations, is $N^q E_{K0}$. The propagation of such pseudo particles along the chain represents collective rotation of the methyl groups and the energy at rest is the activation energy for semiclassical jumping over the potential barrier.

In order to account for tunnelling it is necessary to reconsider the fully periodical Hamiltonian and we suppose that the sine-Gordon excitations are also solutions of Eq. (1), at least to a level of accuracy compatible with the observed life times. Then, the rotational periodicity of the wave functions in Eq. (3) imposes quantization of the kinetic momentum according to Eq. (4)

$$p = t\hbar/r \; ; \quad t = 0, \pm 1, \pm 2, \cdots \tag{8}$$

Furthermore, free translation along the chain takes place only if the kinetic energy is within the tunnelling energy band. The spectrum is then

$$E_{op} \leq \quad E\,(N,n) = \sqrt{N^{2q}E_{K0}^2 + t^2\hbar^2\omega_c^2 L^2/r^2} \quad \leq E_{ip}. \tag{9}$$

Table 1. Comparison of the calculated transitions for the travelling frequencies of N-solitons with observed frequencies

N	$N^q E_{K0}$(meV)	$\nu_{0\to 1}$ (μeV)	
		Calc.	Obs.
22	253.0	540	539[5]
23	264.5	517	514[4]
24	276.0	495	$\approx$ 500[21]
25	287.5	475	472[5]

Numerical values obtained with $L \approx 4r$, according to the crystal structure,[26] are compared to the observed frequencies in Table 1. The bands at 539 and 472 μeV correspond to $N = 22$ and 25, respectively. The frequency calculated for $N = 24$ is close to those reported previously for transient bands. The transient bands at ≈ 514 μeV in the spectra presented in Figure 1 are close to the frequency calculated for $N = 23$. These bands are very close in energy to the breather travelling transition. However, there is no interaction between these dynamics because they correspond to quite different values of the momentum transfer (1 $Å^{-1}$ and 1.5 $Å^{-1}$, respectively). The overall agreement with observations is encouraging and can be certainly improved as E_{ip} and E_{op} do not correspond any longer to observed transitions. This gives more flexibility to the determination of V_0 and V_c.

However, this model does not account for the "hyperfine" structure of the bands at 513.8–510.6 and 470.5–467.4 μeV. Furthermore, whilst the bands at $\approx$ 472 and 514 μeV decay at about the same rates, the band at $\approx$ 500 μeV seems to disappear more rapidly. In Figure 1 it has already vanished after $\approx$ 5 h. These pending problems deserve further investigations.

5. Concluding remarks

Collective rotational dynamics can be represented with pseudo particles that are solutions of the sine-Gordon equation in the continuous limit.

The breather mode represents localized oscillations of the methyl groups with respect to their equilibrium positions. The localized wave form can travel along the chain without dispersion. It can be regarded as a dimensionless pseudo particle. Diffraction by the chain lattice gives rise to quantization of the kinetic momentum.

Tunnelling is a nontrivial problem for infinite chains of coupled methyl groups. Extended states with an energy band structure are in conflict with experiments and transient species must be considered. We suppose that pseudo particles composed of large numbers of kinks or antikinks travelling at the same velocity appear spontaneously as the sample is cooled down. Because of the exclusion principle for fermions, these pseudo particles have spatial extension and are no longer sensitive to the discreteness of the chain lattice. We further suppose that these pseudo particles are also hosted by the fully periodical Hamiltonian. Then, quantization of the kinetic momentum arises from the rotational periodicity of the tunnelling states and free translational occurs for kinetic energy values within the tunnelling energy band. The calculated spectrum fits the observed transitions, but does not account for further splitting of the bands. These pseudo particles disappear naturally at a low temperature, at thermal equilibrium.

We finally conclude that analytical solutions of the sine-Gordon equation are sufficiently close to solutions of the exact Hamiltonian for an isolated infinite chain of coupled methyl groups to account for the spectra.

In the introduction of this talk we referred to nonlinear dynamics giving rise to spatially localized non-dissipative waves in an extended lattice. We suspect that this may be true only in the classical regime. In the quantum regime, non-dispersive excitations behave as dimensionless pseudo particles and localization occurs in momentum space rather than in direct space.

References

1. A. Scott, *Nonlinear Science. Emergence & Dynamics of Coherent Structures* (Oxford University Press, 1999).
2. F. Fillaux and C. J. Carlile, Phys. Rev. B **42**, 5990 (1990).
3. F. Fillaux, C. J. Carlile, and G. J. Kearley, Phys. Rev. B **44**, 12280 (1991).
4. F. Fillaux, C. J. Carlile, J. Cook, A. Heidemann, G. J. Kearley, S. Ikeda, and A. Inaba, Physica B **213&214**, 646 (1995).
5. F. Fillaux, C. J. Carlile, and G. J. Kearley, Phys. Rev. B **58**, 11416 (1998).
6. F. Fillaux, B. Nicolaï, and A. Cousson, *Experimental Studies and Theory of Nonlinear Rotational Dynamics in the Quantum Regime: The Interplay of Structure, Dynamics and Localization in Crystals* (World Scientific Publishing Co., Inc, 2003).
7. R. Dashen, B. Hasslacher, and A. Neveu, Phys. Rev. D **10**, 4130 (1974).
8. R. Rajaraman, *Solitons and instantons. An introduction to solitons and instantons in quantum field theory* (North-Holland, Amsterdam, 1989).
9. A. Scott, F.Chu, and D. McLaughlin, IEEE **61**, 1443 (1973).
10. R. Dashen, B. Hasslacher, and A. Neveu, Phys. Rev. D **11**, 3424 (1975).
11. S. Coleman, Phs. Rev. D **11**, 2088 (1975).
12. M. J. Rice, A. R. Bishop, J. A. Krumhansl, and S. E. Trullinger, Phys. Rev. Letters **36**, 432 (1976).
13. R. Jackiw, Rev. Modern Phys. **49**, 681 (1977).
14. M. B. Fogel, S. E. Trullinger, A. R. Bishop, and J. A. Krumhansl, Phys. Rev. B **15**, 1578 (1977).
15. J. F. Currie, S. E. Trullinger, A. R. Bishop, and J. A. Krumhansl, Phys. Rev. B **15**, 5567 (1977).
16. S. E. Trullinger, Solid State Commun. **29**, 27 (1979).
17. E. Stoll, T. Schneider, and A. R. Bishop, Phys. Rev. Letters **42**, 937 (1979).
18. J. F. Currie, J. A. Krumhansl, A. R. Bishop, and S. E. Trullinger, Phys. Rev. B **22**, 477 (1980).
19. Y. S. Kivshar and B. A. Malomed, Rev. Modern Phys. **61**, 763 (1989).
20. B. Alefeld, A. Kollmar, and B. A. Dasannacharya, J. Chem. Phys. **63**, 4415 (1975).
21. C. J. Carlile, S. Clough, A. J. Horsewill, and A. Smith, Chem. Phys. **134**, 437 (1989).
22. N. LeCalvé, B. Pasquier, G. Braathen, L. Soulard, and F. Fillaux, J. Phys. C: Solid State Phys. **19**, 6695 (1986).
23. R. Hirota, J. Phys. Soc. Jpn **33**, 1459 (1972).
24. P. J. Caudrey, J. D. Gibbon, J. C. Eilbeck, and R. K. Bullough, Phys. Rev. Letters **30**, 237 (1973).
25. M. J. Ablowitz, D. J. Kaup, A. C. Newell, and H. Segur, Phys. Rev. Letters **30**, 1262 (1973).
26. E. K. Morris, Ph.D. thesis, Université d'Orsay (1997).

DISCRETE BREATHERS CLOSE TO THE ANTICONTINUUM LIMIT: EXISTENCE AND WAVE SCATTERING

S. FLACH, J. DORIGNAC AND A. E. MIROSHNICHENKO
Max-Planck-Institut für Physik komplexer Systeme,
Nöthnitzer Str. 38,
D-01187 Dresden, Germany
E-mail: flach@mpipks-dresden.mpg.de

V. FLEUROV
Raymond and Beverely Sackler Faculty of Exact Sciences,
School of Physics and Astronomy,
Tel Aviv University,
Tel Aviv 69978, Israel
E-mail: fleurov@post.tau.ac.il

The anticontinuum limit (i.e. the limit of weakly coupled oscillators) is used to obtain two surprising results. First we prove the continuation of discrete breathers of weakly interacting harmonic oscillators, provided a suitable coupling is chosen. Secondly we derive an analytical result for the wave transmission by a breather of the discrete nonlinear Schrödinger equation at weak coupling. We obtain a resonant full reflection due to a Fano resonance.

1. Introduction

The study of dynamical nontopological localization in nonlinear Hamiltonian lattices is a topic of widespread interest with results ranging from intensive theoretical studies[1] to experiments in such diverse areas as Josephson junction arrays[2], vibrational excitations in crystals[3] and localized excitations in antiferromagnets[4], to name a few. While basic theoretical investigations quickly spread into such directions as e.g. quantum theory and dissipative systems, there are still many interesting questions to be answered in the framework of classical Hamiltonian lattices.

One central class of systems is the model of a chain of interacting oscillators with the Hamiltonian

$$H = \sum_l \left(\frac{P_l^2}{2} + V(X_l) + \epsilon W(X_l - X_{l-1}) \right) \tag{1}$$

where P_l and X_l are pairs of canonically conjugated momenta and displacements of particles at lattice site l satisfying the equations $\dot{X}_l = \partial H / \partial P_l$ and $\dot{P}_l = -\partial H / \partial X_l$. The oscillator potential V and the nearest neighbour interaction W are assumed to be nonnegative functions with $V(0) = W(0) = V'(0) = W'(0) = 0$. The role of the positive constant ϵ is to control the strength of the interaction.

Another important system is the discrete nonlinear Schrödinger equation (DNLS)

$$i\dot{\Psi}_n = C(\Psi_{n+1} + \Psi_{n-1}) + |\Psi_n|^2 \Psi_n \tag{2}$$

where n is an integer labeling the lattice sites, Ψ_n is a complex scalar variable and C describes the nearest neighbour interaction (hopping) on the lattice. The last term in (2) provides with the requested nonlinearity.

2. Weakly interacting harmonic oscillators

According to a proof of existence by Aubry and MacKay[5] system (1) admits time-periodic and spatially localized solutions - discrete breathers - of the form

$$\hat{X}_l(t) = \hat{X}_l(t + \frac{2\pi}{\Omega_b}) \ , \ X_{|l| \to \infty} \to 0 \tag{3}$$

at suitably small coupling ϵ provided none of the multiples of the breather frequency Ω_b resonate with the spectrum of small-amplitude plane waves of (1) which for small coupling implies

$$k\Omega_b \neq \sqrt{V''(0)} \, . \tag{4}$$

In this limit of weak coupling $\epsilon \to 0$ the breather solution is typically assumed to be of a compact form, in the easiest case a single-site excitation $\hat{X}_0(t) \neq 0$ and $\hat{X}_{l\neq 0}(t) = 0$. Then the nonresonance condition (4) at the anticontinuum limit $\epsilon = 0$ can be satisfied only if the oscillator potential V is chosen to be nonisochronous, i.e. $d\Omega_b/dE \neq 0$ where E is the energy of a particle oscillating in V. This excludes from considerations all types of isochronous potentials, especially the well-known harmonic case $V(x) \sim x^2$. This lead to an expectation that for the case of harmonic potentials V there exist no breathers in weakly coupled systems. Below we will show that contrary to this expectation for suitable interaction functions W breathers exist and persist down to zero coupling even for harmonic potentials V.

To proceed we fix the potentials to

$$V(x) = \frac{1}{2}x^2 \, , \; W(x) = \frac{1}{4}x^4 \, . \tag{5}$$

The equations of motion read

$$\ddot{X}_l = -X_l - \epsilon(X_l - X_{l-1})^3 - \epsilon(X_l - X_{l+1})^3 \, . \tag{6}$$

Although the total potential of the system is not a homogeneous function of all coordinates, we may use the time-space separation ansatz[6]

$$X_l(t) = (-1)^l A_l G(t) \tag{7}$$

where A_l is a time-independent amplitude and $G(t)$ is a lattice site independent master function of time. The resulting equations are

$$A_l(\ddot{G} + G) = -G^3 \epsilon \left[(A_l + A_{l-1})^3 + (A_l + A_{l+1})^3 \right] \tag{8}$$

which lead to a differential equation for G

$$\ddot{G} = -G - \kappa G^3 \tag{9}$$

and a difference equation for the amplitudes A_l

$$\frac{\kappa}{\epsilon} A_l = (A_l + A_{l-1})^3 + (A_l + A_{l+1})^3 \tag{10}$$

where $\kappa > 0$ is a separation constant of arbitrary choice. The differential equation (9) corresponds to the problem of an anharmonic oscillator with solutions being periodic in time. The difference equation (10) was studied by Flach[7], where it was proven that localized solutions of the type $A_{|l| \to \infty} \to 0$ do exist by considering (10) as a two-dimensional map and searching for homoclinic trajectories. Thus we arrive at a general existence proof of discrete breathers for the choice (5).

In particular we may choose $\kappa = \epsilon$. Then we easily may continue the system into the zero coupling limit $\epsilon \to 0$ and still keep the breather solution. The time dependence of the solution will correspond to the solution of a harmonic oscillator (cf. (9)). This concludes the proof that discrete breathers can be continued from the uncoupled limit of noninteracting harmonic oscillators by switching on a weak interaction. Note that the interaction has to be an anharmonic function of the coordinates, otherwise the only meaningful solutions would be delocalized plane waves. Note also that at the noninteracting limit the discrete breather solution is not a single site excitation, but is rather a spatially localized excitation of all oscillators. Details of the spatial decay are given by Dey et al[6].

By connecting the existence of breather solutions with tangent bifurcations of band edge plane waves[8], it is actually possible to extend the choice of interaction potentials to a much larger class of functions including also harmonic interaction parts, as well as to consider more general isochronous potentials[9].

Finally we note that it would be interesting to test the variational technique for proving the existence of 'hard' breathers[10] for applicability to the discussed case.

3. Fano resonances

Another important issue concerning discrete breather properties is their scattering impact on small amplitude plane waves. While there exists some literature on that subject[11], there is an ongoing debate concerning the possibility of resonant total reflection of waves by discrete breathers[12], which has been observed numerically in many cases. Again we will use the anticontinuum limit to obtain analytical results. In the following we will consider the case of the DNLS (2). For small amplitude waves $\Psi_n(t) = \epsilon e^{i(\omega_q t - qn)}$ the equation (2) yields the dispersion relation

$$\omega_q = -2C \cos q \ . \tag{11}$$

Breather solutions have the form

$$\hat{\Psi}_n(t) = \hat{A}_n e^{i\Omega_b t} \ , \ \hat{A}_{|n|\to\infty} \to 0 \tag{12}$$

where the time-independent amplitude $\hat{A}_n$ can be taken real valued, and the breather frequency $\Omega_b \neq \omega_q$ is some function of the maximum amplitude $\hat{A}_0$. The spatial localization is given by an exponential law $\hat{A}_n \sim e^{-\lambda|n|}$ where $\cosh\lambda = |\Omega_b|/2C$. Thus the breather can be approximated as a single-site excitation if $|\Omega_b| \gg C$. In this case the relation between the single-site amplitude $\hat{A}_0$ and Ω_b becomes $\Omega_b = -\hat{A}_0^2$. In the following we will neglect the breather amplitudes for $n \neq 0$, i.e. $\hat{A}_{n\neq 0} \approx 0$. In fact $\hat{A}_{\pm 1} \approx \frac{C}{\hat{A}_0} \ll 1$.

We add small perturbations to the breather solution

$$\Psi_n(t) = \hat{\Psi}_n(t) + \phi_n(t) \tag{13}$$

and linearize the equations for $\phi_n(t)$:

$$i\dot{\phi}_n = C(\phi_{n+1} + \phi_{n-1}) - \Omega_b \delta_{n,0}(2\phi_0 + e^{2i\Omega_b t}\phi_0^*) \tag{14}$$

with $\delta_{n,m}$ being the usual Kronecker symbol. The general solution to this problem is given by the sum of two channels

$$\phi_n(t) = X_n e^{i\omega t} + Y_n^* e^{i(2\Omega_b - \omega)t} \tag{15}$$

where X_n and Y_n are complex numbers satisfying the following algebraic equations:

$$-\omega X_n = C(X_{n+1} + X_{n-1}) - \Omega_b \delta_{n,0}(2X_0 + Y_0) , \qquad (16)$$

$$-(2\Omega_b - \omega)Y_n = C(Y_{n+1} + Y_{n-1}) - \Omega_b \delta_{n,0}(2Y_0 + X_0) . \qquad (17)$$

When treating (16,17) as an eigenvalue problem we note that the corresponding matrix is nonhermitian due to the nonzero coupling between the variables X_0 and Y_0. This is a consequence of the fact that the linearized phase space flow around a time-periodic orbit of a general Hamiltonian is characterized by a Floquet matrix which is symplectic. As a result the orbit may be either marginally stable or unstable, i.e. ω may be real or complex. In the following we will however focus on the transmission properties, leaving the issue of linear dynamical stability aside. We note however that in the considered limit of a nearly single-site localized breather solution in the DNLS it is well known that the solution *is* linearly stable[5].

Away from the breather center $n = 0$ Eq. (14) allows for the existence of plane waves with spectrum ω_q. This spectrum will be dense for an infinite chain. As we are interested in the propagation of waves, we will set $\omega \equiv \omega_q$ with some value of q. Then it follows that the X-channel is open and guides propagating waves, while the Y-channel is closed, i.e. its frequency $2\Omega_b - \omega_q$ does not match the spectrum ω_q itself.

Instead of solving (16,17) we will consider a slightly more general set of equations

$$-\omega_q X_n = C(X_{n+1} + X_{n-1}) - \delta_{n,0}(V_x X_0 + V_a Y_0) , \qquad (18)$$

$$-(\Omega - \omega_q)Y_n = C(Y_{n+1} + Y_{n-1}) - \delta_{n,0}(V_y Y_0 + V_a X_0) \qquad (19)$$

which is reduced to (16,17) for $\Omega = 2\Omega_b$ and $V_x = V_y = 2V_a = 2\Omega_b$. We note that for $V_a = 0$ the closed Y channel provides with exactly one localized eigenstate due to a nonzero value of V_y, located at

$$\omega_L^{(y)} = \Omega + \sqrt{V_y^2 + 4C^2} \, . \qquad (20)$$

To compute the transmission coefficient we make use of the transfer matrix method described e.g. by Tong et al[13]. The boundary conditions are:

$$X_{N+1} = Te^{iq} , \; X_N = T , \; Y_{N+1} = D/\kappa , \; Y_N = D , \qquad (21)$$

$$X_{-N-1} = 1 + R , \; X_{-N} = e^{iq} + Re^{-iq} ,$$

$$Y_{-N-1} = F , \; Y_{-N} = \kappa F . \qquad (22)$$

Here T and R are the transmission and reflection amplitudes. F and D describe the exponentially decaying amplitudes in the closed Y-channel, where the degree of localization is connected with the coefficient $\kappa \equiv e^{\lambda}$:

$$\kappa = \frac{1}{2C}\left[-(\Omega - \omega_q) + \sqrt{(\Omega - \omega_q)^2 - 4C^2}\right] . \tag{23}$$

The transfer matrix is a 4×4 matrix, which is defined by (18,19) at $n = 0$. After the corresponding solving of four linear equations we obtain

$$|T|^2 = \frac{4\sin^2 q}{\left(2\cos q - a - \frac{d^2\kappa}{2-b\kappa}\right)^2 + 4\sin^2 q} , \tag{24}$$

$$a = \frac{-\omega_q + V_x}{C} , \; b = \frac{-(\Omega - \omega_q) + V_y}{C} , \; d = \frac{V_a}{C} . \tag{25}$$

This central result allows one to conclude that total reflection is obtained when the condition

$$2 - b\kappa = 0 \tag{26}$$

is realized. It is equivalent to the condition

$$\omega_q = \omega_L^{(y)} , \tag{27}$$

which has a very physical meaning: perfect reflection is obtained when a local mode, originating from the closed Y-channel, is resonating with the spectrum ω_q of plane waves from the open X-channel. The only condition is that the interaction V_a is nonzero. Remarkably, the resonance position does not depend on the actual value of V_a, so there is no renormalization. The existence of local modes which originate from the X-channel for nonzero V_x and possibly resonate with the closed Y-channel is evidently not of any importance. This resonant total reflection is very similar to the Fano resonance effect, as it is unambiguously related to a local state resonating and interacting with a continuum of extended states. The fact that the resonance is independent of V_a is due to the assumed local character of the coupling between the local mode (originating from the Y-channel) and the open channel. If this interaction is assumed to have some finite localization length by itself, then the resonance condition (27) may be renormalized[14].

Returning to the case of a DNLS breather at weak coupling, we insert the values for Ω , V_x ,V_y and V_a into (24,25) and obtain the following expression for the transmission T_b:

$$|T_b|^2 = \frac{4\sin^2 q}{\left(-\frac{2\Omega_b}{C} - \frac{\Omega_b^2}{2C^2}\frac{\kappa_b}{2C - \cos q\kappa_b}\right)^2 + 4\sin^2 q} \tag{28}$$

with

$$\kappa_b = \frac{1}{2C}\left(-(2\Omega_b - \omega_q) + \sqrt{(2\Omega_b - \omega_q)^2 - 4C^2}\right) . \tag{29}$$

The result is that any breather solution of the DNLS close to the anticontinuous limit yields perfect reflection close to $q = \pi/2$ (see also Kim et al[12]). In the very anticontinuous limit perfect reflection is obtained precisely at $q = \pi/2$. Indeed, if we expand (28) in $\frac{C}{\Omega_b}$ we obtain in the lowest order

$$|T_b|^2 \approx \frac{4C^4}{\Omega_b^4} \sin^2 2q \; , \tag{30}$$

provided $\frac{C}{2|\Omega_b|} \ll |\cos q|$. At the same time the DNLS breather is linearly stable, so another conclusion is that total reflection is not related to stability or instability of the scattering periodic orbit. The reason for the appearance of total reflection at the anticontinuous limit is that the closed channel has distance $2\Omega_b$ from the open channel, but the local state of the closed channel is located at the same distance $2\Omega_b$ from the closed channel, leading to a resonance with the open channel.

References

1. A. J. Sievers and J. B. Page, in *Dynamical properties of solids VII phonon physics the cutting edge*, Elsevier, Amsterdam, 1995; S. Aubry, Physica D, **103**, 201 (1997); S. Flach and C. R. Willis, Phys. Rep. **295**, 181 (1998).
2. E. Trias, J. J. Mazo and T. P.Orlando, Phys. Rev. Lett. **84**, 741 (2000); P. Binder, D. Abraimov, A. V. Ustinov, S. Flach and Y. Zolotaryuk, Phys. Rev. Lett. **84**, 745 (2000).
3. B. I. Swanson, J. A. Brozik, S. P. Love, G. F. Strouse, A. P. Shreve, A. R. Bishop and W.-Z. Wang, Phys. Rev. Lett. **82**, 3288 (1999).
4. U. T. Schwarz, L. Q. English and A. J. Sievers, Phys. Rev. Lett. **83**, 223 (1999).
5. R. S. MacKay and S. Aubry, Nonlinearity **7**, 1623 (1994).
6. B. Dey, M. Eleftheriou, S. Flach and G.P. Tsironis, Phys. Rev. E **65**, 017601 (2001).
7. S. Flach, Phys. Rev. E **51**, 1503 (1995).
8. S. Flach, Physica D **91**, 223 (1996).
9. J. Dorignac and S. Flach, in preparation.
10. S. Aubry, G. Kopidakis and V. Kadelburg, Disc. Cont. Dyn. Sys. B **1**, 271 (2001).
11. S. Flach, and C. R. Willis, in: *Nonlinear Excitations in Biomolecules*, ed. by M. Peyrard, Editions de Physique, Springer-Verlag, 1995; S. Kim, C. Baesens, and R. S. MacKay, Phys. Rev. E **56**, R4955 (1997); T. Cretegny, S. Aubry and S. Flach, Physica D **199**, 73 (1998); S. W. Kim, and S. Kim, Phys. Rev. B **63**, 212301 (2001).
12. S. W. Kim and S. Kim, Physica D **141**, 91 (2000); S. Flach, A. E. Miroshnichenko and M. V. Fistul, submitted to CHAOS, cond-mat/0209427.
13. P. Tong, B. Li and B. Hu, Phys. Rev. B **59**, 8639 (1999).
14. S. Flach, V. Fleurov, A. E. Miroshnichenko and M. V. Fistul, in preparation.

A CENTRE MANIFOLD TECHNIQUE FOR COMPUTING TIME-PERIODIC OSCILLATIONS IN INFINITE LATTICES

GUILLAUME JAMES
Laboratoire MIP, UMR 5640,
INSA de Toulouse, Département GMM,
135 avenue de Rangueil, 31077 Toulouse Cedex 4, France.
e-mail : Guillaume.James@gmm.insa-tlse.fr

Time-periodic oscillations in infinite one-dimensional lattices can be expressed in many cases as solutions of an ill-posed recurrence relation on a loop space. We give simple spectral conditions under which all small amplitude solutions lie on an invariant finite-dimensional centre manifold. This result reduces the problem locally to the study of a finite-dimensional mapping and provides an exact version of the rotating-wave approximation method (which can be viewed as a tangent space approximation). We explain how to compute the centre manifold and the reduced mapping at an arbitrary order. In the case of hardening FPU chains, the reduced mapping admits homoclinic orbits corresponding to discrete breathers.

1. An introductive example

For a large class of infinite one-dimensional lattices, time-periodic oscillations can be viewed as solutions of an ill-posed recurrence relation on a loop space. As an example, consider the Fermi-Pasta-Ulam (FPU) lattice

$$\frac{d^2 x_n}{dt^2} = V'(x_{n+1} - x_n) - V'(x_n - x_{n-1}), \quad n \in \mathbb{Z}, \tag{1}$$

which consists in a chain of identical particles nonlinearly coupled to their nearest neighbours. We denote by V a smooth interaction potential satisfying $V'(0) = 0$, $V''(0) = 1$ and by x_n the particle displacement from the trivial equilibrium state. We look for time-periodic solutions of (1) having a given frequency ω. Since V' is locally invertible, one can reformulate (1) using the force variable $y_n(t) = V'(x_n - x_{n-1})(t/\omega)$ (y_n is 2π-periodic in t). This yields the equations

$$\omega^2 \frac{d^2}{dt^2} W(y_n) = y_{n+1} - 2y_n + y_{n-1}, \quad n \in \mathbb{Z}, \tag{2}$$

where $W = (V')^{-1}$. Note that the frequency ω appears in (2) as a bifurcation parameter. For formulating (2) as a mapping in a loop space, we introduce the variable $u_n = (y_{n-1}, y_n) \in D$ where D is a space of 2π-periodic functions of t which has to be precised. Problem (2) takes the form of a mapping

$$u_{n+1} = F_\omega(u_n), \tag{3}$$

where $F_\omega(y_{n-1}, y_n) = (\, y_n\,,\, \omega^2 \frac{d^2}{dt^2} W(y_n) + 2y_n - y_{n-1}\,)$. Here we analyze problem (2) using a discrete *spatial dynamics* approach. This concept originates from the continuous case of partial differential equations. It has been introduced in [9] for studying elliptic partial differential equations in cylindrical domains, formulated as ill-posed evolution problems in the unbounded space coordinate.

We now define appropriate function spaces on which the operator F_ω is acting. We look for $(u_n)_{n\in\mathbb{Z}}$ as a sequence in the following space D

$$D = \{u \in \mathbb{H}^2 \times \mathbb{H}^2, u \text{ even in } t, \int_0^{2\pi} u\, dt = 0\},$$

where $\mathbb{H}^n$ denotes the classical Sobolev space $H^n(\mathbb{R}/2\pi\mathbb{Z})$ (the conditions of eveness in t and zero time-average simplify the problem but are not essential). The recurrence relation (3) holds in

$$X = \{u \in \mathbb{H}^2 \times \mathbb{H}^0, u \text{ even in } t, \int_0^{2\pi} u\, dt = 0\}$$

and the operator $F_\omega : D \to X$ is smooth in a neighbourhood of zero.

By a solution of (3) we mean a sequence $(u_n)_{n\in\mathbb{Z}}$ in D satisfying (3) for all $n \in \mathbb{Z}$. A particular solution is the fixed point $u = 0$ corresponding to the lattice at rest. An important feature of (3) is that the linearized operator $L_\omega = DF_\omega(0)$ is unbounded in X (of domain D) and thus the recurrence relation (3) is ill-posed (in fact there are no solutions of (3) for most initial conditions $u_0 \in D$).

Solutions of (3) homoclinic to $u = 0$ (i.e. satisfying $\lim_{|n|\to+\infty} \| u_n \|_D = 0$) have been extensively studied. They correspond to discrete breather solutions of (2) (spatially localized solutions with time-period 2π). We refer the reader to [1] for a general review on discrete breathers and to [6] for additional references on discrete breathers in FPU lattices. One property making the existence of homoclinic solutions more likely is the invariance $y_n \to y_{-n}$ in (2). Indeed, this invariance implies that (3) is reversible with respect to the symmetry R defined by $R(a, b) = (b, a)$, i.e. if u_n is a solution of (3) then $R\,u_{-n}$ is also a solution. Consequently, if the unstable manifold $W^u(0)$ intersects the fixed set Fix (R) then their intersection also belongs to the stable manifold $W^s(0)$ and homoclinic orbits exist. Note that problem (2) has the invariance $y_n \to y_n(t + \pi)$ and thus F_ω commutes with the symmetry T defined by $(T\,u)(t) = u(t+\pi)$. As a consequence, TR defines an other reversibility symmetry.

In the sequel we shall consider small amplitude solutions of (3), therefore the first step is to examine the linearized problem around $u = 0$. The operator L_ω is given by $L_\omega(y_{n-1}, y_n) = (\,y_n\,,\,(\omega^2 \frac{d^2}{dt^2} + 2)y_n - y_{n-1}\,)$. We denote by σ_k, σ_k^{-1} $(k \geq 1)$ the eigenvalues of L_ω, given by the dispersion relation

$$\sigma^2 + (\omega^2 k^2 - 2)\sigma + 1 = 0 \tag{4}$$

(by convention we denote by σ_k the solution of (4) satisfying $|\sigma_k| \geq 1$ and $\operatorname{Im}\sigma_k \leq 0$). The invariance $\sigma \to \sigma^{-1}$ in (4) is due to reversibility. The invariant subspace under L_ω associated with the pair of eigenvalues σ_k, σ_k^{-1} is spanned by the vectors $(\cos(kt), 0)$ and $(0, \cos(kt))$. When ω is large, the eigenvalues are real negative and lie strictly off the unit circle (both inside and outside). When ω decreases, the σ_k go towards the unit circle. As ω reaches the first critical value $\omega = 2$ (maximal phonon frequency), the eigenvalues σ_1, σ_1^{-1} collide and yield a double (non semi-simple) eigenvalue $\sigma = -1$ (see figure 1). When ω is further decreased, σ_1, σ_1^{-1} rotate on the unit circle and converge towards $\sigma = 1$ as $\omega \to 0^+$. More generally, the pair of eigenvalues σ_k, σ_k^{-1} collide at $\sigma = -1$ when $\omega = 2/k$.

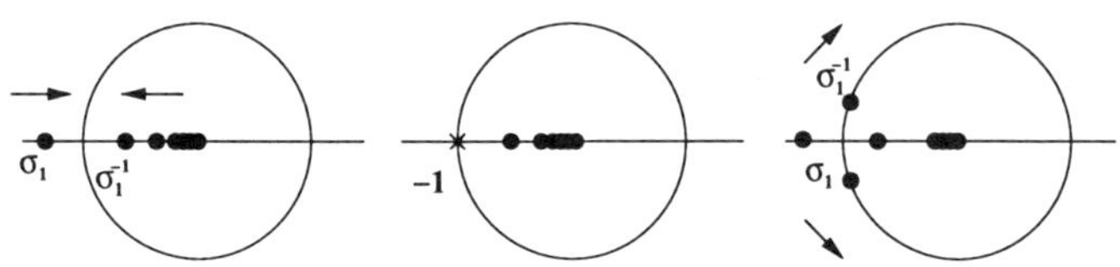

Figure 1. Spectrum of L_ω near the unit circle for $\omega > 2$ (left), $\omega = 2$ (centre), $\omega < 2$ and $\omega \approx 2$ (right).

The above analysis shows that the fixed point $u = 0$ of (3) is hyperbolic when $\omega > 2$. In this case, we shall see that the stable and unstable manifolds $W^s(0)$, $W^u(0)$ can intersect, depending on the properties of V. More precisely, for $\omega \approx 2$ the pair of eigenvalues σ_1, σ_1^{-1} is close to -1 and (3) can admit small amplitude homoclinic orbits to $u = 0$.

In order to perform a weakly non linear analysis, we restrict our attention to the case when $\omega \approx 2$. We introduce the small parameter $\mu = \omega^2 - 4$ and write (3) in the form

$$u_{n+1} = L\,u_n + N(u_n, \mu), \tag{5}$$

where $L = L_2$ is the linearized operator with $\omega = 2$ and

$$N((y_{n-1}, y_n), \mu) = \Big(0\,,\ \frac{d^2}{dt^2} g(y_n, \mu)\Big),$$

with $g(y,\mu) = 4(W(y) - y) + \mu W(y)$. One has $\| N(u,\mu) \|_X = O(\|u\|_D^2 + \mu \|u\|_D) \ll \| L\, u \|_X$ as $(u,\mu) \approx 0$, hence a good starting point for studying (5) is to start from the linear case.

The solutions of

$$u_{n+1} = L\, u_n, \quad u_n \in D, \tag{6}$$

can be computed using Fourier series. Due to the existence of a spectral gap separating $\sigma_1 = -1$ from the remaining (hyperbolic) part of the spectrum, solutions of (6) can be splitted in the following two classes. The first kind of solutions grows at least like $|\sigma_2|^{|n|}$ as $n \to +\infty$ or $-\infty$. The second kind has the form

$$u_n = (y_{n-1}, y_n), \quad y_n = (-1)^n\,(\alpha + \beta\, n)\, \cos t$$

and diverges at most polynomially as $|n| \to +\infty$. Consequently, solutions u_n of (6) which remain bounded in D as $|n| \to +\infty$ necessarily belong to the two-dimensional linear subspace $X_c = \text{Span}\,\{\,(\cos t, 0)\,,\,(0, \cos t)\,\}$ for all $n \in \mathbb{Z}$. The space X_c is denoted as *centre space* and is the invariant space under L associated with the double eigenvalue $\sigma_1 = -1$. The spectral projection π_c on X_c reads $\pi_c\, u = \frac{1}{\pi} \int_0^{2\pi} u(t) \cos t\, dt\, \cos t$.

In the nonlinear case (5), one can locally prove a similar result where the centre space is replaced by a two-dimensional invariant centre manifold[5]. For $\mu \approx 0$, there exists a smooth local manifold $\mathcal{M}_\mu \subset D$ (which can be written as a graph over X_c) invariant under F_ω and the symmetries R, T. One has $\mathcal{M}_\mu = \{u \in D / u = u^c + \psi(u^c, \mu), u^c \in X_c \cap \Omega\}$, where $\psi : X_c \times \mathbb{R} \to (I - \pi_c) D$ is a smooth map satisfying $\psi(u^c, \mu) = O(\|u^c\|^2 + \|u^c\| |\mu|)$ and Ω is a small neighbourhood of 0 in D. More precisely, we have $\psi(\,(a,b)\,\cos t\,,\,\mu) = (\,\varphi(b,a,\mu)\,,\,\varphi(a,b,\mu)\,)\ (a, b \in \mathbb{R})$ where

$$\varphi(a,b,\mu) = -\frac{1}{16} V^{(3)}(0) \cos(2t)\,(ab + \frac{1}{2}a^2 - \frac{7}{2}b^2) + \text{h.o.t.} \tag{7}$$

For $\mu \approx 0$, the centre manifold $\mathcal{M}_\mu$ contains all solutions u_n of (5) staying in Ω for all $n \in \mathbb{Z}$ (small amplitude solutions). Their central component $u_n^c = \pi_c\, u_n$ is given by the two-dimensional mapping

$$u_{n+1}^c = f(u_n^c, \mu), \tag{8}$$

where $f(u^c,\mu) = L\, u^c + \pi_c\, N(u^c + \psi(u^c,\mu), \mu)$. The map $f : X_c \times \mathbb{R} \to X_c$ is smooth in the neighbourhood of 0 and inherits the symmetries of (3) $(f(.,\mu)$

is reversible under R and commutes with T). Setting $u_n^c = (a_n, b_n)\cos t$, the mapping (8) reads

$$\begin{aligned} a_{n+1} &= b_n, \\ b_{n+1} &= -a_n - 2b_n - b_n\mu + c_1 b_n^3 + c_2 a_n b_n^2 + \tfrac{1}{2}c_2 a_n^2 b_n + \text{h.o.t.}, \end{aligned} \tag{9}$$

where $c_1 = \frac{1}{2}V^{(4)}(0) - \frac{17}{16}(V^{(3)}(0))^2$, $c_2 = -\frac{1}{8}(V^{(3)}(0))^2$.

Consequently, the problem of finding small amplitude solutions of (3) for $\omega \approx 2$ reduces to the study of the two-dimensional reversible mapping (9). Note that we are not limited to localized solutions of (3) (i.e. those satisfying $\lim_{|n|\to+\infty} \| u_n \|_D = 0$) since the reduced mapping (9) describes the set of *all* small amplitude solutions when $\mu \approx 0$. For $\mu \approx 0$, each small amplitude solution of (9) corresponds to a solution y_n of (2) given by

$$y_n(t) = b_n \cos t + \varphi(b_{n-1}, b_n, \mu) \tag{10}$$

with $\omega^2 = 4 + \mu$ in (2).

2. Generalization

The above example suggests the following mathematical framework for studying time-periodic oscillations in infinite one-dimensional lattices.

We consider a Hilbert space X and a closed linear operator $L : X \to X$ of domain D (L is in general unbounded). We equip D with the scalar product $\langle u, v\rangle_D = \langle Lu, Lv\rangle_X + \langle u, v\rangle_X$, hence D is a Hilbert space continuously embedded in X. We denote by $\mathcal{U} \times \mathcal{V}$ a neighbourhood of 0 in $D \times \mathbb{R}^p$ and consider a nonlinear map $N \in C^k(\mathcal{U} \times \mathcal{V}, X)$ ($k \geq 2$) satisfying $N(0,0) = 0$, $D_u N(0,0) = 0$. We look for sequences $(u_n)_{n\in\mathbb{Z}}$ in $\mathcal{U}$ satisfying

$$\forall n \in \mathbb{Z}, \quad u_{n+1} = L\,u_n + N(u_n, \mu) \quad \text{in } X, \tag{11}$$

where $\mu \in \mathcal{V}$ is a parameter. In particular, $u = 0$ is a fixed point of (11) when $\mu = 0$. Note that the initial value problem for (11) is in general ill-posed.

We assume that L has the property of spectral separation, i.e. one can find an annulus $\mathcal{A} = \{\, z \in \mathbb{C},\ r \leq |z| \leq R \,\}$ ($r < 1 < R$) such that the only part of the spectrum of L in $\mathcal{A}$ lies on the unit circle (see figure 2). This assumption is essential in the proof of the centre manifold reduction theorem. We shall obviously assume the central ($|z| = 1$) and hyperbolic ($|z| \neq 1$) parts of the spectrum non-empty. We do not require the centre space X_c (invariant subspace under L corresponding to the central part of the spectrum) to be finite-dimensional. However, the centre manifold reduction theorem is more efficient in this case since the local study of (11) is amenable to that of a finite-dimensional mapping. The subspace X_c is finite-dimensional when the spectrum of L on the unit circle consists in a finite number of eigenvalues with finite multiplicities.

The spectral projection π_c on the centre space can be defined in the following way (see e.g. [8])

$$\pi_c = \frac{1}{2i\pi} \int_{\mathcal{C}(R)} (zI - L)^{-1}\, dz - \frac{1}{2i\pi} \int_{\mathcal{C}(r)} (zI - L)^{-1}\, dz,$$

where $\mathcal{C}(r)$ denotes the circle of centre $z = 0$ and radius r (see figure 2). One has $\pi_c \in \mathcal{L}(X, D)$, $X_c = \pi_c X \subset D$ and $\pi_c\, L = L\, \pi_c$. In the sequel we note $\pi_h = I - \pi_c$ and $D_h = \pi_h\, D$.

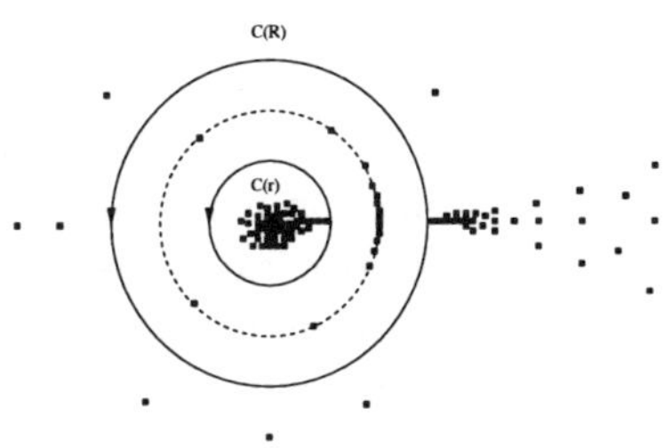

Figure 2. Spectrum of L (dots), unit circle (dashed) and oriented circles $\mathcal{C}(r)$, $\mathcal{C}(R)$.

The condition of spectral separation is satisfied in a large class of infinite one-dimensional lattices where one looks for time-periodic solutions, e.g. in multicomponent Klein-Gordon lattices[6] or diatomic FPU chains[7]. In these examples, the centre space is finite-dimensional and π_c takes a simple form in term of Fourier coefficients. These properties should be preserved in higher-dimensional lattices with appropriate boundary conditions (e.g. periodic) in the transverse directions.

We now state the centre manifold reduction theorem (theorem 2.1 below). The case when L is bounded ($D = X$) has been treated in previous works (see e.g. [10], [2]). The present case when L is unbounded (the one relevant for spatial dynamics in nonlinear lattices) has been treated in [6], to which we refer for the proof. This result can be seen as a discrete analogue of centre manifold reduction theorems for various types of differential equations in Hilbert or Banach spaces ([11], [13], [4], [3]).

Theorem 2.1. *Assume that L has the property of spectral separation. Then there exist a neighbourhood $\Omega \times \Lambda$ of 0 in $D \times \mathbb{R}^p$ and a map $\psi \in C^k(X_c \times \Lambda, D_h)$ (with $\psi(0,0) = 0$, $D_{u^c}\psi(0,0) = 0$) such that for* *all* $\mu \in \Lambda$ *the manifold*

$$\mathcal{M}_\mu = \{ u \in D \,/\, u = u^c + \psi(u^c, \mu), u^c \in X_c \}$$

has the following properties.
i) $\mathcal{M}_\mu$ *is locally invariant under* $L + N(.,\mu)$, *i.e. if* $u \in \mathcal{M}_\mu \cap \Omega$ *then* $Lu + N(u,\mu) \in \mathcal{M}_\mu$.
ii) If $(u_n)_{n\in\mathbb{Z}}$ *is a solution of (11) and* $u_n \in \Omega$ *for all* $n \in \mathbb{Z}$, *then* $u_n \in \mathcal{M}_\mu$ *for all* $n \in \mathbb{Z}$ *and* $u_n^c = \pi_c u_n$ *satisfies the recurrence relation in* X_c

$$\forall n \in \mathbb{Z}, \quad u_{n+1}^c = f(u_n^c, \mu), \tag{12}$$

where $f \in C^k((X_c \cap \Omega) \times \Lambda, X_c)$ *is defined by* $f(.,\mu) = \pi_c\,(L + N(.,\mu)) \circ (I + \psi(.,\mu))$ *and* $f(.,\mu)$ *is locally invertible.*
iii) Conversely, if $(u_n^c)_{n\in\mathbb{Z}}$ *is a solution of (12) such that* $u_n^c \in \Omega$ *for all* $n \in \mathbb{Z}$, *then* $u_n = u_n^c + \psi(u_n^c, \mu)$ *satisfies (11).*

This result reduces the local study of (11) to that of the recurrence relation (12) in the smaller space X_c, which is particularly interesting when X_c is finite-dimensional. In addition, one can show that the invariances of (11) are preserved throughout the reduction procedure. More precisely, if $L + N(.,\mu)$ commutes with a linear isometry $T \in \mathcal{L}(X) \cap \mathcal{L}(D)$ then $\mathcal{M}_\mu$ is invariant under T and $f(.,\mu)$ also commutes with T. If (11) is reversible under a symmetry $R \in \mathcal{L}(D)$, then (under some technical assumptions[6]) $\mathcal{M}_\mu$ is invariant under R and $f(.,\mu)$ inherits reversibility in R.

The centre manifold reduction technique has the advantage of being a constructive method, i.e. the Taylor expansion of the reduction function ψ at $(u^c, \mu) = 0$ (and thus the expansion of f) can be computed at any order. One computes the Taylor expansion of ψ by expanding each side of equation

$$\psi(L\,u^c + \pi_c\,N(u^c + \psi(u^c,\mu),\mu),\mu) = L\,\psi(u^c,\mu) + \pi_h\,N(u^c + \psi(u^c,\mu),\mu) \tag{13}$$

with respect to (u^c, μ) and identifying terms of equal order (equation (13) expresses the fact that $\mathcal{M}_\mu$ is invariant under $L + N(.,\mu)$). This procedure yields a hierarchy of linear problems which can be solved by induction, starting from the lowest order (see (7), (9) for the first terms in the FPU system). Since the method provides explicit expansions of ψ and f, one can in general precisely describe the shape and symmetries of small amplitude bifurcating solutions.

3. Comparison with the rotating-wave approximation

We now compare the centre manifold theorem with the formal rotating-wave approximation (RWA) method[12], using the example of the FPU lattice. The

RWA consists in setting

$$y_n(t) \approx \tilde{\beta}_n \cos t \tag{14}$$

in (2) and neglecting higher harmonics, which yields the following recurrence relation for $\tilde{\beta}_n$ (recall $\omega^2 = 4 + \mu$)

$$\tilde{\beta}_{n+1} + 2\tilde{\beta}_n + \tilde{\beta}_{n-1} = -\mu\,\tilde{\beta}_n + \tilde{B}\,\tilde{\beta}_n^3 + \text{h.o.t.}, \tag{15}$$

where $\tilde{B} = \frac{1}{2}V^{(4)}(0) - \frac{3}{2}(V^{(3)}(0))^2$. Notice that the ansatz (14) describes interaction forces (with zero time-average), whereas mass displacements are considered in [12]. By comparing (14) with (10)-(7), we note that the RWA corresponds to a tangent space approximation (the centre manifold curvature has been neglected), which can lead to misleading results. Indeed, the reduced mapping in normal form reads (set $b_n = \beta_n - \frac{c_2}{12}\beta_n^3$ in (9))

$$\beta_{n+1} + 2\beta_n + \beta_{n-1} = -\mu\,\beta_n + B\,\beta_n^3 + \text{h.o.t.}, \tag{16}$$

with $B = \frac{1}{2}V^{(4)}(0) - (V^{(3)}(0))^2$. Comparison of (15) with (16) shows that the RWA yields a wrong cubic coefficient $\tilde{B} \neq B$ in the reduced mapping when $V^{(3)}(0) \neq 0$ (quadratic terms in (7) consist in a higher harmonic $\cos(2t)$ and contribute to the coefficient B). In the particular case when $V^{(3)}(0) = 0$, the RWA is correct because quadratic terms vanish in (7).

If $B > 0$ and $\mu > 0$ ($\mu \approx 0$), there exist small amplitude solutions of (16) satisfying $\lim_{|n|\to+\infty} \beta_n = 0$ (see [5], [6]). Returning to the FPU system (2), these solutions correspond to discrete breathers having the form $y_n(t) = \beta_n \cos t + \text{h.o.t.}$.

References

1. S Flach and C R Willis. *Phys. Rep.* 295 (1998) p. 181.
2. G Iooss. Bifurcation of maps and applications. *Math. Studies* 36 (1979), Elsevier-North-Holland, Amsterdam.
3. G Iooss. *Nonlinearity* 13 (2000), p. 849-866.
4. G Iooss and K Kirchgässner. *Com. Math. Phys.* 211 (2000), p. 439-464.
5. G James. *C. R. Acad. Sci. Paris.*, 332(1):581–586, 2001.
6. G James. *J. Nonlinear Sci.*, Vol. 13 Issue 1 (2003).
7. G James and P Noble, Breathers on diatomic FPU chains with arbitrary masses. Preprint 2002, laboratory MIP, University of Toulouse.
8. T Kato. Perturbation theory for linear operators. Springer Verlag, 1966.
9. K Kirchgässner. *Journal of Differential Equations* 45 (1982), p. 113-127.
10. J Marsden and M McCracken. The Hopf bifurcation and its applications. Springer Verlag, NY, 1976.
11. A Mielke. *Math. Meth. Appl. Sci.* 10 (1988), p. 51-66.
12. A J Sievers and S Takeno. *Phys. Rev. Lett.* 61 (1988), p.970-973.
13. A Vanderbauwhede and G Iooss. *Dynamics Reported* 1, (C. Jones, U. Kirchgraber and H. Walther, eds) New Series, Springer Verlag (1992), p. 125-163.

ON THE MODULATIONAL STABILITY OF GROSS-PITTAEVSKII TYPE EQUATIONS IN 1+1 DIMENSIONS

Z. RAPTI
Department of Mathematics and Statistics, University of Massachusetts, Amherst MA 01003-4515, USA
E-mail: rapti@math.umass.edu

P. G. KEVREKIDIS
Department of Mathematics and Statistics, University of Massachusetts, Amherst MA 01003-4515, USA
E-mail: kevrekid@math.umass.edu

V. V. KONOTOP
Centro de Física de Matéria Condensada, Universidade de Lisboa, Av. Prof. Gama Pinto, 2, Lisboa 1649-003, Portugal
E-mail: konotop@cii.fc.ul.pt

The modulational stability of the nonlinear Schrödinger (NLS) equation is examined in the cases with linear and quadratic external potential. This study is motivated by recent experimental studies in the context of matter waves in Bose-Einstein condensates (BECs). The linear case can be examined by means of the Tappert transformation and can be mapped to the NLS in the appropriate (constant acceleration) frame. The quadratic case can be examined by using a lens-type transformation that converts it into a regular NLS with an additional linear growth term.

1. Introduction

Intensive studies of Bose-Einstein condensates (BECs) [1] have drawn much attention to nonlinear excitations in them. Recent experiments have revealed the existence of bright [2] and dark [3,4,5] solitons in BECs, as well as topological structures, such as vortices [6] and vortex lattices [7]. An interesting question concerns how such solitary wave structures may arise in this novel context of matter waves in BECs. It is well-known that the dynamics of the wavefunction in BEC is described (at the mean field level, which is an increasingly accurate description, the closer the system is to the zero temperature limit) by the Gross-Pittaevskii (GP) equation which is a variant of well-known nonlinear Schrödinger (NLS) equation [8]. In the NLS context, perhaps the most standard mechanism through which solitons and solitary wave structures appear is through the activation of the modulational instability (MI) of plane waves.

The MI is a general feature of discrete as well as continuum nonlinear wave equations. Its demonstrations span a diverse set of disciplines ranging from fluid dynamics [9] (where it is usually referred to as the Benjamin-Feir instability) and nonlinear optics [10] to plasma physics [11]. One of the early contexts in which its significance was appreciated was the linear stability analysis of deep water waves.

The MI has been examined recently in the context of optical lattices in BECs both in one-dimensional and quasi-one dimensional systems, as well as in multiple dimensions. In the former case, it has been predicted theoretically [12,13] and verified experimentally [14,15] to lead to destabilization of plane waves, and in turn to delocalization in momentum space (equivalent to localization in position space, and the formation of solitary wave structures).

In the present contribution, we discuss the MI conditions for the continuous NLS equation (or equivalently for the GP equation) in (1+1) dimensions (1 spatial and 1 temporal)

$$iu_t + u_{xx} + s|u|^2u + V(x)u = 0. \tag{1}$$

u in this equation describes the slow envelope complex field dynamics (modulating the fast oscillatory dynamics). The subscripts denote partial derivatives with respect to the index, $s \in \{1, -1\}$ illustrates the focusing (+1) or defocusing (−1) nature of the nonlinearity, while $V(x)$ is the external potential.

In section 2.1, we review briefly the results in the absence of the potential (e.g., for $V(x) \equiv 0$). In section 2.2, we examine the case of a linear potential:

$$V(x) = -\alpha x, \tag{2}$$

which is relevant in experimental situations with gravitational [16] (and potentially also electrostatic) fields. In section 2.3, we examine the quadratic potential of the general form:

$$V(x) = -k(t)x^2, \tag{3}$$

which is relevant to contexts in which the (magnetic) trap is strongly confined in the 2 directions, while it is much shallower in the third one [1]. The prefactor $k(t)$ is typically fixed in current experiments, but adiabatic changes in the strength (and in fact even the location of the center) of the trap are experimentally feasible, hence we examine the more general time-dependent case. Finally, in section 3 we summarize our findings and conclude.

2. Analytical Results

2.1. *No potential*

We start by recalling the results for the modulational stability of the NLS (1) without an external potential, i.e. for $V(x) \equiv 0$: To this end we look for perturbed plane wave solutions of the form

$$u(x,t) = (\phi + \epsilon b) \exp[i((qx - \omega t) + \epsilon\psi(x,t))] \tag{4}$$

and analyze the $O(\epsilon)$ terms as

$$b(x,t) = b_0 \exp(i\beta(x,t)), \quad \psi(x,t) = \psi_0 \exp(i\beta(x,t)). \tag{5}$$

Using $\beta(x,t) = Qx - \Omega t$, the dispersion relation connecting the wavenumber Q and frequency Ω of the perturbation (see e.g. [8])

$$(-\Omega + 2qQ)^2 = Q^2(Q^2 - 2s\phi^2) \tag{6}$$

is obtained.

This implies that the instability region for the NLS in the absence of an external potential, appears for perturbation wavenumbers $Q^2 < 2s\phi^2$, and in particular *only for focusing nonlinearities* (to which we will restrict our study from this point onwards).

2.2. *Linear Potential*

The case of a linear potential is relevant to any context of a costant external field (gravitational [16] and electric ones being among the prominent such examples). In this setting, the NLS is well-known to maintain its integrable character [17]. Hence, in some sense, we expect that the modulational results/conditions will not be modified in this case.

The simplest way to illustrate that is by means of the "Tappert transformation" [17] ($\xi = x + \alpha t^2$)

$$u(x,t) = v(\xi,t) \exp(-i\alpha x t - \frac{1}{3} i\alpha^2 t^3) \tag{7}$$

(notice however the difference with the expression proposed in [17]) which brings the Eq. (1) back into the form of the regular NLS equation, without the external potential, in which case the condition of Eq. (6) applies.

Hence, the growth terms will now be $\sim \exp(i(Q\xi - \Omega t))$ with $\Omega = \Omega_r + i\nu$ (when the instability condition is satisfied; $\Omega_r = 2qQ$, cf. Eq. (6)), and hence the instability will be developing according to the spatiotemporal form:

$$u \sim \exp\left(iQ(x + \alpha t^2) - i\Omega_r t + \nu t - i\alpha x t - \frac{1}{3} i\alpha^2 t^3\right). \tag{8}$$

2.3. *Quadratic Potential*

The quadratic potential of Eq. (3) is clearly the most physically relevant example of an external potential in the BEC case, given the harmonic confinement of the atoms by the experimentally used magnetic traps [1]. In particular, to examine the MI related properties in this case, we will use a lens-type transformation [8] of the form:

$$u(x,t) = \ell^{-1} \exp(i f(t) x^2) v(\zeta, \tau) \tag{9}$$

where $f(t)$ is a real function of time, $\zeta = x/\ell(t)$ and $\tau = \tau(t)$. To preserve the scaling we choose [8,18]

$$\tau_t = 1/\ell^2 \tag{10}$$

To satisfy the resulting equations, we then demand that:

$$-f_t = 4f^2 + k(t) \tag{11}$$

$$-\ell_\tau/\ell + 4f\ell^2 = 0. \tag{12}$$

Taking into account (10) the last equation can be solved:

$$\ell(t) = \ell(0) \exp\left(4 \int_0^t f(s) ds\right) \tag{13}$$

what reduces the problem of finding time dependence of the parameters to solution of Eq. (11).

Upon the above conditions, the equation for $v(\zeta, \tau)$ becomes

$$iv_\tau + v_{\zeta\zeta} + |v|^2 v - 2i\lambda v = 0; \tag{14}$$

where

$$f\ell^2 = \lambda, \tag{15}$$

and generically λ is real and depends on time. Thus we retrieve NLS with an additional term, which represents either growth (if $\lambda > 0$) or dissipation (if $\lambda < 0$).

Eq. (14) allows an explicit solution when $v \equiv v(\tau)$ (i.e., a spatially homogeneous solution). The latter is of the form

$$v = \frac{A_0}{\ell(t)} \exp\left(if(t)x^2 + iq\frac{x}{\ell(t)} - iq^2\tau(t)\right) \times$$

$$\exp\left(\Lambda(\tau) + i\theta_0 + iA_0^2 \int_0^{\tau} \exp(2\Lambda(s))ds\right), \tag{16}$$

where $\Lambda(\tau) = 4\int_0^{\tau} \lambda(s)ds$, and A_0 and θ_0 are arbitrary real constants.

A particularly interesting case is that of λ constant. Then from the system of equations (10)-(12) and (14) it follows that k must have a specific form. f,ℓ and τ can then be determined accordingly. In fact, the system (10)-(12) and (14) with λ constant has as its solution

$$k(t) = (t + t^*)^{-2}/16 \tag{17}$$

$$f(t) = (t + t^*)^{-1}/8 \tag{18}$$

$$\ell(t) = 2\sqrt{2\lambda}\sqrt{t + t^*} \tag{19}$$

$$\tau(t) = \ln(\frac{t + t^*}{t^*})/8\lambda. \tag{20}$$

Notice once again, per the assumption of an imaginary phase in the exponential of Eq. (9), that our considerations are valid only for $\lambda \in R$. $t^\star$ in the above equations is an arbitrary constant that essentially determines the "width" of the trap at time $t = 0$ according to Eq. (17).

In this case the modulational condition remains unchanged, but now ω satisfies the dispersion relation $\omega = q^2 - \phi^2 + 2\imath\lambda$, so the growth (if $\lambda > 0$) or dissipation (if $\lambda < 0$) is inherent in equation (4). Moreover, all the terms are modulated by the constant growth (or decay) rate $\exp(2\lambda\tau)$, and the instability (when present) will be developing according to the form $v \sim \exp(i(Q\zeta - \Omega_r\tau) + (\nu + 2\lambda)\tau)$ with $\Omega = \Omega_r + i\nu$. $\Omega_r = 2qQ$.

If $k = k(t)$ is not given by Eq. (17), then λ must be time dependent (e.g., $\lambda \equiv \lambda(t)$). Here one cannot directly perform the MI analysis. However, still in this case, we have converted the explicit spatial dependence into an explicit temporal dependence. An important conclusion that stems from this transformation is that the harmonic potential, viewed in the appropriate frame (of Eq. (9)) can be considered as a form of growth (or dissipation, depending on the sign of λ for $\lambda \in R$) term.

3. Conclusions

In this brief communication, we have examined the problem of modulational instabilities of plane waves in the context of Gross-Pittaevskii equations with an external (linear or quadratic) potential. Both of these cases are directly relevant to current experimental realizations of Bose Einstein condensates.

It was found that the linear (also integrable) case can be reverted to the original NLS setting (wherein the MI conditions are well-known) by an appropriate (Tappert) transformation. This transformation was used to develop the form of the growth of unstable wavenumbers. Notice, however, that these are the wavenumbers of the expansion in a novel spatial coordinate which is essentially the spatial variable in a frame with constant acceleration. Hence, in this case, in the frame moving with a constant acceleration (induced by the constant force), the problem resumes its original NLS format and the MI conditions and resulting growth can be obtained in that frame and then restored in the original frame.

For the case of the quadratic potential, a lens transformation was used to cast the problem in a rescaled space and time frame (in a way very reminiscent of the scaling in problems related to focusing [8,18]). In this rescaled frame, the external potential can be viewed as a form of external growth. For specific forms of temporal dependence of the prefactor of the harmonic term (e.g., for appropriate, non-autonomous quadratic potentials), the resulting prefactor is constant. In such a context once again the MI analysis can be carried through completely, producing similar conditions, but now in the new dynamically rescaled frame/variables (which can be appropriately re-cast in the original variables).

It would be interesting to examine if similar considerations can be generalized to the case of multiple dimensions. Furthermore, in the current scheme of things it seems that the cases of different forms of the potentials need to be treated in different ways (which can be understood in terms of the different physical effects that they represent). Nonethless, it would be very useful, if a general formulation could be developed that could be applied independently of the form of $V(x)$, having as special case limits, the potential forms presented herein.

Finally, it would be worth exploring whether the explicitly demonstrated as modulationally unstable settings presented herein can be used as a means (i.e., initial condition) for directly producing solitary (matter) wave structures in BECs in an alternative fashion to the ones currently used in BEC experiments.

Prof. D.J. Frantzeskakis is gratefully acknowledged for a critical reading of the manuscript and his suggestions. PGK gratefully acknowledges support from a University of Massachusetts Faculty Research Grant, from the Clay Foundation through a Special Project Prize Fellowship and from the NSF through DMS-0204585. VVK gratefully acknowledges support from the European grant COSYC n.o HPRN-CT-2000-00158.

References

1. F. Dalfovo, S. Giorgini, L. P. Pitaevskii, and S. Stringari, Rev. Mod. Phys. **71**, 463(1999).
2. K.E. Strecker *et al.*, Nature **417**, 150 (2002); L. Khaykovich *et al.*, Science **296**, 1290 (2002).
3. S. Burger *et al.*, Phys. Rev. Lett. **83**, 5198 (1999).
4. J. Denschlag *et al.*, Science **287**, 97 (2000).
5. B.P. Anderson *et al.*, Phys. Rev. Lett. **86**, 2926 (2001).
6. M.R. Matthews *et al.*, Phys. Rev. Lett. **83**, 2498 (1999); K.W. Madison *et al.* Phys. Rev. Lett. **84**, 806 (2000); S. Inouye *et al.*, Phys. Rev. Lett.**87**, 080402 (2001).
7. J.R. Abo-Shaeer *et al*, Science **292**, 476 (2001); J.R. Abo-Shaeer, C. Raman and W. Ketterle, Phys. Rev. Lett. **88**, 070409 (2002); P. Engels *et al*, Phys. Rev. Lett. **89**, 100403 (2002).
8. C. Sulem and P.L. Sulem, *The Nonlinear Schrödinger Equation*, Springer-Verlag (New York, 1999).
9. T.B. Benjamin and J.E. Feir, J. Fluid. Mech. **27**, 417 (1967).
10. L.A. Ostrovskii, Sov. Phys. JETP **24**, 797 (1969).
11. T. Taniuti and H. Washimi, Phys. Rev. Lett. **21**, 209 (1968); A. Hasegawa, Phys. Rev. Lett. **24**, 1165 (1970).
12. V.V. Konotop and M. Salerno, Phys. Rev. A **65**, 021602(R) (2002).
13. A. Smerzi, A. Trombettoni, P.G. Kevrekidis, and A.R. Bishop, cond-mat/0207172 (Phys. Rev. Lett., in press).
14. F.S. Cataliotti *et al.*, cond-mat/0207139.
15. M. Kasevich and A. Tuchman (private communication).
16. B. P. Anderson and M. A. Kasevich, Science, **282**, 1686 (1998).
17. H.-H. Chen and C.-S. Liu, Phys. Rev. Lett., **37**, 693 (1976).
18. see e.g., S.I. Siettos, I.G. Kevrekidis and P.G. Kevrekidis, nlin.PS/0204030 and references therein.

ISOCHRONOUS POTENTIALS

S. BOLOTIN

Faculty of Mathematics and Mechanics, Moscow State University, RUSSIA

R.S. MACKAY

Mathematics Institute, University of Warwick, Coventry CV4 7AL, U.K.

A potential $V : \mathbb{R} \to \mathbb{R}$ is said to be *isochronous* if all periodic solutions of $\ddot{x} = -V'(x)$ have the same period, and there is at least one. Harmonic oscillators are examples, but there are many others. In this note, we construct examples of isochronous potentials which are analytic on the whole real axis, and we construct all C^2 isochronous potentials.

1. Introduction

Systems of the form

$$\ddot{x} = -V'(x) \tag{1}$$

are fundamental in mechanics. As is well known, they conserve energy

$$H(x,p) = \frac{1}{2}p^2 + V(x), \tag{2}$$

(where $p = \dot{x}$) and every bounded connected component of a level set $H^{-1}(E) = \{(x,p) \in \mathbb{R}^2 : H(x,p) = E\}$ containing no critical points of H is a periodic orbit. The period is given by

$$T = \sqrt{2}\int_{x_-}^{x_+} \frac{dx}{\sqrt{E - V(x)}}, \tag{3}$$

where $x_- < x_+$ are the extremes of the orbit. We say the potential V is *isochronous* if (i) every periodic orbit has the same period, and (ii) there exists at least one periodic orbit (to rule out trivial cases).

We define *harmonic oscillators* to be those systems for which V is quadratic, $V(x) = V_0 + \frac{1}{2}c(x - x_0)^2$, $c > 0$ (though some people define them to be systems for which every solution is sinusoidal). Every orbit of a harmonic oscillator (except the equilibrium $x = x_0$) is periodic and has the same period $T = 2\pi/\sqrt{c}$. Thus harmonic oscillators are isochronous.

In the physics community, the term "anharmonic" has become a synonym for "non-isochronous". For even potentials this is correct, by the solution of the "tautochrone problem" (the only potentials such that the time to the minimum is independent of starting position are the quadratics) (e.g. §214 of [1]). There are many other isochronous potentials, however. One can take two half-parabolae of different curvatures joined at their minima, or the limiting case of one half-parabola and a hard wall at its minimum. There are smoother examples, like

$$V(x) = x^2 + x^{-2} \tag{4}$$

(communicated to us by S.Flach), which arises for non-zero angular momentum in the rotationally symmetric two-dimensional harmonic oscillator, and $V(x) = 1 + x - \sqrt{1+2x}$ [8], but both of these have a singularity, at $x = 0$ and $x = -\frac{1}{2}$ respectively.

We show below that a C^2 potential is isochronous if and only if its graph arises by horizontally shearing the graph of a parabola, by a shear which preserves monotonicity of the two sides. We construct the general C^2 isochronous potential and a large class of examples which are real analytic on the whole real line.

2. Properties of isochronous potentials

Suppose that V is C^2 and isochronous. Then firstly, periodic orbits of energy E correspond to bounded connected components of $\{x \in \mathbb{R} : V(x) < E\}$, so every periodic orbit encloses at least one local minimum of V, thus V has a local minimum. Secondly, V has no critical points other than local minima, else there would be a family of periodic orbits which come close to one of these other critical points and hence its period goes to infinity so is not constant. By continuity of V', there is precisely one point, or possibly interval, of local minimum. The case of an interval of local minima cannot be isochronous because the period would go to infinity for the family of periodic orbits shrinking onto this interval. If there is a single point of local minimum, it must be non-degenerate, else the period of oscillation around it would again go to infinity as the amplitude goes to zero. We also deduce that $V(x) \to \infty$ as $x \to \pm\infty$, else the period of the family of periodic orbits oscillating about the minimum would go to infinity as E approaches $\min(\lim_{x\to\infty} V(x), \lim_{x\to-\infty} V(x))$, so would not be constant.

We deduce that by shifts of origin and scaling time, we can put any C^2 isochronous potential into what we will call *class A*: the C^2 potentials

V such that $V(0) = 0$, $V'(0) = 0$, $V''(0) = 2$, $V'(x) > 0$ for $x > 0$ and $V'(x) < 0$ for $x < 0$, and $V(x) \to \infty$ as $x \to \pm\infty$. Class A contains the subsets where V is C^r, for $r \in \{3, 4, \ldots, \infty, \omega\}$ (where C^ω denotes real analytic). Then V defines two inverse functions $x_+ : [0, \infty) \to [0, \infty)$ and $x_- : [0, \infty) \to (-\infty, 0]$, such that $V(x_\pm(\phi)) = \phi$ for all $\phi \in [0, \infty)$. By the inverse function theorem, they are as smooth as V on $(0, \infty)$, because $V' \neq 0$ there. This means that if V is C^r for some $r \in \{2, 3, 4, \ldots, \infty, \omega\}$, then so are $x_\pm$ on $(0, \infty)$. Near $\phi = 0$, however, they behave like square roots:

$$x_\pm(\phi) \sim \pm\sqrt{\phi} \text{ as } \phi \to 0, \tag{5}$$

since $V''(0) = 2$.

For a class A potential, every periodic orbit has energy $E \in (0, \infty)$, and every level set in this range is a single periodic orbit. The integral defining the period can be split into its contributions from the left and right-hand sides of zero. Following §214 of [1] (who attributes the idea to Puiseux), or §12 of [4] (we are grateful to R de la Llave for pointing out this reference), we change variables from x to ϕ using the functions $x_\pm$ on the right and left respectively, to obtain

$$T = \sqrt{2} \int_0^E \frac{\Delta'(\phi)}{\sqrt{E - \phi}}\, d\phi, \tag{6}$$

where

$$\Delta(\phi) = x_+(\phi) - x_-(\phi). \tag{7}$$

Write $\phi = Eu$ to obtain

$$T = \sqrt{2E} \int_0^1 \frac{\Delta'(Eu)}{\sqrt{1 - u}}\, du. \tag{8}$$

Differentiating with respect to E shows that V is isochronous iff

$$2\phi\Delta''(\phi) + \Delta'(\phi) = 0 \tag{9}$$

for all $\phi \in (0, \infty)$. This has the general solution $\Delta(\phi) = A\sqrt{\phi} + B$, A, B arbitrary. Equation (5) imposes the restriction $\Delta(\phi) \sim 2\sqrt{\phi}$ as $\phi \to 0$. We deduce that a class A potential V is isochronous if and only if

$$\Delta(\phi) = 2\sqrt{\phi}. \tag{10}$$

Note that for an even potential, $x_- = -x_+$, so $\Delta = 2x_+$, thus it is isochronous iff $x_\pm(\phi) = \pm\sqrt{\phi}$. Hence the only even isochronous potential (up to our class A normalisation conditions) is $V(x) = x^2$.

Returning to the general class A potential, equation (10) is equivalent to

$$x_{\pm}(\phi) = \bar{x}(\phi) \pm \sqrt{\phi} \tag{11}$$

for some function $\bar{x}$ on $[0, \infty)$. Since $x_{\pm}$ come from a class A potential, $\bar{x}$ is C^2 on $(0, \infty)$ and satisfies $\bar{x}(\phi) = o(\sqrt{\phi})$ as $\phi \to 0$ and

$$|\bar{x}'(\phi)| < \frac{1}{2\sqrt{\phi}}, \tag{12}$$

for all $\phi > 0$.

3. Construction of isochronous potentials

Any C^r function $\bar{x}$ on $[0, \infty)$ ($r \in \{2, 3, \ldots, \infty, \omega\}$) satisfying $\bar{x}(0) = 0$ and having $C < 1$ such that

$$|\bar{x}'(\phi)| \leq \frac{C}{2\sqrt{\phi}} \tag{13}$$

generates a C^r isochronous class A potential V. Simply join together the inverse graphs of the functions $x_{\pm}$ defined by Eq. (11), which are strictly monotone and go to $\pm\infty$ by Eq. (13), to construct the graph of V. It remains to check that V is C^r.

For this, it is most convenient to rearrange (11) to express the graph of V as the zero set of the function $K : \mathbb{R} \times \mathbb{R}_+ \to \mathbb{R}$ defined by

$$K(x, \phi) = \phi - (x - \bar{x}(\phi))^2. \tag{14}$$

If $\bar{x}$ is C^r, then so is K. The partial derivative $\partial K/\partial\phi = 1+2(x-\bar{x}(\phi))\bar{x}'(\phi)$. On the zero-set, this takes the value $1 \pm 2\sqrt{\phi}\bar{x}'(\phi)$ which is positive for $\phi > 0$ by condition (13) and is 1 for $\phi = 0$. Thus by the implicit function theorem, the zero-set of K is the graph of a C^r function $V : \mathbb{R} \to \mathbb{R}_+$.

Different functions $\bar{x}$ generate different potentials V. There is an infinite-dimensional set of such functions $\bar{x}$, even of analytic ones. To be concrete, we give one family of examples. Take $\bar{x}(\phi) = \sqrt{1 + \alpha\phi} - 1$ for any $\alpha \in [0, 1)$. This yields the analytic isochronous potentials

$$V(x) = \frac{2 + (1+\alpha)(x^2 + 2x) - 2(x+1)\sqrt{1 + \alpha x(x+2)}}{(1-\alpha)^2}, \tag{15}$$

sketched in Figure 1 for several values of α. Dorignac pointed out that as $\alpha \to 1$ this potential converges on $(0, \infty)$ to the example eq. (4), so it interpolates between this one and the harmonic one.

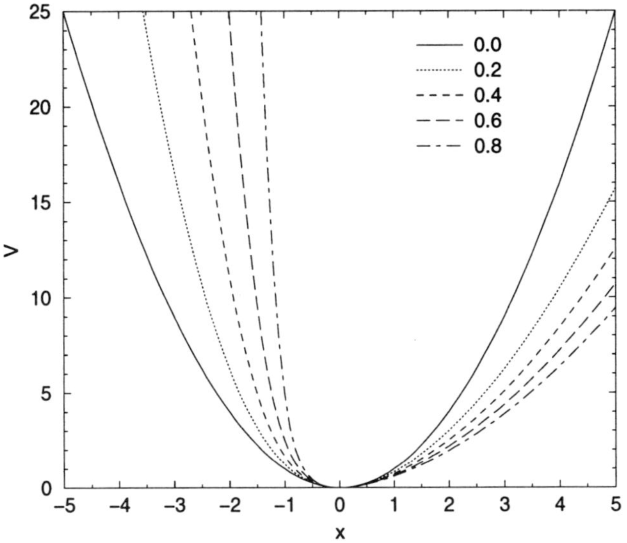

Figure 1. The potentials (15) for α from 0.0 to 0.8 in steps of 0.2.

4. General isochronous potential

The above construction does not quite give all class A isochronous potentials. Although to obtain a C^r potential it is necessary to have $\bar{x} \in C^r$ on $(0, \infty)$ and Eq. (12) (inverse function theorem), it is not necessary to require $\bar{x}$ to be C^r at $\phi = 0$. We will construct the general isochronous class A potential.

It is given by taking

$$\bar{x}(\phi) = \sqrt{\phi} X(\sqrt{\phi}) \tag{16}$$

for any function X which is C^2 on $(0, \infty)$, satisfies

$$|(uX(u))'| < 1 \tag{17}$$

there (where $'$ denotes d/du), and such that $X(u), uX'(u), u^2 X''(u) \to 0$ as $u \to 0$, and $u(\pm 1 + X(u)) \to \pm\infty$ as $u \to \infty$.

To see this, write $\phi = u^2$. Then Eq. (11) becomes

$$x_\pm = u(\pm 1 + X(u)). \tag{18}$$

To make the potential C^2 off zero it is necessary and sufficient that X be C^2 off zero, Eq. (17) hold and $x_\pm \to \pm\infty$ as $u \to \infty$. Denote the inverse functions to Eq. (18) by $u_\pm(x)$ for $x > 0, x < 0$ respectively. They go to zero monotonically through all positive values as x goes to zero. Then

$$V(x) = u^2 = x^2/(\pm 1 + X(u_\pm(x)))^2$$

for $x > 0, x < 0$ respectively. So $V''(0)$ does not exist if $X(u)$ does not go to 0 as $u \to 0$. We deduce that $X(u) \to 0$, and the only possible value for $V''(0)$ is 2.

Differentiating Eq. (18) for $x \neq 0$, we deduce that

$$u'_\pm = (\pm 1 + X + uX')^{-1} \tag{19}$$

and

$$u''_\pm = -\frac{2X' + uX''}{(\pm 1 + X + uX')^3}. \tag{20}$$

Hence

$$V'(x) = 2uu' = \frac{2u}{\pm 1 + X + uX'} \tag{21}$$

goes to 0 as required, because of Eq. (17), and

$$V''(x) = 2u'^2 + 2uu'' = \frac{2(\pm 1 + X - uX' - u^2X'')}{(\pm 1 + X + uX')^3}. \tag{22}$$

This goes to 2 for both signs iff both $X + uX'$ and $X - uX' - u^2X''$ go to zero as $u \to 0$. We already proved that $X \to 0$, so the first condition gives $uX' \to 0$. Using this, the second condition gives $u^2X'' \to 0$. This completes the proof.

An example of a function $\bar{x}$ satisfying all the conditions but which is not C^2 at 0 is generated by $X(u) = \sqrt{u}$.

5. Potentials with the same period function

The approach generalises to answer the question of when two smooth potentials have the same period function (the period function of a family of periodic orbits is the period T as a function of the energy E). By Eq. (6), or §12 of [4], the period function depends only on the function Δ specifying the length of the orbit as a function of energy. Hence any two potentials which are related by just a shear $x \mapsto x + \bar{x}(\phi)$ of their graphs $\phi = V(x)$ have the same period function. By the theory of the Abel transform (e.g. explained in §12 of [4]), different functions Δ give different period functions, so this is the only way to obtain potentials with the same period function.

A simple example is the pair of potentials $V(x) = \tanh^2 x$ (the sech^2 well) and $(1 - e^{-x})^2$ (the Morse potential), for both of which $\Delta(\phi) = 2\tanh^{-1}\sqrt{\phi}$ and which have the same period function $T(E) = \pi\sqrt{2/(1-E)}$ (e.g. differentiate formulae (A2.6 and 7) of [7]).

The issues of how much smoothness is required of the shear to preserve a given amount of smoothness of the potential are the same as in the isochronous case, with identical results in the case of potentials with non-degenerate minima.

6. Relation with previous work

The problem is an old one, dating back at least to Huygens. An equation for Δ in terms of the period function $T(E)$ was given in §12 of [4], assuming V to have a single minimum, which on inserting $T(E) = \text{constant}$, solves the problem up to smoothness questions and the necessity to restrict to class A. Ref. [8] gives a necessary and sufficient condition for an isochronous potential (and [5] a slight extension), but it is not clear from their papers that there are globally defined examples satisfying this. An equation equivalent to the combination of our equations (14) and (16) appears in [6] but without the condition (12) (or (17)) essential for constructing a solution from it. After writing the first version of our paper (April 1999), we became aware of [3], which contains the results of our equations (10) and (12), and gives a criterion for isochronous potentials in terms of "strict involutions", but it is not obvious from this how to make the general C^2 isochronous potential or any examples analytic on the whole of $\mathbb{R}$. There are many other papers dealing with the question of when Hamiltonian systems of more general form than (2) are isochronous, e.g. [2], but they do not answer the question for form (2) specifically.

7. Multidimensional examples

The method allows us to construct many isochronous potentials of more than one degree of freedom: simply take a sum of independent isochronous potentials of the same period. One can also make non-separable rotationally invariant two-dimensional potentials which are isochronous for a particular angular momentum value L: horizontally shear the graph of the radial potential $U(r) = \frac{1}{2}L^2r^{-2} + r^2$ by a function that goes to zero sufficiently fast as the energy goes to infinity, to obtain $\tilde{U}(r) = \frac{1}{2}L^2r^{-2} + V(r)$, and then V will have the required property. The question remains, however, what is the most general multidimensional isochronous potential.

Acknowledgements

Our collaboration was supported by INTAS 10771.

References

1. Appell P, Traite de mécanique rationelle, tome 1 (2nd ed, Gauthiers-Villars, Paris, 1902).
2. Christopher CJ, Devlin J, Isochronous centers in planar polynomial systems, SIAM J Math Anal 28 (1997) 162–177.
3. Cima A, Mañosas F, Villadelprat J, Isochronicity for several classes of Hamiltonian systems, J Diff Eq 157 (1999) 373–413.
4. Landau LD, Lifshitz EM, Mechanics (Pergamon, 1960).
5. Levin JJ, Shatz SS, Nonlinear oscillations of fixed period, J Math Anal Appl 7 (1963) 284–248.
6. Obi C, Analytical theory of nonlinear oscillation VII, The periods of the periodic solutions of the equation $x'' + g(x) = 0$, J Math Anal Appl 55 (1976) 295–301.
7. Percival IC, Richards D, Introduction to dynamics (Cambridge, 1982).
8. Urabe M, Potential forces which yield periodic motions of a fixed period, J Math Mech 10 (1961) 569–578.

BREATHERS ON DIATOMIC FPU CHAINS WITH ARBITRARY MASSES

GUILLAUME JAMES†, PASCAL NOBLE‡

†*Laboratoire MIP, Département GMM, INSA de Toulouse,*
135 avenue de Rangueil, 31077 Toulouse Cedex 4, France.
e-mail : Guillaume.James@gmm.insa-tlse.fr

‡*Laboratoire MIP, Université Paul Sabatier,*
118 route de Narbonne, 31 062 Toulouse Cedex, France.
e-mail : noble@mip.ups-tlse.fr

We give a proof for the existence of breathers in diatomic FPU chains with arbitrary mass ratios. This completes an existence result proved by Livi, Spicci and MacKay for large mass ratios. The proof doesn't use the concept of anticontinuous limit and is based on a discrete centre manifold reduction. For interaction potentials satisfying a hardening condition, we find breathers with frequencies slightly above the optic band, or in the gap slightly above the acoustic band. When the potential satisfies the opposite softening condition, we obtain breathers with frequencies in the gap slightly below the optic band.

1. introduction

A Fermi-Pasta-Ulam (FPU) chain consists in a one-dimensional chain of particles connected by anharmonic springs. We consider a diatomic FPU chain with alternating masses m_1, m_2 ($m_1 \leq m_2$) and look for spatially localized and time periodic solutions ("breather" solutions). The first breather existence result for this kind of system has been proved by Livi, Spicci and MacKay [7] by continuation from the anticontinuous limit. Their result is valid for hardening potentials and large mass ratio m_2/m_1 (see [2] for numerical continuation results up to $m_1 = m_2$). The case $m_1 = m_2$ has been studied with different approaches. For hardening potentials, the existence of breathers has been proved using a variational technique in [1] and a centre manifold reduction in [3], [4]. In the latter case, the equations are viewed as a recurrence relation in a loop space and the problem can be locally reduced to the study of a finite-dimensional mapping (breathers are obtained as homoclinic orbits to 0). In addition, Kiselev, Bickham and Sievers [6] have obtained a numerical evidence of the existence of breathers for $m_1 = \frac{1}{2}m_2$ (mass ratio corresponding to KBr crystals) and a soft potential.

In this paper, we prove a breather existence result for arbitrary mass ratio using a general centre manifold reduction theorem [5] which completes previous results obtained via the anticontinuous limit [7]. Moreover, we prove the existence of breathers for softening potentials, which confirms the numerical results of Kiselev et al [6]. The outline of the paper is as follows. We first write the equations of motion as a mapping in a loop space. Then we perform a spectral study of the linearized operator around the trivial equilibrium state. For some critical values of the parameters, we have collision of pairs of simple eigenvalues at +1 or −1 and homoclinic bifurcations can occur. We analyze these bifurcations and prove the existence (or non-existence) of small amplitude breathers (homoclinic orbits to 0), depending on the properties of the potential. This is done by reducing the problem to a reversible mapping on a 2-dimensional centre manifold.

2. Formulation of the mathematical problem

The diatomic FPU chain is described by the equations

$$\begin{aligned} \frac{d^2}{dt^2}x_{2n} &= V'(x_{2n+1}-x_{2n}) - V'(x_{2n}-x_{2n-1}), \\ m\frac{d^2}{dt^2}x_{2n+1} &= V'(x_{2n+2}-x_{2n+1}) - V'(x_{2n+1}-x_{2n}), \end{aligned} \tag{1}$$

where $m = m_1/m_2 \in (0,1)$ and V is a smooth interaction potential normalized by $V''(0) = 1$ ($V'(0) = 0$). We look for solutions of (1) with frequency ω. To cut off the invariance of (1) under translations, we use the rescaled force variable $y_n = V'(x_n - x_{n-1})(\frac{t}{\omega})$. Note that y_n is 2π-periodic in time. By integrating (1) one observes that the time average of y_n is independent of n. Since we are interested in spatially localized solutions, we fix $\int_0^{2\pi} y_n(t)dt = 0$ in the sequel.
Using variables y_n, problem (1) leads to the new system:

$$\begin{aligned} m\omega^2\frac{d^2}{dt^2}(W(y_{2n+1})) &= y_{2n+2} - (m+1)y_{2n+1} + my_{2n}, \\ m\omega^2\frac{d^2}{dt^2}(W(y_{2n})) &= my_{2n+1} - (m+1)y_{2n} + y_{2n-1}, \end{aligned} \tag{2}$$

where $W(y) = (V')^{-1}(y)$ is the local inverse of V' and satisfies $W(y) = y + W_2y^2 + o(y^2)$. Breather solutions of (1) with frequency ω correspond to 2π-periodic solutions of (2) satisfying $\lim_{|n|\to\infty} \|y_n\|_{L^\infty} = 0$. We now formulate (2) as a first order recurrence relation in a loop space. For this purpose, we define $u_n = y_{2n}$, $v_n = y_{2n-1}$ and $Y_n = (u_n, v_n)$. Problem (2) can be rewritten

$$Y_{n+1} = F_{m,\omega}(Y_n) \tag{3}$$

with

$$F_{m,\omega}(u,v) = \begin{pmatrix} D(\frac{1}{m}(D(u)-v)) - mu \\ \frac{1}{m}(D(u)-v) \end{pmatrix} \tag{4}$$

and $D(u) = m\omega^2 \frac{d^2}{dt^2} W(u) + (m+1)u$. Now (m,ω) play the role of bifurcation parameters.

We now define appropriate function spaces on which $F_{m,\omega}$ is acting. We look for $(u_n, v_n) \in \mathbb{D}$ where $\mathbb{D} = \{(u,v) \in \mathbb{H}^4 \times \mathbb{H}^2, u, v \text{ even}, \int_0^{2\pi} u dt = \int_0^{2\pi} v dt = 0\}$ and $\mathbb{H}^n$ denotes the classical Sobolev space $H^n(\mathbb{R}/2\pi\mathbb{Z})$. The recurrence relation (3) holds in $\mathbb{X} = \{(u,v) \in \mathbb{H}^0 \times \mathbb{H}^2, u, v \text{ even}, \int_0^{2\pi} u dt = \int_0^{2\pi} v dt = 0\}$. The operator $F_{m,\omega} : \mathbb{D} \to \mathbb{X}$ is smooth in a neighbourhood $\mathcal{U}$ of $Y = 0$ in $\mathbb{D}$. Note that the fixed point $Y = 0$ of $F_{m,\omega}$ corresponds to the lattice at rest. We now examine the symmetry properties of equation (3). On the one hand, equation (3) is invariant under the symmetry $TY = Y(\cdot + \pi)$ i.e. $F_{m,\omega}$ commutes with T. Moreover, if y_n is a solution of (2) then $\tilde{y}_n = y_{-n-1}$ also satisfies (2). This implies that $F_{m,\omega}$ is reversible with respect to the symmetry $R(u,v) = (v,u)$ i.e. if $Y \in \mathcal{U}$ and $F_{m,\omega}(RY) \in \mathcal{U}$, one has $(F_{m,\omega} \circ R)^2 Y = Y$. We now study the spectrum of the linearized operator at $Y = 0$.

3. Spectral properties of the linearized operator

The linearized operator $L_{m,\omega} = DF_{m,\omega}(0)$ reads

$$L_{m,\omega}(u,v) = \begin{pmatrix} A(\frac{1}{m}(Au - v)) - mu \\ \frac{1}{m}(Au - v) \end{pmatrix},$$

where $Au = m\omega^2 \frac{d^2 u}{dt^2} + (m+1)u$. The operator $L_{m,\omega} : \mathbb{D} \subset \mathbb{X} \to \mathbb{X}$ is unbounded in $\mathbb{X}$ (of domain $\mathbb{D}$) and closed. We now examine the spectral properties of $L_{m,\omega}$ (see [5] for the proof).

Lemma 3.1. *For all $m \in (0,1)$ and $\omega > 0$, the spectrum of $L_{m,\omega}$ is unbounded, discrete and can be written $\sigma(L_{m,\omega}) = \{0\} \cup \Sigma_{m,\omega}$, where 0 belongs to the essential spectrum and $\Sigma_{m,\omega}$ consists in non-zero eigenvalues. The set $\Sigma_{m,\omega}$ is contained in the union of the real axis and the unit circle, and invariant under $z \to \bar{z}$, $z \to z^{-1}$. The eigenvalues form sequences $(z_k)_{k \geq 1}$ and $(z_k^{-1})_{k \geq 1}$ (with $|z_k| \geq 1$ and $Im\, z_k \geq 0$) determined by the equation*

$$\frac{1}{2}(z_k + z_k^{-1}) = 1 + k^2\omega^2 (\frac{m}{2}k^2\omega^2 - m - 1), \quad k \geq 1. \tag{5}$$

We now study the variations of the spectrum of $L_{m,\omega}$ as we vary the parameter $(\omega^2, m) \in S$, $S = (0, +\infty) \times (0, 1)$. The eigenvalues z_k, z_k^{-1} belong to the positive real axis when $k^2\omega^2 \geq 2(1 + \frac{1}{m})$ and collide at $z = +1$ when $k^2\omega^2 = 2(1 + \frac{1}{m})$. Note that all the eigenvalues belong to the positive real axis (and lie outside the unit circle) when $\omega^2 > 2(1 + \frac{1}{m})$. For $\frac{2}{m} \leq k^2\omega^2 \leq 2(1 + \frac{1}{m})$ or $k^2\omega^2 \leq 2$, the eigenvalues z_k, z_k^{-1} belong to the unit circle. They lie on the negative real axis when $2 \leq k^2\omega^2 \leq \frac{2}{m}$ and collide at $z = -1$ when $k^2\omega^2 = \frac{2}{m}$ or $k^2\omega^2 = 2$. One can see that z_k does not vary monotonically for $\omega^2 \in [\frac{2}{k^2}, \frac{2}{k^2 m}]$ and reaches its minimum $z_k = -1/m$ at the mid value $\omega^2 = \frac{m+1}{k^2 m}$. The above analysis leads us to consider the curves $\Gamma_k^+ : \omega^2 = \frac{2}{k^2}(1 + \frac{1}{m}), \Gamma_k^- : \omega^2 = \frac{2}{k^2 m}, \Gamma_k^a : \omega^2 = \frac{2}{k^2}$ in the parameter space S. The infinite collection of curves Γ_k^+, Γ_k^-, Γ_k^a $(k \geq 1)$ divides S in different regions corresponding to different numbers of eigenvalues on the unit circle (these curves are depicted in figure 1 for $k = 1, 2$).

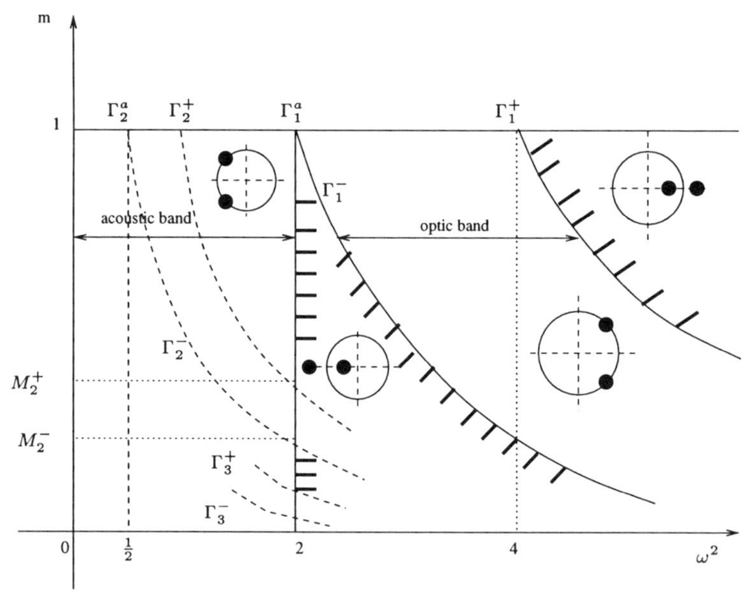

Figure 1. Spectrum of $L_{m,\omega}$ near the unit circle in some parameter regions. In the shaded regions, z_1, z_1^{-1} are the only eigenvalues close to the unit circle and the fixed point $Y = 0$ is hyperbolic.

The set of possible bifurcation scenarios is broad. With the aim of finding discrete breathers as homoclinic orbits to $Y = 0$, we shall restrict ourselves to the simplest situation when z_1, z_1^{-1} are the only eigenvalues close to the unit circle and the fixed point $Y = 0$ is hyperbolic. The corresponding regions in the parameter space S are depicted in figure 1 (shaded regions). One of these regions is situated in the neighbourhood of Γ_1^+, at the right side (i.e. above the top of the optic band). An other region is in the neighbourhood of Γ_1^-, at the left side (i.e. below the bottom of the optic band). The situation is more complicated when (ω^2, m) is close to Γ_1^a and at the right side (i.e. above the top of the acoustic band), since the curves Γ_1^a and $\Gamma_k^{\pm}$ $(k \geq 2)$ intersect at the point $(\omega^2, m) = (2, M_k^{\pm})$ with $M_k^+ = \frac{1}{k^2-1}, M_k^- = \frac{1}{k^2}$. We shall only consider the case when $m \in I_k = (M_{k+1}^+, M_k^-)$ $(k \geq 1)$ is fixed and $\omega^2 \approx 2$, since z_1, z_1^{-1} are the only eigenvalues close to the unit circle. In the following lemma, we compute the centre space $\mathbb{X}_c$ i.e. the generalized eigenspace corresponding to the eigenvalues lying on the unit circle.

Lemma 3.2. *If $(\omega^2, m) \in \Gamma_1^+$, $z = +1$ is a double non-semi-simple eigenvalue of $L_{m,\omega}$. If $(\omega^2, m) \in \Gamma_1^-$ or $(\omega^2, m) \in \Gamma_1^a$ with $m \in I_k$ $(k \geq 1)$, $z = -1$ is a double non-semi-simple eigenvalue of $L_{m,\omega}$. Moreover, the centre space $\mathbb{X}_c$ is the same for all these parameter values and is spanned by the vectors $V_u = (\cos t, 0)$ and $V_v = (0, \cos t)$. The spectral projection Π_c on $\mathbb{X}_c$ reads $\Pi_c(u, v) = (P_c u, P_c v)$, where $P_c u = \frac{1}{\pi} \int_0^{2\pi} u(t) \cos t \, dt \, \cos t$.*

We now fix $m \in (0,1)$ and set $\omega^2 = \omega_c^2 + \mu$, where $(\omega_c^2, m) \in \Gamma_1^+, \Gamma_1^-$ or Γ_1^a (in this case we further assume $m \in I_k$) and $\mu \approx 0$ is a small bifurcation parameter. Problem (3) can be written $Y_{n+1} = L_{m,\omega_c} Y_n + \mathcal{N}(Y_n, \mu)$ where $\mathcal{N}(0, \mu) = 0$, $D\mathcal{N}(0,0) = 0$.

4. Centre manifold and normal form computation

Using the centre manifold theorem for quasilinear mappings [4], one obtains for $\mu \approx 0$ the existence of a smooth local manifold $\mathcal{M}_\mu$ invariant under $F_{m,\omega}$ and the symmetries R, T (see [5] for more details). This manifold can be written $\mathcal{M}_\mu = \{Y \in \mathbb{D}/Y = Y^c + \Psi(Y^c, \mu), Y^c \in \mathbb{X}_c \cap \Omega\}$ where $\Psi(., \mu) : \mathbb{X}_c \to (I - \Pi_c)\mathbb{D}$ is a smooth map satisfying $\Psi(Y, \mu) = O(||Y||^2 + ||Y|||\mu|)$ and Ω is a small neighbourhood of 0 in $\mathbb{D}$. For $\mu \approx 0$, the centre manifold $\mathcal{M}_\mu$ contains all solutions Y_n staying in Ω for all $n \in \mathbb{Z}$ (small amplitude solutions). Their central component $Y_n^c = \Pi_c Y_n$ is given by the two dimensional mapping

$$Y_{n+1}^c = f(Y_n^c, \mu), \tag{6}$$

where $f(Y^c, \mu) = L_{m,\omega_c} Y^c + \Pi_c \mathcal{N}(Y^c + \Psi(Y^c, \mu), \mu)$. The map $f : \mathbb{X}_c \to \mathbb{X}_c$ is smooth, reversible under R and commutes with T.

4.1. *Centre manifold computation*

Setting $Y^c = (a,b)\cos(t)$, we identify $\mathbb{X}_c$ with $\mathbb{R}^2$. We now compute the Taylor expansion of Ψ. The symmetry properties of Ψ implies that $T\Psi(a,b,\mu) = \Psi(-a,-b,\mu), R\Psi(a,b,\mu) = \Psi(b,a,\mu)$. Consequently the Taylor expansion of Ψ has the form:

$$\Psi(a,b,\mu) = \begin{pmatrix} \Psi_{011}a\mu + \Psi_{101}b\mu + \Psi_{020}a^2 + \Psi_{110}ab + \Psi_{200}b^2 \\ \Psi_{101}a\mu + \Psi_{011}b\mu + \Psi_{200}a^2 + \Psi_{110}ab + \Psi_{020}b^2 \end{pmatrix} + \text{h.o.t.}$$

Setting $Y_n^c = (a_n, b_n)\cos(t)$, coefficients Ψ_{pqr} are obtained by identification of the terms $a_n\mu$, $b_n\mu$, a_n^2, b_n^2 and a_nb_n in the Taylor expansion of the equation:

$$\Psi(\Pi_c F_{m,\omega}(Y_n^c + \Psi(Y_n^c,\mu)),\mu) = (I - \Pi_c)F_{m,\omega}(Y_n^c + \Psi(Y_n^c,\mu)) \tag{7}$$

(this relation is due to the invariance of $\mathcal{M}_\mu$ under $F_{m,\omega}$, see [5] for more details). The identification of quadratic terms in (7) leads to a linear system of ODE for Ψ_{pqr}. Solving this system yields $\Psi_{011} = \Psi_{101} = 0$ and the functions $\Psi_{020}, \Psi_{200}, \Psi_{110}$ are colinear to $\cos(2t)$ (their exact form depends on the considered value of ω_c and is given in the appendix).

4.2. *Normal form computation*

In the following lemma, we write the reduced mapping (6) in *normal form*, i.e. we perform a change of variables which only keeps its essential terms (see [5] for a detailed proof). In what follows we use the notation $Y_n^c = (a_n, b_n)\cos(t)$.

Lemma 4.1. *There exists a smooth local diffeomorphism h_μ defined on a neighbourhood of $(a,b) = 0$ which transforms (6) into the following mapping:*

$$A_{n+1} \pm 2A_n + A_{n-1} = g_\mu(A_n, A_n \pm A_{n-1}) \tag{8}$$

$$B_{n+1} = A_{n+1} \pm A_n \tag{9}$$

where $(A_n, B_n) = h_\mu(a_n, b_n)$ ($h_\mu(0) = 0$) and

$$\begin{aligned} g_\mu(A_n, B_n) = &-(m + 1 - \omega_c^2)\,(2\mu A_n - \tfrac{B\omega_c^2}{8}A_n^3) \\ &+o(\|(A_n,B_n)\|^3 + \mu\|(A_n,B_n)\|), \end{aligned} \tag{10}$$

with a $-$ sign in (8), (9) for $(\omega^2, m) \in \Gamma_1^+$ and a $+$ sign for $(\omega^2, m) \in \Gamma_1^-$ or Γ_1^a ($m \in I_k, k \geq 1$). The coefficient B in (10) is given by (for $V''(0) = 1$)

$$B = \frac{1}{2}V^{(4)}(0) - (V^{(3)}(0))^2. \tag{11}$$

As a conclusion, the mapping (6) can be locally transformed into the simpler second order scalar recurrence relation (8) via the diffeomorphism h_μ. The form of h_μ depends on the considered value of ω_c and is given in the appendix.

5. Homoclinic solutions of the reduced mapping

We now consider the recurrence relation (8) for $\mu \approx 0$ and study the existence of small homoclinic orbits to $A_n = 0$ (these orbits correspond to small amplitude breather solutions of (2)). In the case when $(\omega_c^2, m) \in \Gamma_1^-$ or $(\omega_c^2, m) \in \Gamma_1^a$ we use the variable $U_n = (-1)^n A_n$. For $(\omega_c^2, m) \in \Gamma_1^+$ we shall fix $U_n = A_n$ in the sequel. Equation (8) yields

$$U_{n+1} - 2U_n + U_{n-1} = \pm g_\mu(U_n, U_n - U_{n-1}), \tag{12}$$

with a + sign for $(\omega_c^2, m) \in \Gamma_1^+$ and a − sign for $(\omega_c^2, m) \in \Gamma_1^-$ or $(\omega_c^2, m) \in \Gamma_1^a$. In the sequel, we choose (ω^2, m) in the shaded regions of figure 1 (then $U_n = 0$ is an hyperbolic equilibrium). We denote by hardening case the situation when $B > 0$ and softening case the opposite situation when $B < 0$. If $\mu B > 0$, (12) admits homoclinic orbits to 0 (see [4], [5] for a proof) and there exist small amplitude breather solutions of (2). This result heavily relies on the fact that (12) is reversible (due to the reversibility of f). In the hardening case, this yields the existence of breathers with frequencies $\omega^2 \gtrsim 2$ (for $m \in I_k$) and $\omega^2 \gtrsim 2(1 + \frac{1}{m})$. In the softening case, we obtain discrete breathers with frequencies $\omega^2 \lesssim \frac{2}{m}$. If $\mu B < 0$ in (12), the *local* stable and unstable manifolds $W^s(0)$ and $W^u(0)$ do not intersect in the neighbourhood of 0 and (2) admits no small amplitude breather solution in the corresponding parameter range. We sum up the situation in figure 2.

Appendix

This appendix gives the leading order terms in the definition of the centre manifold and the change of variables h_μ. In the sequel we note $W_2 = -\frac{1}{2}V^{(3)}(0)$. In the case when $(\omega_c^2, m) \in \Gamma_1^+$ one has

$$\Psi_{110} = \frac{1}{8m} W_2 \cos(2t), \quad \Psi_{020} = -\frac{1}{16}\frac{8m-1}{m} W_2 \cos(2t),$$
$$\Psi_{200} = \frac{1}{16m} W_2 \cos(2t)$$

and $h_\mu(a,b) = (\,-(\alpha+2)\,a - \alpha\, b,\ -2(1+\alpha)\,(a+b)\,) + o(||(a,b)||)$.

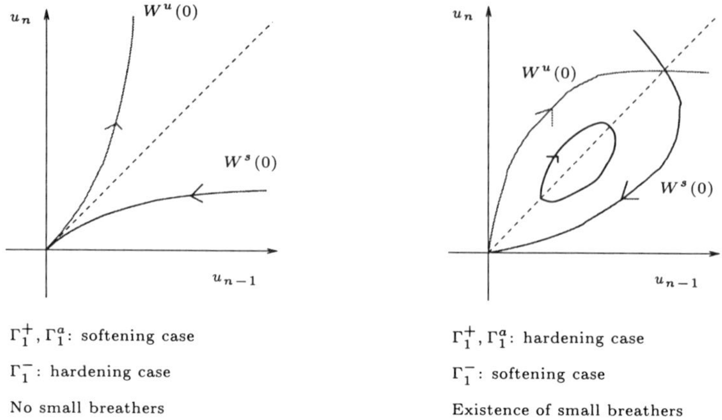

Figure 2. Local stable and unstable manifolds $W^s(0)$ and $W^u(0)$ in (12).

For $(\omega_c^2, m) \in \Gamma_1^-$ we have

$$\Psi_{110} = \frac{1}{8}\frac{m^2+2m-3}{m(m-3)} W_2\cos(2t), \quad \Psi_{200} = \frac{1}{16}\frac{m^2-6m-3}{m(m-3)} W_2\cos(2t),$$
$$\Psi_{020} = -\frac{1}{16}\frac{7m^2-34m+3}{m(m-3)} W_2\cos(2t)$$

and $h_\mu(a,b) = (\,(\alpha-2)\,a + \alpha\, b\,,\, -2(1-\alpha)\,(a+b)\,) + o(\|(a,b)\|)$.

The case when $(\omega_c^2, m) \in \Gamma_1^a, m \in I_k, k \geq 1$ yields

$$\Psi_{110} = \frac{1}{8}\frac{5m^2-6m+1}{m^2(3m-1)} W_2\cos(2t), \quad \Psi_{200} = \frac{1}{16}\frac{3m^2+6m-1}{m^2(3m-1)} W_2\cos(2t),$$
$$\Psi_{020} = -\frac{1}{16}\frac{24m^3+5m^2-6m+1}{m^2(3m-1)} W_2\cos(2t)$$

and $h_\mu(a,b) = (\,\alpha\, a + (2-\alpha)\, b\,,\, 2(\alpha-1)\,(a-b)\,) + o(\|(a,b)\|)$.

References

1. S. Aubry, G. Kopidakis and V Kadelburg, Discrete and Continuous Dynamical Systems-Series B, Vol 1, Number 3, 2001, pp 271-298.
2. T. Cretegny, R. Livi and M. Spicci, Physica D 119, 1998, pp 88-98.
3. G. James, C.R.Acad.Sci.Paris, t. 332, Série I (2001), pp 581-586.
4. G. James, *Centre manifold reduction for quasilinear discrete systems*, to appear in Journal of Nonlinear science (submitted 2002).
5. G. James and P. Noble, *Breathers on diatomic FPU chains with arbitrary masses*, preprint 2002, laboratory MIP, University of Toulouse.
6. S.A. Kiselev, S.R. Bickham and A.J. Sievers, Phys Rev B 50, number 13, 1994, pp 9135.
7. R. Livi, M Spicci and R.S. MacKay, Nonlinearity 10, 1997, pp 1421-1434.

DYNAMICS OF DISCRETE BREATHERS IN FLEXIBLE CHAINS

J.M. SANCHO AND M. IBAÑES,

Department d'Estructura i Constituents de la Matèria
Universitat de Barcelona
Diagonal 647, E–08028 Barcelona, Spain
E-mail: jmsancho@ecm.ub.es; marta@ecm.ub.es

G. P. TSIRONIS

Department of Physics
University of Crete and
Foundation for Research and Technology-Hellas
P.P. Box, 2208
71003 Heraklion, Crete. Greece
E-mail:gts@physics.uoc.gr

We present the study of discrete breather dynamics in curved polymer-like chains that consist of masses connected via nonlinear springs. For chains with strong angular rigidity, we find that breather motion is strongly affected by the presence of bended regions of the polymers, while the breathers themselves show a very strong resilience and remarkable stability in the presence of geometrical changes. For more flexible chains modeled via second neighbor interactions we find that chain geometry typically does not destroy the localized breather states but, contrary to the angularly rigid chains, it induces some small but constant energy loss. Furthermore, we find that a curved segment acts as an active gate reflecting or refracting the incident breather and transforming its velocity to a value that mostly depends on the DB frequency. We analyze the physical reasoning behind these seemingly general breather properties.

Intrinsic localized modes (ILM's) or discrete breathers (DB's) are space localized lattice oscillation modes whose main ingredients are nonlinearity in the interactions and lattice discreteness [1]. DB properties, including generation [1], rigorous existence[2], dynamics and mobility[3] are nowadays well established.

The work to be presented here attempts to deviate from simple one dimensional lattice models by introducing the possibility of a different spatial structure and as a consequence that of a chain bending. We will thus be concerned here with polymeric chains of masses coupled with springs that can move in principle in the whole (x, y) plane and are characterized by local and global elastic properties. Our basic objective is how energy localization in the form of DB's interplays with single chain polymer elasticity and shape. More specifically, the main question we want to address here is if a breather can propagate in curved portions of a polymer and what are the features of its kinetics as it traverses straight and curved segments of it.

Our minimum parameter model will be an arbitrarily shaped chain of equal masses coupled via nonlinear springs involving only two body polynomial-type interactions without on-site potentials. The chain is thus able to move on the plane according to interaction potentials between first and second neighboring masses. Thus, the chain consists of N molecular units interacting through pair-wise two body interactions. Each mass unit in the chain is labeled by an index n, while its location is specified through the pair (x_n, y_n) denoting its location on the plane with respect to an absolute Cartesian system. Since we will use first and second neighbor interaction potentials we need to introduce the following two Euclidean distances:

$$d_n = \left[(x_n - x_{n-1})^2 + (y_n - y_{n-1})^2\right]^{1/2}, \tag{1}$$

$$e_n = \left[(x_n - x_{n-2})^2 + (y_n - y_{n-2})^2\right]^{1/2}. \tag{2}$$

We note that d_n, e_n are simply the distances on the plane between the n-th unit and the $n-1$-th and $n-2$th units respectively. The polymer chain plasticity as well as rigidity is controlled by the ensemble of first and second neighbor constant equilibrium distances $\{a_n\}$ and $\{b_n\}$ respectively. The constant a_n is the equilibrium oscillator distance between units n and $n-1$ while b_n is that between the n-th and $n-2$-th units. The explicit configuration of these two set of constants fixes the desired equilibrium geometry of the polymer chain [5].

The neighbor interactions are generated through Fermi-Pasta-Ulam (FPU) type potentials:

$$V(d_n) = K_1 \frac{(d_n - a_n)^2}{2} + \beta_1 \frac{(d_n - a_n)^4}{4}, \tag{3}$$

$$W(e_n) = K_2 \frac{(e_n - b_n)^2}{2} + \beta_2 \frac{(e_n - b_n)^4}{4}. \tag{4}$$

Although this type of linear chains with only first neighbor interaction are in principle highly degenerate, the inclusion of a second neighbor interaction will reduce substantially the degeneracy and there is no need for other additional external constraints.

Accordingly the Hamiltonian for the planar polymer chain can be then written as:

$$H = \sum_n \frac{m}{2}(\dot{x}_n^2 + \dot{y}_n^2) + \sum_n \{V(d_n) + W(e_n)\} \tag{5}$$

where the index n runs over all polymer masses. Then the resulting equations of motion are:

$$m\,\ddot{x}_n = -\frac{\partial}{\partial x_n}U_n \qquad m\,\ddot{y}_n = -\frac{\partial}{\partial y_n}U_n, \tag{6}$$

which are numerically integrated with a Runge–Kutta algorithm of fourth order.

Although the modeling presented here is general, in the following we will restrict ourselves to the case $a_n = a\ \forall n$. b_n depends on the geometrical structure we want to study (see Fig. 1).

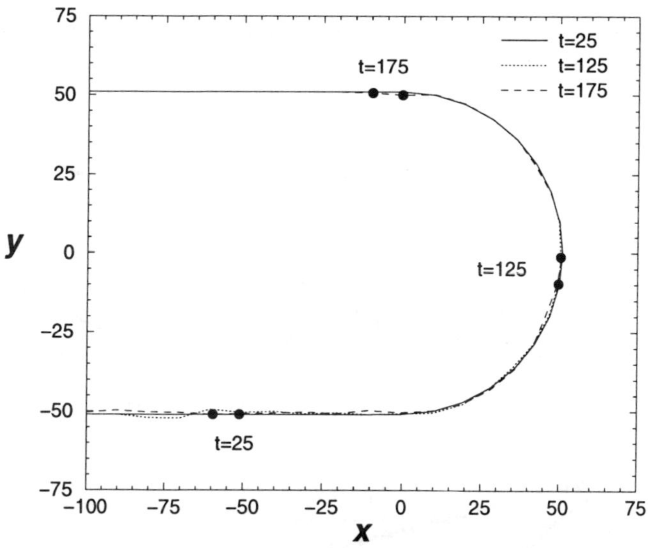

Figure 1. Chain geometry and breather position at different times for a chain with first and second neighbor interactions ($K_1 = \beta_1 = 1$ and $K_2 = \beta_2 = 0.3$). The center, which occupies two sites, of the DB is marked with the black circles. The hairpin geometry is given by $\alpha = \pi/16$ and $a = 10$. x and y are in units of a in all figures. $T_b = 2.122$

The first neighbor nonlinear interaction, $V(d_n)$, is mostly responsible for the local longitudinal dynamics whereas the second neighbor interaction for the geometric and angular rigidity of the chain. The second neighbor interaction is thus playing a role similar to the angular constraint of the restricted model of Ref. [6], while, on the other hand permitting transversally both flexibility and focusing. Since interactions decay with distance, first neighbor interactions have to be taken stronger than second neighbor interactions, i.e. $K_1 > K_2$ and $\beta_1 > \beta_2$.

The basic setup of our study is a mobile breather generated in a straight part of a chain which moves towards a curved portion of the chain. For the numerically exact generation of one dimensional DB's we used the accurate yet approximate algebraic method [4]. Furthermore, for rendering the DB mobile we used a simple variant of the pinning mode excitation method [3].

We have considered a hairpin structure, launch a longitudinal DB in the straight region while removing any linear sound modes possibly induced initially on the lattice as a result of the breather kick. The propagation of the DB in a hairpin shaped chain with first and second interactions is shown in Fig. 1. We observe that the spatial structure of the chain is stable due to the inclusion of second neighbor interactions. In the absence of them the chain is unstable [6]. Other possible scenarios that mimic rigid chains are possible, see for example [7,8,?].

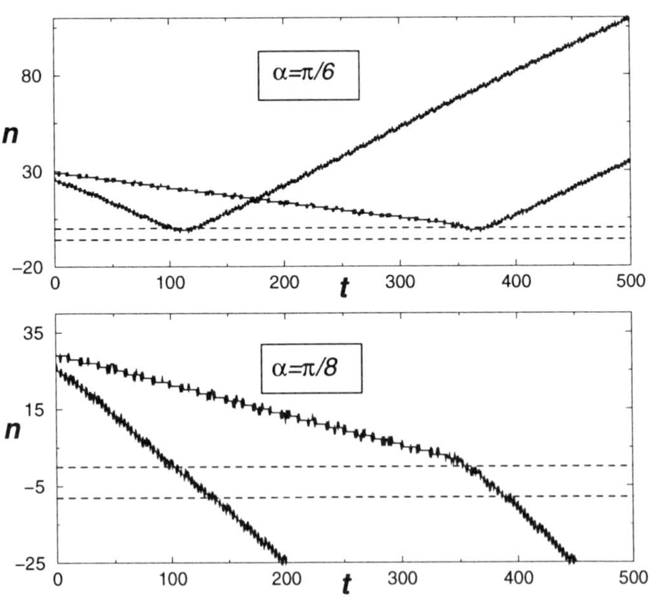

Figure 2. Position of the DB center as a function of time for two initial velocities and two hairpin curvatures given by $\alpha = \pi/6$ (top) and $\alpha = \pi/8$ (bottom). Depending on the curvature, the DB enters into the curved region or rebounces. Parameter values: $K_2 = \beta_2 = 0.3$ and $T_b = 2.122$.

The most interesting results can be appreciated in Fig. 2. DB enters into the curved region for low curvatures and rebounces for high curvatures. In all cases there is an energy loss when reaching the curved region. Moreover a fine tuning of the second neighbor interaction values K_2 and β_2 can reduce energy losses when the breather enters into the curve.

In addition, we found that DB's rebounce or exit the curved region always with approximately the same asymptotic velocity, v_F, independently of the initial velocity and the curvature. If the initial velocity of the DB is lower than v_F, the DB increases its velocity when entering into the curved region or while rebouncing (Fig. 2). If the initial velocity is above v_F, the DB decelerates when traversing the curved region . Thus, the hairpin is transparent for DB's with a velocity near v_F. Nevertheless the final velocity does not seem to depend very much on the harmonic or anharmonic nature of the second neighbor interactions neither on the strengths of the parameters with the exception of breathers frequency. In Fig. 3 the explicit dependence of v_F on the DB frequency ω_b is shown.

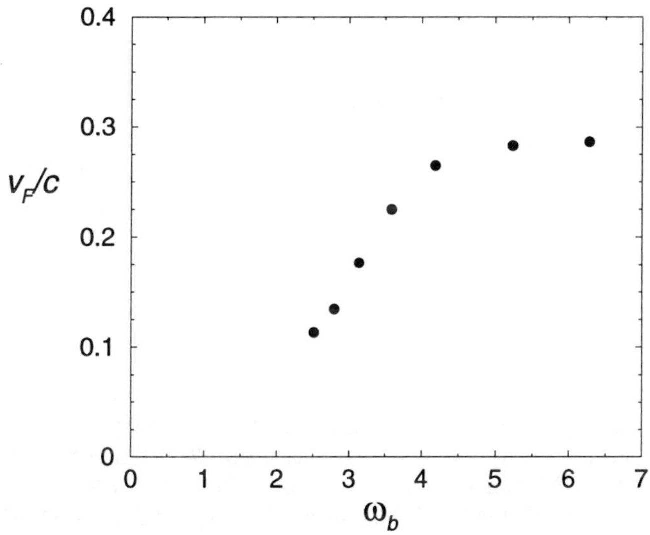

Figure 3. Ratio between the asymptotic velocity and the sound velocity ($c = \sqrt{K_1 + 4K_2}$) of an initially static DB in a straight chain with second neighbor interactions and with a transverse perturbation applied on it, versus DB frequency. The phonon band is below $\omega \approx 2$.

Moreover for an initially stationary breather ($v_i = 0$) with transversal perturbation we find that the breather begins moving and reaches shortly the terminal velocity v_F. We thus find that an initial random perturbation in the transversal breather direction has effects very similar to those of the geometric bend. A similar result is observed if, for instance, the transversal perturbation is not included initially in the breather shape but is effected after it is launched at some other spatial region. Thus the features we have described are due to the flexibility of the chain, which can move in two dimensions.

Thus the bend region acts as an active gate selecting a DB according to its frequency. Preliminary work on a biopolymer in three dimensions and an $\alpha - helix$ spatial structure is in agreement with these main conclusions. Further work is in progress.

As a result of this study the discrete breather emerges as an efficient energy transfer agent in more complex geometries. It is seen as able to be generated as a local depository of energy, transport it across chain segments with different local geometric properties, survive local environment changes and adapt to the local strain requirements.

Acknowledgments

This work has been supported partially by the European Union under the RTN project LOCNET (HPRN-CT-1999-00163) and by the Dirección General de Enseñanza Superior e Investigación Científica (Spain) under project BFM2000-0624. This research was also partially supported by US Department of Energy under contract W-7405-ENG-36. This work is dedicated to Prof. Domingo Gonzalez, from the University of Zaragoza, on the occasion of his retirement.

References

1. A. J. Sievers and S. Takeno, *Phys. Rev. Lett.* **61**, 970 (1988).
2. R. S. MacKay and S. Aubry, *Nonlinearity* **7**, 1623 (1994).
3. Ding Chen, S. Aubry and G. P. Tsironis, *Phys. Rev. Lett.* **77**, 4776 (1996).
4. M. Ibañes, J. M. Sancho and G. P. Tsironis, *Phys. Rev.* **E65**, 041902 (2002).
5. G. P. Tsironis, M. Ibañes, J. M. Sancho, *Europhysics Letters* **57**, 697 (2002).
6. G. P. Tsironis, *J. Phys. A: Math. Gen.* **35**, 951 (2002).
7. R. Reigada, J. M. Sancho, M. Ibañes and G. P. Tsironis, *Journal of Physics* **A 34**, 8465 (2001.
8. J. Cuevas, F. Palmero, J.F.R. Archilla and F.R. Romero, Phys. Lett. A**299**, 221(2002).
9. S.F.Mingaleev, Y.B.Gaididei, P.L.Christiansen, and Y.S.Kivshar, Europhysics Letters **59**, 402-409 (2002).

CRITICAL DYNAMICS OF DNA DENATURATION *

N. THEODORAKOPOULOS

Theoretical and Physical Chemistry Institute,
National Hellenic Research Foundation,
Vasileos Constantinou 48,
116 35 Athens, Greece
E-mail: nth@eie.gr

M. PEYRARD AND T. DAUXOIS

Laboratoire de Physique, UMR-CNRS 5672,
ENS Lyon, 46 Allée d'Italie,
69364 Lyon Cédex 07, France

We present detailed molecular dynamics results for the displacement autocorrelation spectra of the Peyrard-Bishop model of thermal DNA denaturation. As the phase transition is approached, the spectra depend on whether the wavelength is smaller than, or exceeds the correlation length. In the first case, the spectra are dominated by a single peak, whose frequency approaches the bare acoustic frequency of the harmonic chain, and whose linewidth approaches zero as $T_c - T$. In the second case, a central peak (CP) feature is dominant, accounting for most of the weight; the linewidth of the CP appears to be temperature-independent. We also present force autocorrelation spectra which may be relevant for analyzing the statistical properties of localized modes.

1. Introduction

The thermal denaturation of DNA, i.e. the separation of the two strands upon heating, is a typical thermodynamic instability. It can be modelled along the lines of other thermodynamic instabilities (e.g. wetting, solid-on-solid adsorption), by associating a single, one-dimensional coordinate with the distance of a base pair[1]. Details can be found in this volume[2] and in the original literature cited there.

The equilibrium properties of the system near the phase transition are characterized by a divergent correlation length ξ, and a discontinuity in the specific heat; in other words, this is a second-order transition; the feature

*Work partially supported by EU contract HPRN-CT-1999-00163 (LOCNET network).

which sets it apart from other structural, or order-disorder transitions is that, as the transition temperature is approached from below, the order parameter diverges, i.e. the low-temperature phase becomes continuously unstable.

In this paper, we present results for the dynamical correlations of the order parameter, obtained by numerical simulation. At low and intermediate temperatures, the spectra appear to be dominated by the properties of localized anharmonic motion ("discrete breathers"). As the critical temperature is approached from below, the spectra depend solely on whether the wavelength is smaller or larger than the correlation length. In the first case, they reflect the dynamics of "islands" of the high temperature phase. In the second case, they are dominated by a strong central peak, whose width appears to depend on the wavevector, but not on the temperature.

The paper is structured as follows: Section 2 introduces the notation and numerical procedure. Section 3 presents the main results, and an analysis along the lines of relaxational/oscillational phenomenology. Section 4 is a sketch of a tentative, alternative theory, along the lines of the Mori-Zwanzig projection operator formalism. Section 5 presents a brief summary and discussion.

2. Notation and numerical procedure

We consider the "minimal" Hamiltonian model of homogeneous DNA denaturation proposed by Peyrard and Bishop[1] (PB),

$$H(y) = \sum_n \left[\frac{p_n^2}{2} + \frac{1}{2R}(y_n - y_{n-1})^2 + V(y_n) \right] \quad , \tag{1}$$

where y_n, p_n are dimensionless, canonically conjugate coordinates and momenta of the nth base pair transverse to the chain, and $V(y) = (1-e^{-y})^2$. R is a dimensionless parameter which describes the relative strength of on-site vs. elastic interactions; here $R = 10.1$.

The thermodynamic properties of (1) have been reviewed in Ref. 2. This work describes the spectra of dynamical correlations

$$S_{AA}(q,\omega) = \frac{1}{N} \int_{-\infty}^{+\infty} \frac{dt}{2\pi}\, e^{-i\omega t} \sum_{m,n} e^{iqa(n-m)} < A_n(t) A_m(0) > \tag{2}$$

where, in this paper, mostly $A_n = y_n$. The integral of (2) over all frequencies,

$$S_{AA}(q) \equiv \int_{-\infty}^{\infty} d\omega\; S_{AA}(q,\omega) = \frac{1}{N} \sum_{m,n} e^{iqa(n-m)} < A_n A_m > \tag{3}$$

can be computed exactly using the transfer-integral result for the equal time correlations (cf. Eq. 55 of Ref. 2). It is expedient to consider normalized spectral functions

$$\hat{S}_{AA}(q,\omega) = \frac{S_{AA}(q,\omega)}{S_{AA}(q)} \quad . \tag{4}$$

The angular brackets in Eqs. (2)-(3) denote canonical ensemble averages. Typically, we implemented this by repeated molecular dynamics (MD) simulations of the system for many different initial conditions, Fourier-transforming the spatiotemporal correlations obtained from each run, and averaging over all runs to obtain the final result.

The equations of motion,

$$\ddot{y}_n = \frac{1}{R}(y_{n+1} + y_{n-1} - 2y_n) - V'(y_n) \quad ; \quad n = 1, 2, ...N \tag{5}$$

with periodic boundary conditions, $y_0 = y_N$, $y_{N+1} = y_1$, and typical system size $N = 1024$, were numerically integrated for an interval $T = 410$, using a 4-th order Runge-Kutta algorithm, with a time step equal to 0.02. Initial conditions were "canonical", in the sense that (i) the velocities $\dot{y}_n$ were Gaussian-distributed, and (ii) the positions y_n were random variables distributed according to the potential energy part of the Hamiltonian (1); in addition, the system was "thermalized" for a certain time, using a Nosé procedure[3].

3. Spectra: Phonons vs. central peak

At intermediate temperatures, the spectra are characterized by an anharmonic phonon component and a strong, low-frequency intensity (cf. Fig. 1a); this low-frequency component becomes even more pronounced at lower values of the wavevector. There is considerable residual structure in the spectrum; in particular, a secondary peak at lower frequencies appears to be a consistent feature.

As the temperature increases, and the instability approaches, the structure becomes significantly simpler. The decisive quantity is the correlation length ξ. If the wavelength is shorter than ξ, i.e. $q\xi/(2\pi) > 1$, the spectrum in effect probes the "droplets" of the high-temperature phase, of typical size ξ, which are present in the low-temperature phase; consequently, the main feature of Fig. 1b is a peak, from the acoustic phonons[3]. At the smallest values of the ratio $q\xi/(2\pi)$, a central peak (CP) feature begins to grow, and eventually dominates the spectrum at values $q\xi/(2\pi) << 1$; this is the case in Fig. 2.

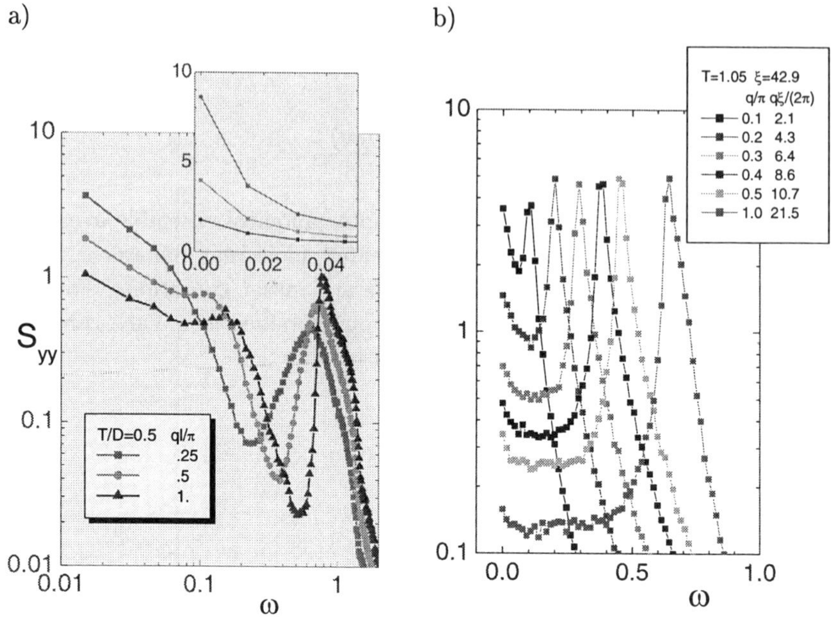

Figure 1. Normalized dynamical correlation spectra. Panel (a) presents the results at $T = 0.5$, for selected values of q. The inset shows the zero-frequency details (linear frequency scale). Panel (b) presents results at $T = 1.05$, for a variety of q values. The inset shows the ratio of the correlation length ξ and the wavelength. If the wavelength is smaller than the correlation length, the spectrum is dominated by the phonon peak. At this length scale, the spectra probe the "droplets" of the high-temperature phase present in the system. Note the gradual buildup of a central peak at the lowest values of q.

A first attempt to analyze the data can be made in terms of a phenomenological relaxation/oscillation spectral function, similar to the one used in analyzing structural phase transitions[4], i.e.

$$\hat{S}_{yy}(q,\omega) = \frac{1}{\pi\omega} Im \frac{\omega_0^2}{\omega_0^2 - \omega^2 - i\omega\Gamma} \tag{6}$$

where $\Gamma = \Gamma_0 + \delta^2/(\gamma - i\omega)$ is a relaxational memory kernel, and the q-dependence of all the parameters has been suppressed. If $\Gamma_0 \ll \delta^2/\gamma$, it is possible for the spectrum (6) to split into phonon-like

$$\hat{S}_{yy}(q,\omega) \approx \frac{1}{\pi}(1-\rho)\frac{\omega_\infty^2 \Gamma_0}{(\omega^2 - \omega_\infty^2)^2 + (\omega\Gamma_0)^2} \quad , \quad \omega \gg \gamma \tag{7}$$

and CP

$$\hat{S}_{yy}(q,\omega) \approx \frac{1}{\pi}\rho\frac{\gamma'}{\omega^2 + \gamma'^2} \quad , \quad \omega \ll \gamma \tag{8}$$

contributions, where $\omega_\infty^2 = \omega_0^2 + \delta^2$, $\rho = \delta^2/\omega_\infty^2$, and $\gamma' = \gamma(1-\rho)$.

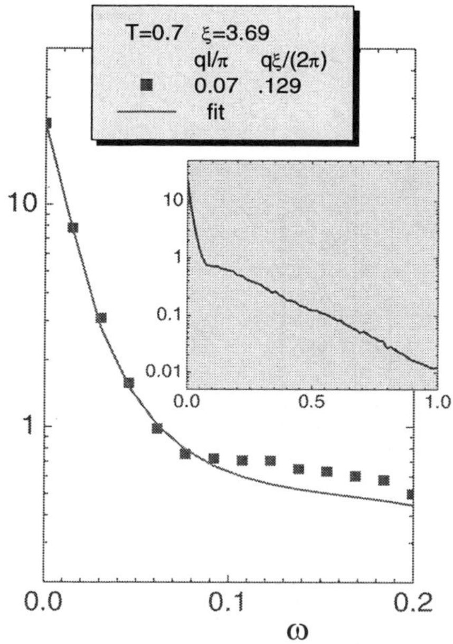

Figure 2. Normalized dynamical correlation spectra at $T = 0.7$, and $q/\pi = 0.07$. At long-wavelengths (compared to ξ) the spectrum is dominated by the CP feature. The inset shows that the tail of the spectra drops off with a different slope. The fit has been obtained using Eq. (6).

Fig. 3a shows that the CP linewidth γ' is largely independent of temperature; it does however depend on q, roughly linearly, as long as $q\xi/(2\pi)$ does not exceed unity, i.e. as long as the CP is appreciable.

Returning to the phonon peak, it is possible to follow the decreasing phonon linewidth as the transition is approached and the dynamics - with the exception of the very long wavelengths $q\xi/(2\pi) \leq 1$ - evolves towards the harmonic limit(Fig. 3b). Our data is consistent with a linear critical slowing down, i.e. $\Gamma_{ph} \propto T_c - T$.

4. Force autocorrelations

It is instructive to consider the spectrum of the force autocorrelations, i.e. $A_n \equiv f_n = -V'(y_n)$ in Eqs. (2) - (4). At low temperatures one might try to describe the spectra in terms of a crude model of independent local modes (ILM), i.e. site-independent solutions of the $R \to \infty$ (anticontinuum) limit

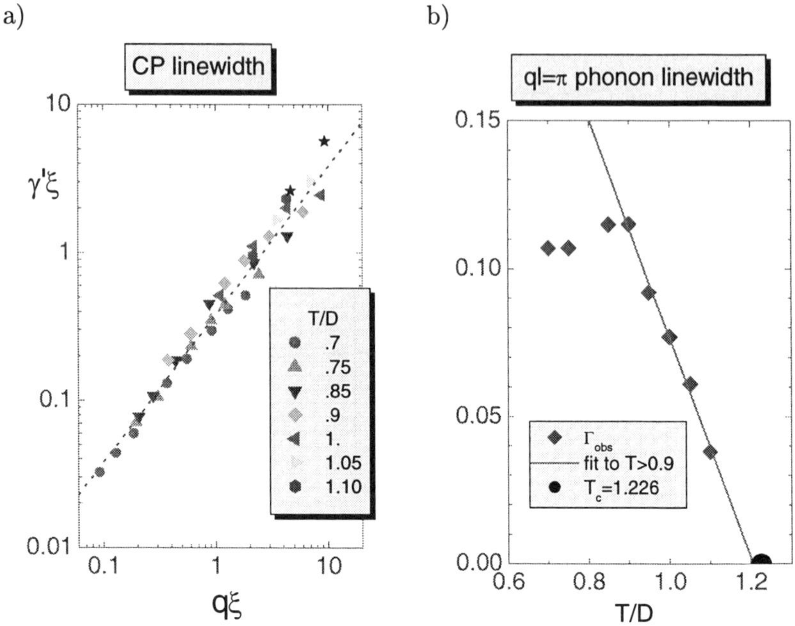

Figure 3. Linewidths. Panel (a) presents the linewidth of the central peak as a function of wavelength. The dashed line has unit slope. Panel (b) presents the linewidth of the $q = \pi$ phonon as a function of temperature. The linewidth extrapolates to zero at a temperature not far from T_c.

of (5). The explicit form of the second time derivative is

$$f_{\lambda,\delta}(t) = \lambda^2 \epsilon_\lambda^{1/2} \frac{\cos(\lambda t + \delta) - \epsilon_\lambda^{1/2}}{\left[1 - \epsilon_\lambda^{1/2} \cos(\lambda t + \delta)\right]^2} \tag{9}$$

where $0 < \lambda < \lambda_{max} = \sqrt{2}$, $0 < \delta < 2\pi$ depend on the initial conditions and $\epsilon_\lambda = 1 - \lambda^2/2$ is the energy of the ILM. Since $R \gg 1$, the above form should be a good approximation to exact one-site discrete breathers[6]. The canonical average of autocorrelations of (9) is

$$< f(t) f(0) >= \int_0^{2\pi} \frac{d\delta}{2\pi} \int_0^{\lambda_{max}} d\lambda \; Q(\lambda) \; f_{\lambda,\delta}(t) f_{\lambda,\delta}(0) \tag{10}$$

where $Q(\lambda) = Z^{-1} \exp(-\epsilon_\lambda / T)$ and Z is determined from the normalization condition $\int d\lambda \; Q(\lambda) = 1$. The MD force spectra at $T = 0.2$ are shown in Fig. 4a, along with a numerical Fourier transform of (10). Overall agreement is satisfactory, except for the very low frequency part of the spectrum.

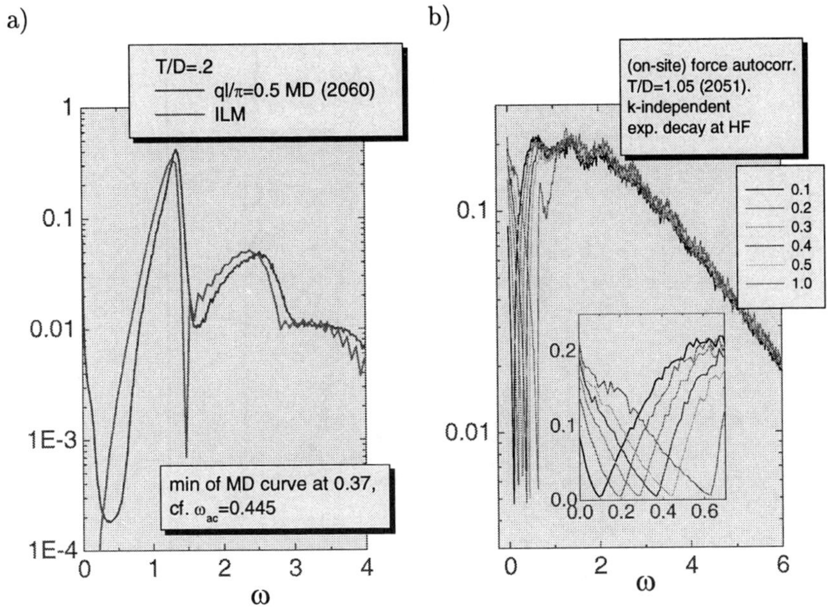

Figure 4. Normalized spectra of force autocorrelations. Panel (a) shows the results at low temperatures. The wavy curve is a theoretical estimate based on an independent localized mode picture, i.e. the normalized spectrum of (10). Panel (b) presents results at higher temperatures. Note (i) the extreme dip, almost to zero intensity, which occurs almost exactly at the acoustic phonon frequencies (inset), and (ii) the exponential decay at higher frequencies.

At higher temperatures, force autocorrelation spectra exhibit the following features: (i) a very pronounced dip occurring almost exactly at the bare acoustic frequencies $\hat{\omega}_q = (2/\sqrt{R})\sin(q/2)$, characteristic of the high-temperature phase, and (ii) a roughly q-independent decay at higher frequencies. The observed form of the force spectra motivates an improved version of (6), with $\omega_0^2 \equiv T/S(q)$, as demanded by the Mori-Zwanzig projection operator formalism[5], and $Re\Gamma = \Gamma_0 \exp(-\omega/\omega_c) + \Gamma_1$, where $\Gamma_1 = a(1-\omega/\hat{\omega}_q)^{3/2}$ if $\omega < \hat{\omega}_q$, and $\Gamma_1 = b(\omega - \hat{\omega}_q)^{1/2}\exp(-\omega/\omega_c)$ if $\omega > \hat{\omega}_q$. Preliminary fits obtained with this improved Ansatz for the memory function (approximating $Im\Gamma$ by a constant, equal to its value at the peak, and setting ω_c equal to the value obtained by the exponental decay constant of force spectra, cf. above and Fig. 4b) are shown in Fig. 5; they seem to reproduce the MD data much better, using the same number of adjustable parameters.

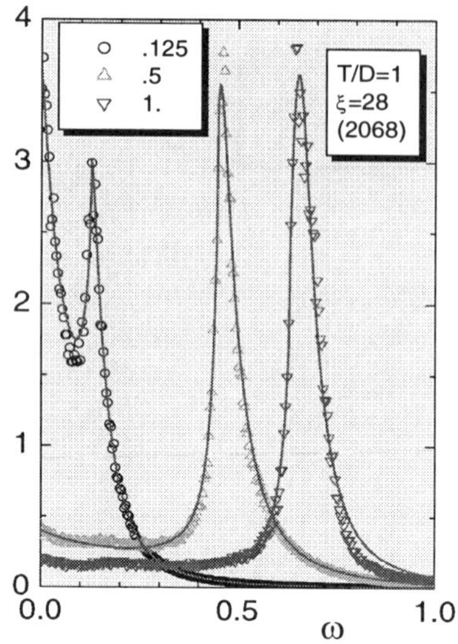

Figure 5. The spectra S_{yy} at $T = 1.0$. Fits are obtained with the improved Mori-Zwanzig Ansatz, which incorporates a memory kernel with a dip, similar to the one shown in Fig. 4b.

5. Concluding remarks

The MD data presented show that the critical dynamics of the Peyrard-Bishop model of DNA thermal denaturation can be thought of as follows: At length scales shorter than the correlation length ξ, which correspond to "droplets" of the high temperature phase, the system reflects the properties of the unstable phase; oscillatory dynamics of the soft, acoustic phonons is the result. The linewidth of these phonons appears to vanish linearly as $T_c - T$ ("critical slowing down"). At length scales longer than the correlation length, the dynamics is dominated by the central peak. Fluctuations are stronger, as evidenced from the divergence of the static structure factor $S(q)$; the typical time scales of these fluctuations appears however to be non-critical. It remains a challenge to the theory to establish whether these "non-critical", q-dependent dynamics can be associated with localized excitations. The preliminary analysis performed at low temperatures suggests that a picture of independent localized modes provides a reasonable description of the force autocorrelations - with the important exception of the very low frequency regime. Perhaps a more detailed theory of discrete breather statistical mechanics can improve our understanding of this part of the spectra.

References

1. M. Peyrard and A.R. Bishop, *Phys. Rev. Lett.* **62**, 2755 (1989).
2. N. Theodorakopoulos, *this volume.*
3. T. Dauxois, M. Peyrard and A. R. Bishop, *Phys. Rev. E* **47**, R44 (1993).
4. F. Schwabl, *Phys. Rev. Lett.* **28**, 500 (1972).
5. D. Forster, *Hydrodynamic Fluctuations, Broken Symmetry and Correlation Functions*, Benjamin, New York (1976).
6. J. L Marin and S. Aubry, *Nonlinearity* **9**, 1501 (1996).

SCATTERING AND CONFINEMENT OF DISCRETE BREATHERS IN INHOMOGENEOUS FPU CHAINS

I. BENA, A. SAXENA*, M. IBAÑES AND J. M. SANCHO

Department d'Estructura i Constituents de la Matèria, Universitat de Barcelona, 08028 Barcelona, Spain, E-mail: ioana@ecm.ub.es

* *Theoretical Division, Los Alamos National Laboratory, Los Alamos, New Mexico 87545, USA*

We investigate the scattering of a moving discrete breather (DB) on a junction (or pair of junctions) in a Fermi-Pasta-Ulam (FPU) chain consisting of two (or three) segments with either different masses of the particles or different interaction parameters. We consider three distinct cases: (i) an abrupt junction (ii) an up ramp, and (iii) a rectangular trap. The latter is a three-segment structure which can lead to DB confinement. Depending on the mass or parameter difference at the junction and DB characteristics (frequency and velocity), the DB can either reflect from, or transmit through, or get trapped at the junction or on the ramp. For a heavy-light mass junction, the DB can even split at the junction into a reflected and a transmitted DB. The latter can subsequently split into two or more DBs.

Moving discrete breathers (DBs) in homogeneous nonlinear chains have been studied extensively in recent years[1,2]. Disorder introduced by impurities can render such chains inhomogeneous and the scattering of a DB by impurities has also been explored[3]. However, the scattering of a DB in the presence of "engineered disorder", or "ordered inhomogeneity", is an important issue which has only recently been addressed in the context of an FPU chain with a junction[4]. Here we consider three different kinds of ordered inhomogeneities in FPU chains (Fig. 1) and discuss peculiar DB reflection, transmission, trapping, splitting and confinement penomena. Our results have important implications for realistic situations such as junctions between different electron-phonon coupled chains and optical fibers with variable refractive indices.

The FPU model represents a one-dimensional (1D) chain of particles with no on-site potential (i.e., an acoustic chain), with the Hamiltonian

$$H = \sum_n \left[\frac{m\dot{x}_n^2}{2} + \frac{\alpha}{2}(x_{n+1} - x_n)^2 + \frac{\beta}{4}(x_{n+1} - x_n)^4 \right] ,$$

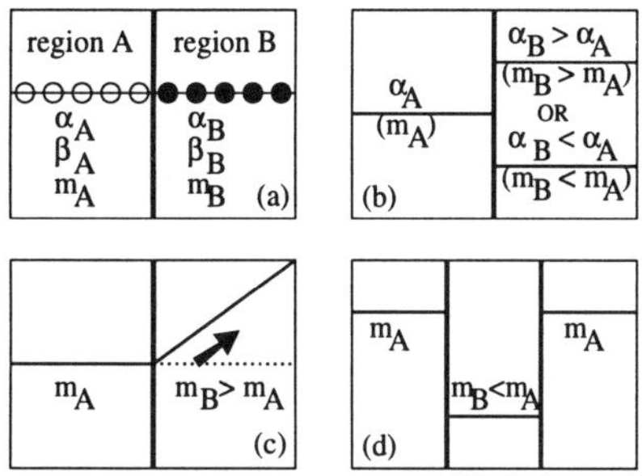

Figure 1. Schematic of an inhomogeneous FPU chain: (a) on the two sides of the junction (thick vertical line) the chain has different parameters; (b) an abrupt junction, either in the harmonic interaction constant α or in the mass m; (c) an up mass ramp; and (d) a rectangular mass trap.

where α and β denote, respectively, the strengths of the linear and nonlinear nearest-neighbor interactions; x_n is the elongation at the n-th particle, and m is the mass of the particles. For simplicity, all these quantities are expressed in *dimensionless units.*

The FPU lattice admits DB-like solutions[1,2] with periods T_{DB} that are smaller than the minimum period of the phonon spectrum, i.e., $T_{DB} < \pi\sqrt{m/\alpha}$. In Fig. 2 we depict the interaction of a DB (coming from region A) with an abrupt α-junction. Depending on the values of α_B, the DB can either reflect or transmit (with different velocities) through the junction. Sometimes the DB can also get trapped in region B. Analogous behaviors are obtained for a mass- and a β-type of junctions. In Fig. 3 an initial

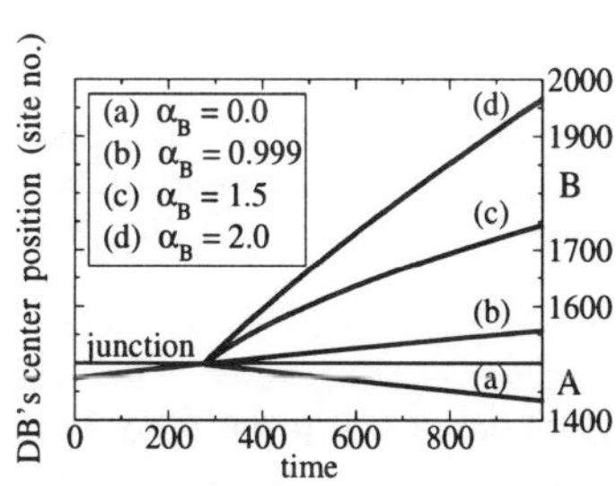

Figure 2. Typical behavior of a DB at an α-type abrupt junction [see Fig. 1(b)]. Here $T_{DB} = 2.1$ for the initial DB, $\alpha_A = 1$, $\beta_A = \beta_B = 1$, and $m_A = m_B = 1$.

DB splits (at an abrupt mass junction) into a reflected and a transmitted DB. Later on, the transmitted DB furhter splits into two other DBs (the circles on the figure indicate the regions of the splittings). When the mass difference between regions A and B is even larger, the transmitted DB (that

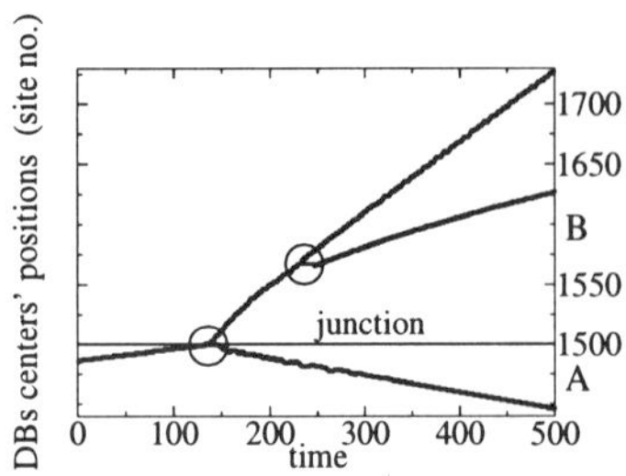

Figure 3. DB splitting at an abrupt heavy-light mass junction [see Fig. 1(b)]. $T_{DB} = 2.1$, $\alpha_A = \alpha_B = 1$, $\beta_A = \beta_B = 1$, $m_A = 1$, while $m_B = 0.5$.

has progressively less energy) might split into three or even four smaller DBs. Decreasing m_B furhter leads practically to the disappearance of the transmitted DB, and to a substantial amount of phonon creation. A mass ramp is considered in Fig. 4. The point where the DB is reflected on the ramp depends on the slope of the ramp, the initial velocity of the DB, as well as DB's frequency–the case illustrated in this figure. Sometimes the DB can get trapped on the ramp. Based on the results obtained for a

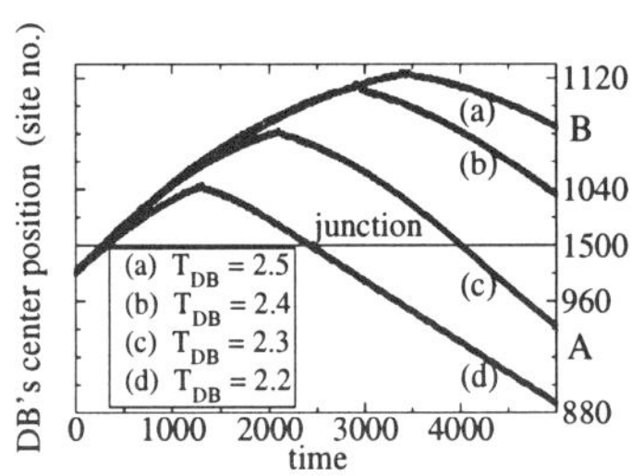

Figure 4. Typical behavior of a DB on an up mass ramp [see Fig. 1(c)]. Here $T_{DB} = 2.1$ for the initial DB, $\alpha_A = \alpha_B = 1$, $\beta_A = \beta_B = 1$, $m_A = 1$, while the mass of the particles in part B of the chain increases as $m_B(k) = 1 + 2 \times 10^{-5}\, k$, where k is the position label of the particle in the B part.

single junction, one can envision the possibility that a DB gets confined in the middle region of an $A - B - A$ mass 'sandwich structure' of Fig. 1(d). Such a configuration is ideal for studying (multiple) collisions of DBs. For the trap depicted in Fig. 5, (a) when the size of the trap (here 40 particle sites) is larger than a typical size of the DB, then both incident DBs (from the two A segments) can get into the trap and continue to collide with each other and the walls of the trap, with a certain (small) energy loss–evident from the fact that both DBs slow down in their movement. (b) When the size of the trap (here 5 particle sites) is "too small" to contain both DBs,

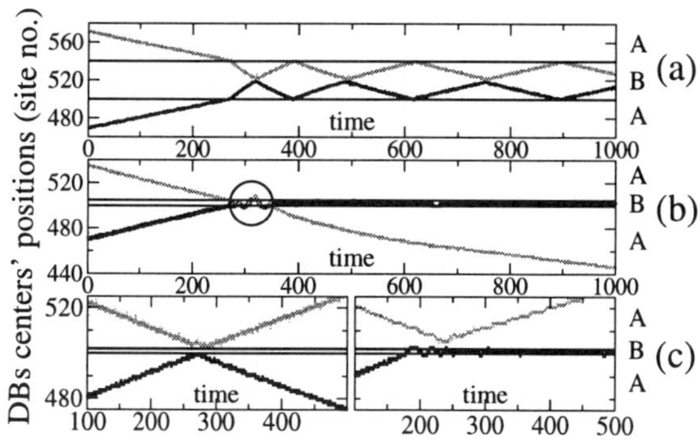

Figure 5. Examples of collisions between two *identical* DBs (here $T_{DB} = 2.1$) that arrive symmetrically at a trap of the type $A-B-A$ mass 'sandwich structure' [see Fig. 1(d)]. $\alpha_A = \alpha_B = 1$, $\beta_A = \beta_B = 1$, $m_A = 1$, and $m_B = 0.9$.

one of them is trapped and the other DB is expelled from the trap after a short coexistence time (denoted by a circle). Note that the final DBs are slightly different (in frequency) from the initial ones. (c) Finally, it is possible to have a "head-on collision" between the DBs while reaching simultaneously the two opposite walls of a very narrow trap (here 2 particle sites). Then both DBs are reflected back to their respective segments (left panel). If there is a DB that arrives first at this narrow trap, it is trapped; when a second one arrives, it is directly reflected (right panel).

Some of the observed DB behavior can be rationalized in terms of estimating the Peierls-Nabarro barrier for inhomogeneous chains[4]. We are currently exploring different shapes of the trap such as a wedge or a continuously varying smooth profile. A quantitative understanding of the cases considered here remains an open question for further study.

We are indebted to G. P. Tsironis for discussions. This work has been supported by the European Union under the RTN project LOCNET (HPRN-CT-1999-00163) and by the U.S. Department of Energy. A.S. gratefully acknowledges a fellowship from Iberdrola (Spain).

References

1. S. Flach and C. R. Willis, *Phys. Rep.* **295**, 181 (1998).
2. P. G. Kevrekidis, K. Ø. Rasmussen, and A. R. Bishop, *Intern. J. Mod. Phys.* **B15**, 2833 (2001).
3. See, e.g., K. Forinash, M. Peyrard, and B. Malomed, *Phys. Rev.* **E49**, 3400 (1994); J. Cuevas, F. Palmero, and J. F. R. Archilla, *arXiv:nlin/020326* and references therein.
4. I. Bena, A. Saxena, and J. M. Sancho, *Phys. Rev.* **E66**, 036617 (2002) and references therein.

RESONANT FLUXONS IN JOSEPHSON WINDOW JUNCTIONS : A NUMERICAL AND ANALYTICAL STUDY

A. BENABDALLAH

Institut für Physik, TU Chemnitz, D-09107 Chemnitz, Germany.
E-mail: benabdallah@physik.tu-chemnitz.de

J. G. CAPUTO

Laboratoire de Mathématique, INSA de Rouen B.P. 8, 76131 Mont-Saint-Aignan cedex, France.
E-mail: caputo@lmi.insa-rouen.fr

We present a numerical and analytic study of the influence of the passive region on fluxon dynamics in a window junction. We examine the effect of the extension of the passive region and its electromagnetic characteristics, its surface inductance L_I and capacitance C_I. When the velocity in the passive region v_I is equal to the Swihart velocity (1) a one dimensional model describes well the operation of the device. When v_I is different from 1, the fluxon adapts its velocity to v_I. In both cases we give simple formulas for the position of the limiting voltage of the zero field steps.

1. Electrodynamics of the window Josephson junction

The present design of low temperature superconducting devices based on semi-conductor technology enables to integrate Josephson junctions into superconducting strip-lines to realize complex devices. A simple example is the window design shown in Figure 1 where a single rectangular junction is surrounded by a uniform passive region.

The evolution of the phase inside the junction Ω_J and passive region $\Omega\backslash\Omega_J$ is given by the two-dimensional coupled sine-Gordon waves equation [2]

$$\frac{\partial^2\phi}{\partial t^2} - \Delta\,\phi + \sin\phi + \alpha\frac{\partial\phi}{\partial t} = 0 \quad \text{in} \quad \Omega_J\,,$$

$$C_I\frac{\partial^2\psi}{\partial t^2} - \frac{1}{L_I}\Delta\,\psi = 0 \quad \text{in} \quad \Omega\backslash\Omega_J\,, \tag{1}$$

together with the coupling for the phase and it's derivative, the current through the interface conditions

$$\psi = \phi \quad \text{and} \quad \frac{1}{L_I}\frac{\partial\psi}{\partial n} = \frac{\partial\phi}{\partial n} \quad \text{on} \quad \partial\Omega_J$$

and the external boundary conditions which physically indicates a lateral injection of current or/and an external magnetic field

$$\frac{1}{L_I}\nabla\psi \cdot n = J_{ext} \text{ on } \partial\Omega ,$$

In the Eq. (1) the parameter α is a damping, C_I and L_I are respectively, the capacitance per unit surface and sheet inductance and the unit of space is the Josephson penetration depth λ_J and the unit of time is the plasma frequency in the junction [3].

Here we consider a rectangular window of length $l = 10$ and width $w = 1$ embedded in a rectangular passive region of extension w' as shown in Figure 1. We will not consider the influence of an external magnetic field and will assume the external current feed to be of overlap type. We assumed throughout the study a small damping $\alpha = 0.01$ which is typical of under damped Josephson junctions.

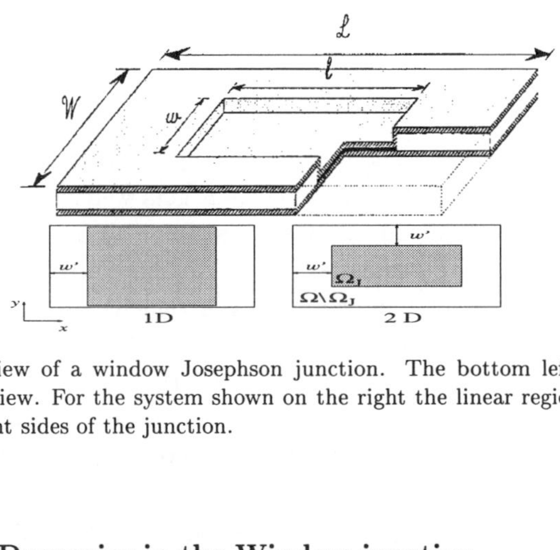

Figure 1. A view of a window Josephson junction. The bottom left panel shows a schematic top view. For the system shown on the right the linear region exists only on the left and right sides of the junction.

2. Fluxons Dynamics in the Window junction

The propagation of fluxons in Josephson junctions is the base of their use as high frequency oscillators. As the fluxons shuttles back and forth in the junction, it gives voltage impulse at every collision with the junction boundaries. In the absence of the passive region the fluxon velocity is limited by one so the current voltage characteristic shows equally spaced resonances called "Zero field steps" (ZFS) for the (multi)fluxons solutions. The propagation of the fluxons in window junction is affected by the presence of the passive region. To estimate the time average of the voltage for the window junction we compute the velocity in the junction using the

McLaughlin-Scott soliton perturbation equations [5] and get for the fluxon velocity $v = \frac{1}{\sqrt{1+(\frac{4\alpha}{\pi\gamma})^2}}$, and in the passive part the velocity $v_I = \frac{1}{\sqrt{L_I C_I}}$, so that the average of the voltage in time in the limit of large current can be approximated by

$$V_{1D} \equiv < \varphi_t > = \frac{\Delta\varphi}{\Delta t} = \frac{2\pi}{2w'/v_I + l/v}, \tag{2}$$

where the denominator is just the time take by the fluxon (phase jump of 2π) to travel across the device. We now confirm this estimation first when the electric properties of the device are homogeneous and then when they are different in the junction and the passive region so that $v_I \neq 1$. First we consider a situation where the electric properties are homogeneous in the whole device so that ($L_I = C_I = 1$) and change the extension of the passive region w'.

In Fig. 2 (left panel) we plot the (IV) characteristic curves for four values of $w' = 0, 1, 2$ and 3. This figure show that the position of the ZFS moves toward the left when w' increases. This agrees with formula (2). For large value of w' (typically $w' = 12$), the fluxon becomes unstable. This seems to be due to the fixed points that exist on each junction/passive region interface [1]. The fluxon can be trapped easier by one of them as w' increases because the driving force due to the current gets weaker. The second case tested is when the velocity in the passive region $v_I \neq 1$. To simplify things we fix w' to 2 in all the presented runs. We show in Fig. 2(right panel) the IV characteristics for four values of the velocity in the passive region $v_I = 0.33, 0.5, 1$ and 2. Notice that the positions of the ZFS of the 1D effective model follow (2). This figure also shows that the position of the ZFS moves from the right to the left as v_I decreases from 2 to 0.33. In 2D case it well known that dynamics of the fluxon in a pure Josephson junction can be reduce it to 1D effective problem. For example in the dynamical case Eilbeck and al [4] found that when ($I/8\mathrm{L} \ll 1$) then the phase is uniform in the y direction so that the reduction is possible. For a window junction the situation is different because we have a passive region around the junction which has electrical properties, a capacity and inductance which fix the velocity v_I. As shown in the Fig. 3 (left panel) the comparison between 2D and 1D was done for the two cases of homogeneous electrical properties and different geometries and a fixed geometry and different electrical properties. In Fig. 3 (left panel) one can see the good agreement between the 1D and the 2D calculations. Notice that the limiting values of the ZFS computed numerically and the estimates from (2) in excellent agreement.

When the velocity in the passive region $v_I \neq 1$ then the formula (2) is

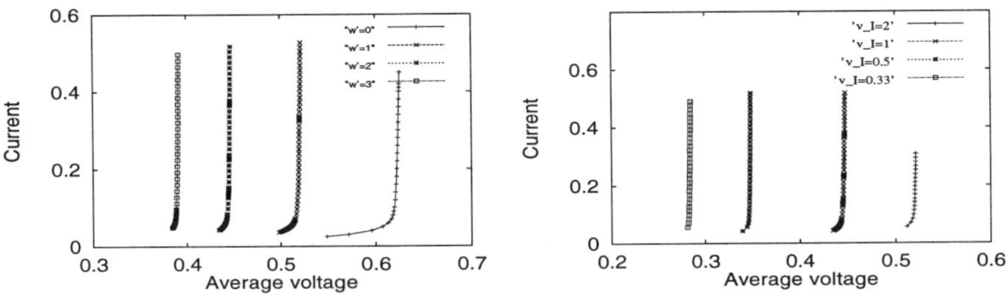

Figure 2. Left panel: The 1D current voltage characteristic for $L_I = C_I = 1$ and various extensions of the passive region w' (left) and for $w' = 2$ and various velocity in the passive region (right).

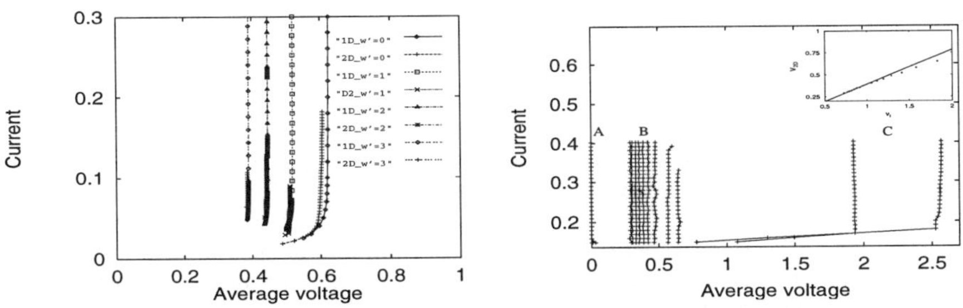

Figure 3. Left panel: IV curves for a homogeneous window junction and various extensions of the passive region w'. Both 1D and 2D calculations are presented for comparison. Right panel: IV curves for $0.1 \leq L_I \leq 3$, $C_I = 1$ and $w' = 3$. The insert shows the fluxon velocity in region B as a function of the linear wave speed v_I in the passive region.

completely off for the 2D calculations and the reduction of the 2D window junction problem to 1D effective problem is impossible. By using the interface condition [2] one can give an estimate for the average voltage $V_{2D} = \frac{2\pi}{2w'/v_I + l/v_I}$. This gives a very good approximation of the limiting voltage (see Fig. 3 right panel right panel region B).

References

1. A. Benabdallah, J. G. Caputo and N. Flytzanis, *Physica D* **161**, 79 (2002).
2. A. Benabdallah and J. G. Caputo, *J. Appl. Phys.* **92**, 3853 (2002).
3. A. Barone and G. Paterno, *Physics and Applications of the Josephson effect*, J. Wiley, (1982).
4. J. C. Eilbeck, P. S. Lomdahl, O. H. Olsen and M. R. Samuelsen,*J. Appl. Phys.* **57** 861 (1985).
5. D.W. McLaughlin and A.C.Scott, *Phys. Rev. A.* **18** 1652 (1978).

INTERACTION OF MOVING BREATHERS WITH AN IMPURITY

J CUEVAS, F PALMERO, JFR ARCHILLA AND FR ROMERO
Nonlinear Physics Group of the University of Sevilla, Department of Applied Physics I
ETSI Informática, Avda Reina Mercedes s/n, 41012, Sevilla, Spain
E-mail: jcuevas@us.es

We analyze the influence of an impurity in the evolution of moving discrete breathers in a Klein–Gordon chain with non-weak nonlinearity. Three different behaviours can be observed when moving breathers interact with the impurity: they pass through the impurity continuing their direction of movement; they are reflected by the impurity; they are trapped by the impurity, giving rise to chaotic breathers. Resonance with a breather centred at the impurity site is conjectured to be a necessary condition for the appearance of the trapping phenomenon.

1. Introduction

The interaction of nonlinear localized oscillations with impurities in a system can play an important role in its transport properties. This problem has been studied during the last decades within different frameworks, e.g. the scattering of kinks with impurities in the continuous sine-Gordon and ϕ^4 models [1] and in the Frenkel–Kontorova model [2]. The interaction of a moving discrete breather with an impurity in a Klein–Gordon chain has been considered by Forinash et al [3]. In this case, it is assumed that the system has weak nonlinearity. Here, we are interested in the study of the features of the interaction of moving discrete breathers with an impurity at rest in a Klein–Gordon chain of oscillators with non-weak nonlinearity. We also establish a hypothesis for the appearance of trapping of a breather by an impurity.

2. The Model

We consider a Klein–Gordon chain with nearest neighbours attractive interactions with Hamiltonian given by:

$$H = \sum_{n=1}^{N} \left(\frac{1}{2}\dot{u}_n^2 + V_n(u_n) + \frac{1}{2}C(u_n - u_{n-1})^2 \right), \qquad (1)$$

where $V_n(u_n) = D_n(e^{-u_n} - 1)^2$ is the substrate potential at the n-th site. The inhomogeneity is introduced assuming a different well depth at only one site, i.e., $D_n = D_o(1 + \alpha\delta_{n,0})$, then we refer to the particle located at $n = 0$ as an impurity. $\alpha \in [-1, \infty)$ is a parameter which tunes the magnitude of the inhomogeneity.

This Hamiltonian leads to the dynamical equations which have stationary and moving localized solutions (i.e., stationary and moving breathers). The former are calculated using the methods based in the anti–continuous limit [5] and the latter are calculated using the marginal mode method [4].

The dynamical equations can be linearized if the amplitudes of the oscillations are small. These equations have $N - 1$ non-localized solutions (*linear extended modes*) and one localized solution, (*linear impurity mode*). Their frequencies, ω_E and ω_L, respectively, are given by:

$$\omega(q,\alpha) = \sqrt{\omega_o^2 + 4C\sin^2\frac{q(\alpha)}{2}}, \quad \omega_L^2 = \omega_o^2 + 2C + \mathrm{sign}(\alpha)\sqrt{\alpha^2\omega_o^4 + 4C^2}, (2)$$

where $q \in (0, \pi]$ if $\alpha < 0$ and $q \in [0, \pi)$ if $\alpha > 0$. Figure 1 shows the dependence on α.

3. Numerical simulations

We have studied the behaviour of moving breathers when they interact with an impurity varying the value of the inhomogeneity parameter α. We have found four different regimes, separated by critical values of the parameter α [6]:

- *Barrier.* The impurity acts as a potential barrier. It occurs either with $\alpha > 0$ or $\alpha \in (-1, \alpha_1)$ with $\alpha_1 < 0$. If $\alpha \gtrsim 0$, the breather can pass through the impurity provided the translational velocity is high enough [7].
- *Excitation.* The impurity is excited and the breather is reflected. It occurs for $\alpha \in (\alpha_1, \alpha_2)$. This behavior is shown in figure 2.
- *Trapping.* The breather is trapped by the impurity. It occurs in the interval $\alpha \in (\alpha_2, \alpha_3)$. When the moving breather is close to the impurity, it becomes trapped while its center oscillates between the

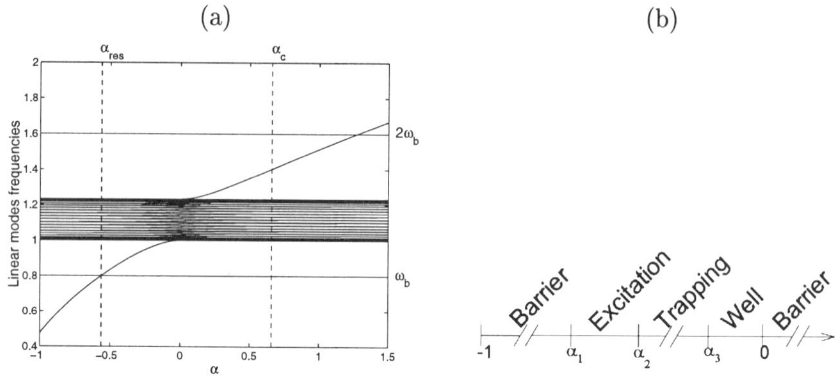

Figure 1. (a) Frequencies of the linear modes versus the parameter α. At $\alpha = \alpha_{res}$ and $\alpha = \alpha_c$, two different bifurcations occur, being the first one due to the resonance between the impurity mode and the breather. (b) Different regimes in the interaction of a moving breather with an impurity introduced as an inhomogeneity in the potential well depth.

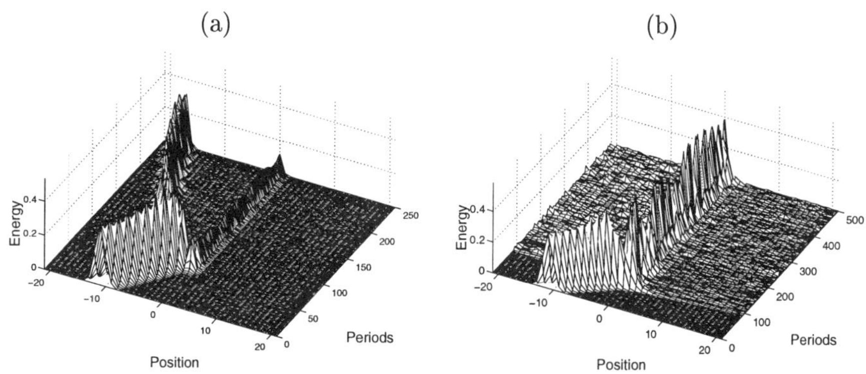

Figure 2. (a) Interaction of a breather with an impurity for $\alpha = -0.52$, which corresponds to the impurity excitation case. (b) Evolution of the moving breather for $\alpha = -0.3$, which corresponds to the trapping case. The moving breather becomes trapped by the impurity; afterwards, the breather emits phonon radiation and its energy centre oscillates between the sites adjacent to the impurity.

neighbouring sites, as figure 2 shows. The trapped breather emits a great amount of phonon radiation and seems to be chaotic.

- *Well.* The impurity acts as a potential well. It occurs for $\alpha \in (\alpha_3, 0)$ and consists of an acceleration of the breather as it approaches to the impurity, and a deceleration after the impurity has been passed through.

4. Discussion

It is observed that the breather bifurcates with the zero solution at $\alpha = \alpha_{res}$. That is, for α smaller than this value, no impurity breather exists. At $\alpha = \alpha_{res}$, the frequency of the impurity mode coincides with the moving breather frequency, i.e., in (2), $\omega_{\rm L} = \omega_{\rm b}$.

The scenario for the trapped breathers when $\alpha < 0$ is the following: the impurity mode has $q = 0$, and also all the particles of the impurity breather vibrate in phase; this vibration pattern indicates that the impurity breather bifurcates from the impurity mode and it will be the only localized mode that exists when the impurity is excited for $\alpha > \alpha_{res}$. Thus, when the moving breather reaches the impurity, it can excite the impurity mode. For $\alpha < \alpha_{res}$, the moving breather is always reflected. In addition, the impurity breather does not exist. Therefore, there might be a connection between both facts, i.e., the existence of the impurity breather seems to be a necessary condition in order to obtain a trapped breather.

If $\alpha > 0$, the impurity mode has $q = \pi$ but the impurity breather's sites vibrate again in phase, that is, the impurity breather does not bifurcate from the impurity mode. There are two different localized excitations: the tails of the (linear) impurity mode and the impurity breather. Thus, if the moving breather reaches the impurity site, it will excite these localized excitations. Therefore, we conjecture that the existence of both linear localized entities at the same time may be the reason why the impurity is unable to trap the breather when $\alpha > 0$.

Trapping hypothesis: *The existence of an impurity breather for a given value of α is a necessary condition for the existence of trapped breathers. However, if there exists an impurity mode with a vibration pattern different from the impurity breather one's, the trapped breather does not to exist.*

References

1. M Salerno, MP Soerensen, O Skovgaard and PL Christiansen, *Wave Motion* 5:49 1983. Z Fei, YuS Kivshar, and L Vázquez. *Phys. Rev. A*, 45:6019, 1992. *Phys. Rev. A*, 46:5214, 1992.
2. OM Braun and YuS Kivshar. *Phys. Rev. B*, 43:1060, 1991. *Phys. Rep.*, 306:1, 1998.
3. K Forinash, M Peyrard, and BA Malomed. *Phys. Rev. E*, 49:3400, 1994.
4. Ding Chen, S Aubry, and GP Tsironis. *Phys. Rev. Lett.*, 77:4776, 1996.
5. JL Marín and S Aubry. *Nonlinearity*, 9:1501–1528, 1996.
6. J Cuevas, F Palmero, JFR Archilla, and FR Romero. *Jour. Phys. A: Math. Gen.*, 2002. In press.
7. J Cuevas, F Palmero, JFR Archilla, and FR Romero. *Phys. Lett. A*, 299:221, 2002.

A NONRECIPROCAL FREQUENCY DOUBLER OF ELECTROMAGNETIC WAVES BASED ON A NONLINEAR PHOTONIC CRYSTAL

V. KUZMIAK*
Department of Physics and Astronomy, University of California Irvine,
4129 Frederic Reines Hall, Irvine, 92697-4575, USA
E-mail: vkuzmiak@lucy.ps.uci.edu

V. V. KONOTOP
Departamento de Física Universidade de Lisboa
Av. Prof. Gama Pinto 2, Lisbon P-1649-003, Portugal
E-mail: konotop@cii.fc.ul.pt

We have shown that an electromagnetic wave corresponding to the second harmonic transmitted through a dual periodic photonic crystal possessing $\chi^{(2)}$ nonlinearity exhibits a profound difference in the output intensity depending on the direction of the propagation and, therefore, such a system appears to be suitable in the design of a device that corresponds to non-reciprocal frequency doubler that operates at the second harmonic and allows propagation of the EM wave in one direction and forbids the propagation in the opposite direction.

1. Introduction

The theory of the propagation of the electromagnetic(EM) wave in photonic crystals was developed on the basis of close analogy with the quantum particle dynamics in atomic crystals [1]. Recently, the noreciprocal transmission has been numerically studied in the one-dimensional(1D), nonlinear, anisotropic photonic band gap material with a spatial gradation in the linear refractive index [2]. We describe an alternative underlying principle that constitutes the operational mechanism for a nonreciprocal frequency doubler which also can be viewed as a basic element of a light diode. We consider a TE electromagnetic wave incident from the left on a 1D multilayer stack which consists of two kinds of periodic media as it is shown in Fig. 1 (a). The left system shown in Fig. 1 (b) has band gaps that can be

*permanent address: iree as,prague, czech republic

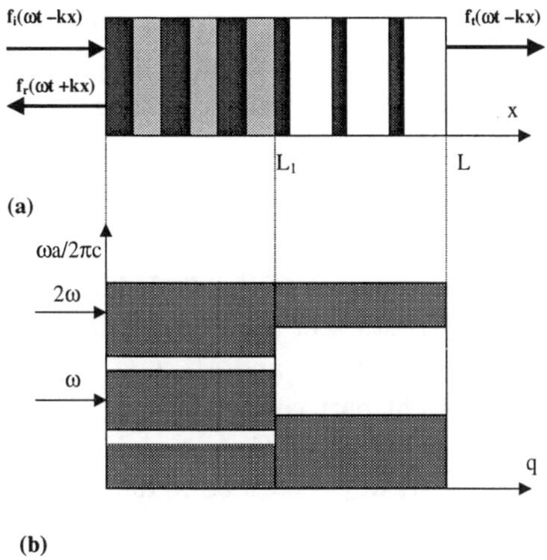

Figure 1. (a) Multilayered 1D stack composed of two periodic substructures. (b) Schematic picture of the photonic band structure of the left and right parts of the dual 1D stack. Shadowed regions indicate the allowed bands while empty regions correspond to the stop bands.

arbitrarily small and it satisfies the resonant conditions for the second harmonic generation(SHG)[3]. Simultaneously, we require for the left structure to possess $\chi^{(2)}$ nonlinearity, while the right structure is assumed not to support strong SHG. We assume TE electromagnetic wave $\mathbf{E} = (0, 0, E(x,t))$ incident on the stack from the left(see Fig. 1 (a)) with the frequency ω that belongs to an allowed zone of the left structure and falls into the gap of the right structure. The second harmonic of the incident wave, with the frequency $\omega_2 = 2\omega_1$ and the wave vector $q_2 = 2q_1 + Q$ (where Q is the vector of the reciprocal lattice of the left structure), which is generated in the left structure, belongs to an allowed zone of the both subsystems.

2. Theory

The wave equation for the amplitude E can be written in the form

$$-\frac{\partial^2 E(x,t)}{\partial x^2} + \frac{1}{c^2}\frac{\partial^2}{\partial t^2}\int_{-\infty}^{\infty} \epsilon(x, t-t')E(x,t')\,dt' = -\frac{4\pi}{c^2}\frac{\partial^2}{\partial t^2}\chi^{(2)}(x)[E(x,t)]^2,$$

where

$$\epsilon(x,t) = \begin{cases} \epsilon_1(x,t) & 0 < x < L_1 \\ \epsilon_2(x,t) & L_1 < x < L_2, \end{cases}$$

and the functions $\epsilon_j(x,t)$ $(j = 1, 2)$ are periodic: $\epsilon_j(x+a,t) = \epsilon_j(x,t)$. The similar requirement is imposed on the $\chi^{(2)}$. We seek the solution of the wave equation in the form $E(x,t) = \sum_{n=0}^{\infty}[A_n(x)\cos(n\omega t) + B_n(x)\sin(n\omega t)]$. To solve the problem we substitute this expansion in the wave equation (1) and obtain the set of the simultaneous equations for the Fourier coefficients $A_n(x)$ and $B_n(x)$.

3. Results

To calculate the transmission properties of the diode-like structure shown in Fig. 1 we consider a finite 1D stack of the length L where the left part consists of four alternating layers of GaAs and $Al_{0.5}Ga_{0.5}As$ with the filling fraction $f_1 = 0.5$, while the right part consists of four alternating layers of GaAs and vacuum with $f_2 = 0.3$. Both GaAs and $Al_{0.5}Ga_{0.5}As$ possess the second-order nonlinearity with $\chi^{(2)} = 1.68 \times 10^{-10}$m/V and $\chi^{(2)} = 1.73 \times 10^{-10}$m/V, respectively. Energy of the transmitted wave for the

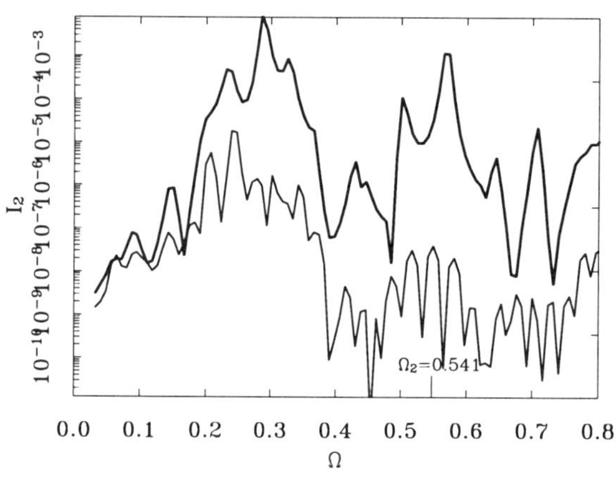

Figure 2. Output energy of the second harmonic signal I_2 vs. normalized frequency $\Omega = \omega a/2\pi c$ evaluated for the whole 1D dual structure for the positive (solid line) and negative (thin line) incidence of the electromagnetic wave when $A_1(0) = 1.0\text{E}{+}03$ V/m.

wave incident from the left side is defined as $T = \sum_{n=1}^{\infty} I_n(L)/\sum_{n=1}^{\infty} I_n(0)$ where $I_n = A_n^2 + B_n^2$. When an incident light contains only the fundamental harmonic, the first and second harmonics are dominant and thus $T \approx T_1 + T_2$ (where $T_1 = I_1(L)/I_1(0)$ is the transmittance of the first harmonic and the transformation coefficient $T_2 = I_2(L)/I_1(0)$ describes the energy transfer from the fundamental to the second harmonic). We seek the solution for the monochromatic cosine incident wave propagating through the structure of the finite length possessing weak nonlinear nonlinearity and, therefore, the system of Eqs. for the coefficients A_n and B_n can be reduced to two lowest modes [5] and we solve a set of four coupled ODE's of the first order by using the shooting method. The integrals $A'_n(L)$, $A_n(L)$, $B'_n(L)$, and $B_n(L)$ are evaluated by the Runge-Kutta method. The nonreciprocity the transmission spectra the second harmonic signal propagating through the dual structure shown in Fig. 1(a) is is demonstrated in Fig. 2 which displays a profound difference between the intensity of the second harmonic I_2 for the positive and negative incidence of the electromagnetic wave.

4. Discussion and Conclusion

We proposed a new concept of a frequency doubler based on the nonlinear 1D photonic crystal that exhibits nonreciprocal properties with respect to the direction of propagation of the incident wave. The nonreciprocity is achieved due to the SHG in the left part of the dual periodic structure that possesses $\chi^{(2)}$ nonlinearity and Bragg reflection of the fundamental harmonic in the right part. We have shown that for an appropriate frequency of the fundamental signal such a device can act as a light diode that operates at the frequency of the second harmonic.

Acknowledgments

The work has been supported by the NATO Linkage grant No. PST.CLG.978177. VVK acknowledges partial support from the Programme "Human Potential-Research Training Networks", contract No. HPRN-CT-2000-00158.

References

1. For review see *Photonic Band Gaps and Localization*, edited by C.M. Soukoulis (Plenum, New York, 1993); *J. of Lightwave Tech.* vol.17, Nov. 1999.
2. M. Scalora et al., J. Appl. Phys. **76**, 2023(1994); M. D. Tocci et al., Appl. Phys. Lett.**66**, 2324(1995).
3. V. V. Konotop and V. Kuzmiak, J. Opt. Soc. Am. B **16**,1370 (1999); *ibid* **17**,1874 (2000).
4. *Handbook of Optical Constants*, edited by E. D. Palik (Academic, New York, 1985)
5. We note that although the generation of harmonics higher than the second one is neglected the reduced system of equations does take into account an exhaustion of the intensity of the incident fundamental radiation (i.e. the energy decay of the of the first harmonic due to the transfer into the second one).

STATIONARY ENERGY TRANSPORT IN NONLINEAR LATTICES

S. LEPRI [1, *] R. LIVI[2,*, †] A. POLITI [3,*]

[1] *Dipartimento di Energetica "S. Stecco", via S. Marta 3 I-50139 Firenze (Italy)*

[2] *Dipartimento di Fisica, via G. Sansone 1 I-50019 Sesto Fiorentino (Italy)*

[3] *Istituto Nazionale di Ottica Applicata, L.go e. Fermi 6 I-50125 Firenze (Italy)*

Low-dimensional systems often display unusual transport properties. We focus on heat transport in crystals with reference to simple mathematical models consisting of classical oscillators coupled on a lattice. In one and two dimensions the finite-size thermal conductivity diverges with the volume and the corresponding linear constitutive relation (Fourier's law) breaks down. The universality of the divergence law is emphasized.

The theoretical and numerical study of simple models for lattice thermal conductivity was initiated already in the sixties. The hope was to get a deeper insight and possibly provide an alternative (non–perturbative) computational approach to the traditional ones (e.g. relaxation-time approximation of the Boltzmann-Peierls equation). However, it was only after the publication of the first convincing numerical evidence of a diverging thermal conductivity in anharmonic chains [1] that this subject attracted a renovated interest within the theoretical community. A fairly complete overview is given in Ref. [2] where the role of lattice dimensionality on the breakdown of Fourier's law has been discussed in detail. Indeed, the analysis of several examples clarified that anomalous conductivity should occour generically whenever an acoustic phonon dispersion is present (with some remarkable exceptions however [3]).

The study of toy models would not be only useful for a more complete theoretical understanding but should be also relevant for real materials. For instance, recent molecular dynamics results with realistic carbon potentials indicate already an unusually high conductivity of Single-Walled Nanotubes (SWNT) [4] and a power-law divergence with the tube length with an ex-

---

*istituto nazionale di fisica della materia - udr firenze

†istituto nazionale di fisica nucleare - sezione di firenze

ponent very close to the one observed for simple 1d models [5]. On the other hand, simulations of such complicated interactions may rapidly reach the available computational limits. Dealing with computationally simpler models should instead allow for a more accurate quantitative prediction of scaling laws.

A physically meaningful class of models is represented by arrays of point atoms interacting with their neighbours through simplified force laws. Limiting the discussion to 1d, we let m_n and x_n be respectively the mass and the position of the n-th atom and consider the equation of motion

$$m_n\ddot{x}_n = -F_n + F_{n-1} \quad ; \qquad F_n = -V'(x_{n+1} - x_n)\,, \tag{1}$$

where V is the interaction potential. The microscopic expression of the heat current is

$$J = \sum_n \left[\frac{1}{2}(x_{n+1} - x_n)(\dot{x}_{n+1} + \dot{x}_n)\,F_n + \dot{x}_n h_n\right]\,. \tag{2}$$

where h_n is a suitably defined local energy [2]. For small oscillations (compared to the lattice spacing a), the second term can be neglected and $x_n - x_{n-1} \simeq a$, so that Eq. (2) is well approximated by

$$J \simeq \frac{a}{2}\sum_n(\dot{x}_{n+1} + \dot{x}_n)\,F_n \tag{3}$$

The customary way to evaluate the thermal conductivity κ is through the Green-Kubo (GK) formula (with $L = Na$ being the chain length)

$$\kappa_{GK} = \frac{1}{k_B T^2}\lim_{t\to\infty}\int_0^t d\tau \lim_{L\to\infty} L^{-1}\langle J(\tau)J(0)\rangle \quad .$$

A crucial (and sometimes overlooked [7]) point is that such formulae look formally identical in different ensembles but the definition of J is different since the "systematic" parts associated with other conservation laws must be subtracted out [8]. For instance, expression (3) is correct in the microcanonical ensemble with zero total momentum, while in the canonical ensemble (for large N)

$$J = \frac{a}{2}\sum_n(\dot{x}_{n+1} + \dot{x}_n)\,F_n - av\langle\sum_n F_n\rangle$$

with v being the center–of–mass velocity. This choice insures that the autocorrelation of J vanishes for $t \to \infty$.

An alternative approach to transport coefficients is nonequilibrium molecular dynamics (NEMD). It amounts to simulate a system between two reservoirs operating at different temperatures T_+ and T_-. To this aim, several methods have been proposed in terms of both deterministic and stochastic dynamics [2]. Regardless of the actual thermostating scheme, after a short transient, an off-equilibrium stationary state sets in with a net heat current flowing through the lattice. The finite-size thermal conductivity $\kappa(N)$ is estimated as the ratio between the time-averaged flux and the overall temperature gradient $(T_+ - T_-)/L$. Notice that κ represents an effective transport coefficient including both boundary and bulk scattering mechanisms.

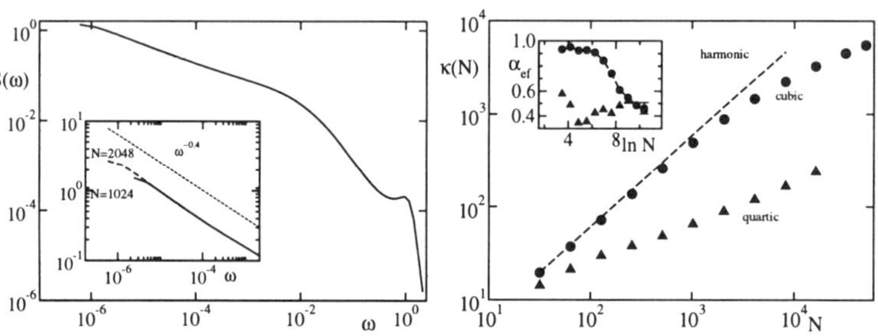

Figure 1. Fermi-Pasta-Ulam (FPU) model: power-law divergence in the power spectrum of J (left) and finite-size conductivity for $T_+ = 1.1, T_- = 0.09$ (right). Results for the cubic and quartic FPU potentials are compared with the harmonic chain. The inset contains the effective exponent $\alpha_{ef} = \Delta \ln \kappa / \Delta \ln N$ versus $\ln N$.

Anomalous behaviour manifests itself as: *(i)* an nonintegrable algebraic decay of equilibrium correlations of the heat current; *(ii)* a divergence of $\kappa(N)$ in the thermodynamic limit $N \to \infty$. This is very much reminiscent of the problem of long-time tails in fluids: transport coefficients in low spatial dimensions may *not exist at all*, thus implying a breakdown of usual hydrodynamics. For example, in 1d (see Fig. 1)

$$\kappa(N) \propto N^{\alpha} \qquad (\alpha > 0) \quad , \qquad \langle J(t)J(0)\rangle \propto t^{-\beta} \qquad (\beta < 1)$$

Consistency of the two approaches is verified by assuming that $\kappa(N)$ can be estimated by extending the integral in the GK formula up to the "transit time" L/v_s (v_s being the sound speed) thus obtaining $\kappa \propto N^{1-\beta}$ (the latter exponent is the one reported in the last column of Table 1). The available data (see Table 1) strongly suggest that α is a universal exponent and the observed values are somehow in between the estimates coming from mode-coupling theories [2] ($\alpha = 2/5$) or recent renormalization group calculations [12] ($\alpha = 1/3$).

Besides the case of perfect lattices, where irreversible behaviour arises only from anharmonicity, some work has been devoted to the role of disor-

Table 1. Survey of the estimates of the divergence exponent α for different 1d models. For diatomic chains r indicates the mass ratio.

Model	Reference	α (NEMD)	α (GK)
FPU-β	Lepri et al. [2]	0.37	0.37
FPU-α	Lepri [9]	$\lesssim 0.44$	-
Diatomic FPU r=2	Vassalli [10]	0.43	compatible
Diatomic Toda r=2	Hatano [11]	0.35-0.37	0.35
	Vassalli [10]	0.39	compatible
Diatomic Toda r=8	Vassalli [10]	0.44	compatible
Diatomic hard points	Hatano [11]	0.35	-
	Grassberger et al. [6]	0.32	0.34
SWNT (Tersoff)	Maruyama [5]	0.32	-

der. Impurity scattering should decrease the conductivity but simulations support the opposite conclusion [2]. This is even more relevant in view of the joint role of disorder and nonlinearity in enhancing the energy transport [13].

Acknowledgments

This work is part of the EC network LOCNET, Contract No. HPRN-CT-1999-00163 and of COFIN00 project *Caos e localizzazione in meccanica classica e quantistica.*

References

1. S. Lepri, R. Livi, A. Politi, Phys. Rev. Lett. 78 (1997) 1896.
2. S. Lepri, R. Livi, A. Politi, cond-mat/0112193.
3. C. Giardinà, R. Livi, A. Politi, M. Vassalli, Phys. Rev. Lett. 84 (2000) 2144.
4. S. Berber, Y. Kwon, D. Tomanek, Phys. Rev. Lett., 84 (2000) 4613.
5. S. Maruyama, Physica B 323 (2002) 193 .
6. P. Grassberger, W. Nadler, L. Yang Phys. Rev. Lett. 89 (2002) 180601
7. T. Prosen and D.K. Campbell, Phys. Rev. Lett. 84 (2000) 2857.
8. M. S. Green, Phys. Rev. 119 (1960) 829.
9. S. Lepri, Eur. Phys. J. B 18 (2000) 441.
10. M. Vassalli, Diploma Thesis, University of Florence, 1999.
11. T. Hatano, Phys. Rev. E 59 (1999) R1.
12. O. Narayan, S. Ramaswamy, cond-mat/0205295
13. G. Kopidakis, S. Aubry, Phys. Rev. Lett. 84 (2000) 3236.

QUANTUM TARGETED ENERGY TRANSFER

P. MANIADIS, G. KOPIDAKIS *AND S. AUBRY
Laboratoire Léon Brillouin (CEA-CNRS), CEA Saclay
91191-Gif-sur-Yvette Cedex, France
E-mail: maniadis@llb.saclay.cea.fr

Quantum manifestations of TET are investigated in systems which exhibit classical Targeted Energy Transfer. It is found that TET persists in the quantum case but with some differences. First, TET is not complete but is limited by the zero point quantum energy (if any) of the quantum oscillators. Second, close to TET threshold (if any), quantum effects tend to spread much faster the energy wave packet between the two oscillators.

When two coupled harmonic oscillators are resonant, any amount of energy initially injected on one oscillator, oscillates between them with a frequency proportional to their coupling. However, when the weakly coupled oscillators are anharmonic, resonance does not persist in general, since the oscillator frequency depends on the amplitude. It has been found that when the nonlinear oscillators are appropriately tuned and under some well determined conditions, resonance may be preserved throughout a complete energy exchange. This selective transfer between two nonlinear oscillators, the donor and the acceptor or, more generally, between Discrete Breathers has been termed Targeted Energy Transfer (TET). Important ramifications in physics, chemistry, and especially in biological physics and in the design of devices have been suggested [1,2]. The classical TET theory has been illustrated in Discrete Nonlinear Schrödinger (DNLS) [1,2] and Klein-Gordon oscillator models [3] and applied to the problem of ultrafast electron transfer [4]. In this paper we briefly describe how TET persists in purely quantum systems. Details are given elsewhere [3].

The classical system is described (in the action angle-representation) by the Hamiltonian

$$\mathcal{H} = H_D(I_D) + H_A(I_A) + H_V(I_D, I_A, \theta_D, \theta_A) \tag{1}$$

*also at department of physics, university of crete, p.o. box 2208, 71003 heraklion, crete, greece where H_D and H_A is the donor and acceptor Hamiltonian, respectively, with I_D, θ_D and I_A, θ_A the respective conjugate action-angle variables. The small coupling between donor and acceptor is described by the term H_V which is consider to be proportional to a small parameter λ. In this system, a selective transfer of energy may appear if certain conditions are fulfilled [1,2]. There must exist some value of action I_T and energy E_T such that $E_T = H_D(I_T) = H_A(I_T)$. In addition, the two oscillators must remain in resonance during the transfer:

$$\omega_D(I) = \left.\frac{dH_D}{dI}\right|_I = \left.\frac{dH_A}{dI}\right|_{I_T - I} = \omega_A(I_T - I) \tag{2}$$

When these conditions are fulfilled, it has been shown that if energy equal to E_T (and corresponding action I_T) is initially on the donor, then this energy will be transferred to the acceptor in a time which is inverse proportional to the coupling parameter λ. This process is very selective in the sense that a small variation in the total energy of the system is sufficient to destroy the resonance condition and therefore block the transfer. It has also been shown that an important function for the study of TET is the detuning function, which is defined as $\epsilon_T(I) = E_T - H_D(I) - H_A(I_T - I)$. In order for TET to occur, this detuning function must be bounded by the maximum (w.r.t. the slow variable $\theta = \theta_D - \theta_A$) and the minimum of the coupling function V (which is defined as the average of H_V over the fast variable $\theta_0 = \theta_D + \theta_A$), $\langle V \rangle_{min}(I) < \epsilon_T(I) < \langle V \rangle_{max}(I)$. Since the coupling function is proportional to λ we can define a critical coupling λ_{cr} such that TET appears for $\lambda > \lambda_{cr}$ [1,2].

In order to solve the purely quantum problem for a pair of weakly coupled donor-acceptor described by Hamiltonian (1), we first consider the quantum DNLS dimer:

$$\hat{H} = \omega_D \hat{N}_D + \frac{\chi_D}{2}\hat{N}_D^2 + \omega_A \hat{N}_A + \frac{\chi_A}{2}\hat{N}_A^2 - \lambda(a_D^+ a_A + a_A^+ a_D) \tag{3}$$

where a_i^+ and a_i^+ are boson creation and annihilation operators ($i = D, A$ for donor and acceptor, respectively) with normal commutation relations $[a_i, a_j] = 0$ and $[a_i, a_j^+] = \delta_{i,j}$. $\hat{N}_i = a_i^+ a_i$ are the corresponding number operators (we consider that $\hbar = 1$). The resonance condition for the classical dimer leads to the relation $\omega_D - \omega_A = I_T(\chi_A - \chi_D)/2$ and $\chi_A = -\chi_D$. It is found that the same condition is valid for the quantum dimer [3]. The total number of bosons $n = < \hat{N} > = < \hat{N_D} + \hat{N_A} >$ corresponds to the classical action and it is conserved. In this case, all eigenvalues of the Hamiltonian corresponding to total number of bosons equal to $n = I_T$ are degenerate and equal to E_T (in the limit of $\lambda = 0$) forming a quantum path for the transfer of n bosons, initially on the donor, to the acceptor. The states of the Hamiltonian can be found using the states of the number operator $|\psi_i\rangle = \sum_{j=0}^{n} c_{i,j} |\mathbf{j}\rangle$, where $|\mathbf{j}\rangle = |j\rangle |n-j\rangle$ are the states of the number operator corresponding to n bosons with j bosons on the donor and $n-j$ bosons on the acceptor and $c_{i,j} = \langle \mathbf{j} | \psi_i \rangle$. Any initial state of the system (which is not an eigenstate of the Hamiltonian), will have a time evolution given by the expression $|\Psi(t)\rangle = \sum_{i=0}^{n} b_i e^{-iE_i t} |\psi_i\rangle$ where $b_i = \langle \psi_i | \Psi(0) \rangle$.

In order to study the quantum manifestation of TET, we choose as an initial condition the eigenstate $|\Psi(0)\rangle = |\mathbf{n}\rangle = |n\rangle |0\rangle$, which corresponds to n bosons on the donor and zero on the acceptor. This is not an eigenstate of the Hamiltonian and therefore it will have a time evolution. To view this time evolution we measure the average of $\hat{N}_D$ on this state, $\left\langle \hat{N}_D \right\rangle = \langle \Psi(t) | \hat{N}_D | \Psi(t) \rangle = \sum_{j=1}^{n} j |\alpha_j(t)|^2$ where $\alpha_j(t) = \sum_{i=1}^{n} b_i c_{i,j} \exp(-iE_i t)$. As can be seen in figure (1a), when the coupling is smaller than the critical value, only a small part of the bosons initially on the donor are transferred to the acceptor. The time evolution of the classical dimer is also plotted (dashed line) for comparison. When the coupling is larger than the critical value, after some time all bosons are transferred to the acceptor (figure 1b). When the detuning function is strictly zero everywhere, the quantum and the classical behavior of the system is the same. In this plot the parameters are such that the detuning function is small but nonzero during the transfer. Due to the nonzero values of the detuning function, a quantum dispersion appears and, as a result, the behavior of the quantum system resembles the classical one only for a short time. After the first transfer, the wave-packet looses its coherence and the bosons, equally spread on both donor and acceptor.

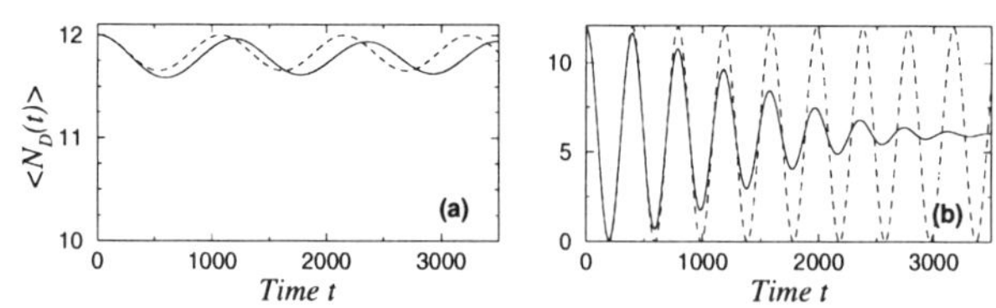

Figure 1. Boson number $< \hat{N}_D(t) >$ on the donor versus time in the quantum case (full line) with $n = 12$ bosons compared to the time evolution of the corresponding classical action $I_D(t)$ on the donor (dashed line). Fig.*(a)* corresponds to $\lambda = 0.0005$ and fig. *(b)* corresponds to $\lambda = 0.0015$.

Next, we study a system of two coupled Klein-Gordon oscillators described by the Hamiltonian $H_{KG} = H_D + H_A + H_V$ where $H_i = \frac{1}{2}p_i^2 + V_i(x_i)$

is the Hamiltonian of the uncoupled oscillators ($i = D, A$ for donor and acceptor, respectively) and $H_V = -\lambda x_D x_A$ is the coupling. The donor and acceptor potentials are chosen to be sixth order symmetric polynomials in order to have bound states and discrete spectrum when we solve the Schrödinger equation, $V_D(x_D) = \frac{\omega_D^2}{2}x_D^2 + \frac{\alpha_D}{4}x_D^4 + \frac{\beta_D}{6}x_D^6$ and $V_A(x_A) = \frac{\omega_A^2}{2}x_A^2 + \frac{\alpha_A}{4}x_A^4 + \frac{\beta_A}{6}x_A^6$. It has been shown in [3] that for a given donor and for given E_T or I_T it is possible to tune the parameters of the acceptor potential in order to have TET between these two oscillators in the classical limit with relatively small critical coupling and detuning function. For the investigation of the quantum TET in this dimer it is necessary to solve the Schrödinger equation for each individual oscillator as well as for the coupled system. For convenience in the numerical calculations, $\hbar$ is considered as a parameter given by the semiclassical quantization rule $\hbar = I_T/(l_D + l_A + 1)$ where l_D, l_A are the excitation levels of donor and acceptor, respectively. For the numerical computation, it is necessary to introduce the standard harmonic oscillator creation and annihilation operators for the donor and the acceptor, $\alpha_D^+ = \sqrt{\frac{\omega_D}{2\hbar}}x_D + i\sqrt{\frac{1}{2\hbar\omega_D}}p_D$, $\alpha_D = \sqrt{\frac{\omega_D}{2\hbar}}x_D - i\sqrt{\frac{1}{2\hbar\omega_D}}p_D$, $\alpha_A^+ = \sqrt{\frac{\omega_A}{2\hbar}}x_A + i\sqrt{\frac{1}{2\hbar\omega_A}}p_A$, $\alpha_A = \sqrt{\frac{\omega_A}{2\hbar}}x_A - i\sqrt{\frac{1}{2\hbar\omega_A}}p_A$ and substitute x_D, p_D, x_A, p_A in the Hamiltonian using the above expressions. Using similar calculations like in the quantum DNLS dimer it is found that TET survives in the quantum limit. Two differences exist between these two models. In the Klein-Gordon dimer the ground state energy of the donor and of the acceptor is not zero and therefore not all energy can be transferred from donor to acceptor. The second difference is that this dimer is not integrable and the total number of bosons is not conserved. Thus, quantum dispersion is stronger in this case. In addition, the detuning function cannot be zero but it can take very small values. For the investigation of quantum TET in this dimer we calculate the average of the donor Hamiltonian over the wave function $\Phi(t)$ which is chosen such that at $t = 0$ all energy is on the donor.

In figure (2), the time evolution of $< H_D(t) >$ is plotted. As initial condition, a state of the uncoupled system is chosen with the donor excited in the $l = 8$ level and the acceptor in the ground state. As can be seen in figure (2a), when the coupling is smaller than the critical value, the energy remains mainly at the donor. When the coupling is larger than the critical value, all but the ground state energy is transferred to the acceptor, as shown in figure (2b). For comparison, the evolution of $H_D(t)$ in the classical system is plotted with doted line.

It has been shown here that TET exists not only in classical but also

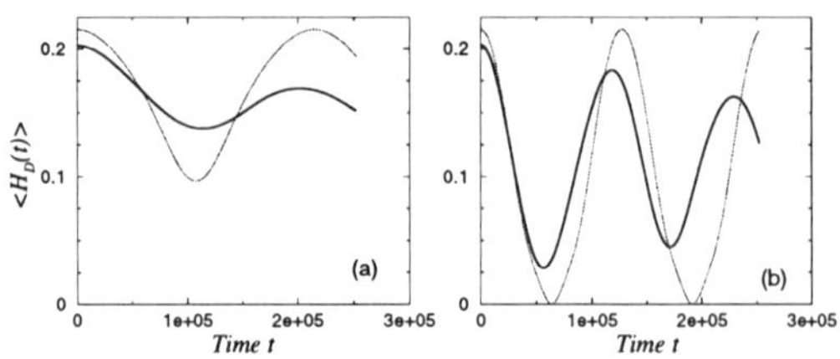

Figure 2. Energy $< H_D(t) >$ on the donor versus time in the quantum case (full line) with $l = 8$ bosons compared to the time evolution of corresponding classical energy $E_D(t)$ on the donor (doted line). Fig. *(a)* corresponds to $\lambda = 0.00002$ and fig. *(b)* corresponds to $\lambda = 0.00006$.

in quantum systems. The resonance conditions for the quantum transfer are the same as for the classical one. In the quantum system, the number of bosons n corresponds to the classical action I_T. When the dimer is tuned for a value I_T the quantum levels corresponding to n bosons are almost degenerate forming a quantum path for the wavepacket to transfer from the donor to the acceptor (quantum detuning function). When the classical detuning function of the system is strictly zero everywhere during the transfer, the quantum behavior is exactly the same as the classical one (DNLS dimer). When the detuning function is not zero everywhere, quantum dispersion appears making the quantum evolution different from the classical one after the first transfer (DNLS and Klein-Gordon dimer). In the Klein-Gordon dimer there are extra differences between classical and quantum due to the nonzero ground state energy of each oscillator. Only part of the total energy can be transferred. The time needed for the first transfer is also modified since smaller amount of energy is transferred and therefore less time is needed.

Acknowledgments

This work has been supported by the European TMR program LOCNET HPRN-CT-1999-00163.

References

1. S. Aubry, G. Kopidakis, A. M. Morgante and G. P. Tsironis, Physica B **296**, 222 (2001).
2. G. Kopidakis, S. Aubry and G. P. Tsironis, Phys. Rev. Lett. **87**, 165501 (2001).
3. P. Maniadis, G. Kopidakis and S. Aubry, submitted to Physica D.
4. S. Aubry and G. Kopidakis, preprint cond-mat/0210215 and in preparation.

BREATHER SCATTERING IN THE DISSIPATIVE DRIVEN FRENKEL-KONTOROVA MODEL

MATTHIAS MEISTER

Dpto. Física de la Materia Condensada and Instituto de Biocomputación y Física de Sistemas Complejos, Universidad de Zaragoza, 50009 Zaragoza, Spain.
E-mail: matthias@posta.unizar.es

The dissipative and driven Frenkel-Kontorova model supports moving breather solutions as attractors of the dynamics in a certain range of the system parameters. It is natural, then, to study the behaviour of such localised excitations in head-on collisions. The final state of a scattering process depends very sensitively on the initial conditions and can be qualitatively changed by very weak perturbations. At fixed values of the system parameters a larger number of initial conditions is used and from them several different bound states of two breathers are found. Bound states can be characterised by their velocity and by the distance between the breather cores. It turns out that certain values of this distance are preferred and that there also is a clear relation between the distance and the velocity.

1. The model

We consider the standard Frenkel-Kontorova model (discrete sine-Gordon model) describing a system of harmonically coupled masses each of which is subject to a periodic on-site potential. The (dimensionless) Hamilton function of the model is

$$H = \sum_n \left\{ \frac{1}{2} p_n^2 + \frac{1}{(2\pi)^2} \left[1 - \cos(2\pi u_n) \right] + \frac{C}{2} \left(u_{n+1} - u_n \right)^2 \right\}, \tag{1}$$

with p_n the momentum of the n-th particle, u_n its deviation from the equilibrium position and C the strength of the coupling. We are considering the case that there is one particle per period of the potential in the ground state of the chain. The equations of motion derived from (1), if damping and an external driving are also taken into account, read

$$\frac{d^2}{dt^2} u_n = -\frac{1}{2\pi} \sin\left(2\pi u_n\right) + C\left(u_{n+1} + u_{n-1} - 2u_n\right) - \alpha \frac{d}{dt} u_n + F \sin(\omega t). \tag{2}$$

Here α determines the strength of the damping and F gives the amplitude of the harmonic driving force, ω its (angular) frequency. Equations (2) support breather solutions which are attractors of the system. They can be found by a simple continuation procedure making use of this attractor property: Starting from the anticontinuum limit, the coupling C is increased by a small step and the system is given time to relax, then C is increased again, and so on. Along this continuation path several bifurcations are encountered [1], of particular interest to the following is that in a certain range of the coupling C the breathers are mobile.

2. Scattering

The mobile breathers can be used for numerical scattering experiments. In particular, the simulations have been done on a ring of 1000 particles, with $\alpha = 0.02, F = 0.02, \omega = 0.2\pi$, and $C = 0.890$. We created 110 initial two-breather configurations from two single mobile breather configurations. The second of these two breathers was created from the first one by letting the latter evolve in time for $1, 2, 3, \ldots, 110$ full periods of the driving force and then creating the inverted configuration $(u_j^{inv}, \dot{u}_j^{inv}) = (u_{N+1-j}, \dot{u}_{N+1-j})$, which represents a breather with a velocity opposite to the original one. The inverted and the original (before the time evolution !) breather were then pasted into a two breather configuration in such a way that the distance between the breathers is the same for all initial configurations. Note that because of the quasiperiodicity of their time evolution, the two breathers in the pair are not mirror-images of each other.

Four possible types of final state of a scattering process are found: The breathers can rebounce, one or both of the breathers can be destroyed, or the breathers can form a bound state (see below). The interaction between the breathers is mediated by the phonons continuously emitted by both breathers. Altogether we have found 56 bound states out of 110 initial conditions; this number includes the bound states formed in collisions subsequent to previous reflections of the breathers on the ring. However, it turns out that the final state of a scattering process depends very sensitively on the initial and/or boundary conditions: using the same initial conditions but free boundary conditions instead of periodic ones or a larger system can lead to qualitatively different results, e.g. annihilation instead of bound state formation. Our simulations did not reveal any systematics that would allow to predict the outcome of a collision from the initial conditions. Gaussian white noise coupled to the system can also change the final state, even if the noise is so weak that neither evidence for diffusion of the single breathers (on the scale set by the system size) nor fluctuations of the energy distribution in the system (on the scale set by the breather cores) are visible. This indicates that the final state of a scattering process depends very sensitively on the internal degrees of freedom of the breather, the quasiperiodic time evolution of which shows in the lack of systematics in the results of our simulations. Once formed, however, the bound states prove to be more stable against perturbations. For the sake of completeness we mention that symmetric initial conditions can lead to (immobile) bound states. These specific bound states are extremely sensitive to symmetry-breaking perturbations (like noise) and therefore rather unphysical.

3. Bound states

When a bound state is formed in a collision, the two colliding breathers are locked together by the phonon oscillations between them. Two obvious characteristics of a bound state are the distance between the breather cores and the velocity at which the bound state propagates through the system. From a distance-velocity diagram (see Fig. 1) it is evident that there are preferred values of the distance and that there is a clear relation between distance and velocity. The values of the velocity are distributed around the velocity of a single breather ($v \approx 0.01858$) at the given values of the system parameters. The distance values are approximately equally spaced, as can be seen from the positions of the points in the diagram relative to the vertical dashed lines. In particular we recognize three groups of distance-velocity pairs (the three 'lines' of negative slope) within which equidistance between the distance values is even more pronounced, whereas there are some slight shifts between the three groups.

The distribution of distance-velocity pairs in Fig. 1 is connected to the phonon oscillations between the two breather cores. For the phonons in the damped system there is a frequency-range where the damping of the phonons is small [1]. This range is analogous to the phonon band of the undamped system. Due to the translation of the breathers in a bound state, emitted phonons are subject to Doppler-shifts. Only phonons whose frequency after the shift is in the low-damping range can propagate over significant distances. For the given system parameters, 3ω is the only harmonic of the driving force that meets this requirement. Note that in our case the Doppler-effect creates two frequencies out of one: the emission from the breather cores into the region of the chain between them is in the direction of the translation for one of the breather cores and in the opposite direction for the other. This leads to a beat frequency and a beat wavelength. Preliminary results indicate that this beat wavelength plays a part in the determination of the distance value at a given velocity. The situation, however, is complicated by the fact that in a reference frame comoving with the bound state there is not only the frequency 3ω, but also frequencies $3\omega + 2\pi v n$, where v is the translation velocity and n

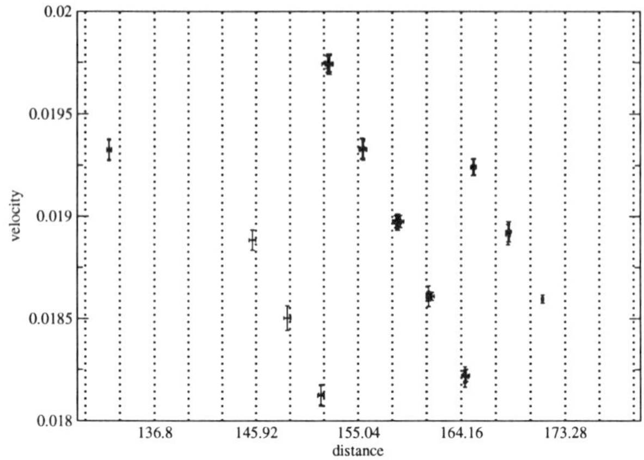

Figure 1. Distance-velocity chart for bound states. The values were obtained from runs over 50000 time units. Positions of the breathers were taken every 5000 time units. So we obtained 11 (10+initial) values for the distance and 10 values for the velocity. Averages were taken and the error bars correspond to one standard deviation. Bound states with positive and negative velocities were found. The chart shows the absolute value of the velocities. Note that many of the 'points' shown are actually clusters of points.

is an integer. These frequencies are combination frequencies of 3ω with $2\pi v$, a frequency arising due to the propagation of the bound state along a system of periodicity 1. These frequencies are also Doppler-shifted when observed from the rest frame of the chain, giving rise to many possible beat frequencies and wavelengths. The mechanism of the interplay of all these frequencies/wavelengths has not been clarified yet. Furthermore, it must be taken into account that the phonons between the breather cores may well be describable by a linear theory, whereas the breather cores themselves have to be described by the full nonlinear equations. The breather cores are not pointlike, but have some width, which will also contribute to the distance.

Acknowledgments

This work was made possible by a fellowship from the European Commission within the Research Training Network LOCNET (HPRN-CT-1999-00163).

References

1. J. L. Marín, F. Falo, P. J. Martínez, and L. M. Floría, Phys. Rev. E **63**, 066603 (2001)

TRANSVERSE DC MAGNETIZATION FOR AC DRIVEN SPINS *

A. E. MIROSHNICHENKO, S. FLACH AND A. A. OVCHINNIKOV

Max-Planck-Institut für Physik komplexer Systeme
Nöthnitzer Strasse 38, D-01187 Dresden, Germany

We study zeroth harmonic generation for spins $s = \frac{1}{2}$ under the influence of external dc and ac magnetic fields in the $x - z$ plane by means of symmetry analysis. In particular we obtain a dc magnetization in the *transverse* y direction.

1. Introduction

A spin system under the influence of an external ac field is a well known object of study using NMR techniques. The second harmonic generation due to magnetic resonance was studied earlier[1]. Here we focus on the *zeroth*-harmonic generation for a spin $s = \frac{1}{2}$ in the presence of a dc magnetic field in z direction and an ac field in the $x - z$ plane. By means of a symmetry analysis we found that it is possible to observe a *transverse* dc magnetization in y direction if the angle between the applied ac field and the x axis is nonzero. For a single spin $s = \frac{1}{2}$ the effect is rather small[2]. To enlarge its value we have considered two interacting spins $s = \frac{1}{2}$. The presence of exchange coefficients does not change symmetry considerations. But for some particular values of these coefficients there is an avoided-crossing of energy levels which leads to a strong enhancement of the induced dc magnetization.

2. Density matrix approach

Our quantum system is described by the Hamiltonian $H = H_0 + H_1(t)$, where $H_1(t) = H_1(t + T)$ is periodic in time. In order to describe the time-dependent statistical evolution of such a system we use the quantum

*This work is supported by Deutsche Forschungsgemeinschaft FL200/8-1

Liouville equation for the density matrix[2] $\rho(t)$

$$\frac{\partial\rho}{\partial t} = i[H,\rho] - \nu(\rho - \rho_e) \tag{1}$$

where $[A,B] = AB - BA$. By means of the second term in the rhs of Eq.(1) we assume that our spin system is coupled to an environment, where ν represents the inverse relaxation time. ρ_e is an equilibrium state, which in the absence of $H_1(t)$ should be the Gibbs distribution

$$\rho_e(H_0) = \frac{1}{Z_0}e^{-\beta H_0} \ , \ Z_0 = Tr(e^{-\beta H_0}) \ . \tag{2}$$

The value $\bar{A}(t)$ of an observable corresponding to the operator A is defined by $\bar{A}(t) = Tr(A\rho(t))$ and its time average is abbreviated by $\tilde{A} = \frac{1}{T}\int_0^T \bar{A}(t)\,dt$.

3. The case of a single spin

We consider one spin $s = \frac{1}{2}$ in the presence of a constant magnetic field in z-direction and an ac magnetic field in the $x - z$ plane. This system can be described by the following Hamiltonian

$$H(t) = h_0 S_z + h(t)(\alpha S_x + \gamma S_z), \tag{3}$$

where $\alpha = \sin\phi$ and $\gamma = \cos\phi$. We assume that the field $h(t) = h(t+T)$ has zero mean $\int_0^T h(t)\,dt = 0$. The spin component operators are given by the Pauli matrices: $S_{x,y,z} = \frac{1}{2}\sigma_{x,y,z}$.

We can rewrite Eq.(1) in terms of the observables $\bar{S}_{x,y,z}$

$$\begin{aligned} \dot{\bar{S}}_x &= (h_0 + \gamma h(t))\bar{S}_y - \nu\bar{S}_x \ , \\ \dot{\bar{S}}_y &= \alpha h(t)\bar{S}_z - (h_0 + \gamma h(t))\bar{S}_x - \nu\bar{S}_y \ , \\ \dot{\bar{S}}_z &= -\alpha h(t)\bar{S}_y - \nu(\bar{S}_z + C) \ , \end{aligned} \tag{4}$$

where $C = \frac{1}{2}\tanh(\beta h_0/2)$. It is a special case ot the Bloch equations[4].

Due to its linearity, the system (4) possesses only one time-periodic attractor. All spin components on this attractor can be expanded into a Fourier series

$$\bar{S}_{x,y,z}(t) = A_{0\ x,y,z} + \sum_{n\neq 0} e^{i\omega nt} A_{n\ x,y,z} \ . \tag{5}$$

After averaging over the period T, we obtain $\tilde{S}_{x,y,z} = A_{0\ x,y,z}$.

The symmetries which leave Eq.(4) invariant and preserve $\bar{S}_z$ and thus ρ_e are the following:

1) $\nu = 0$, $\gamma = 0$, $h(-t) = -h(t)$:

$$\bar{S}_x \to -\bar{S}_x \ , \ \bar{S}_y \to \bar{S}_y \ , \ \bar{S}_z \to \bar{S}_z \ , \ t \to -t,$$

2) $\nu = 0$, for all γ , $h(-t) = h(t)$:

$$\bar{S}_x \to \bar{S}_x \ , \ \bar{S}_y \to -\bar{S}_y \ , \ \bar{S}_z \to \bar{S}_z \ , \ t \to -t,$$

3) $\nu \neq 0, \gamma = 0, \ \ h(t + T/2) = -h(t)$:

$$\bar{S}_x \to -\bar{S}_x \ , \ \bar{S}_y \to -\bar{S}_y \ , \ \bar{S}_z \to \bar{S}_z \ , \ t \to t + \frac{T}{2} \ .$$

Therefore we conclude that for case 1) $\tilde{S}_x = 0$, for case 2) $\tilde{S}_y = 0$ and for case 3) $\tilde{S}_x = \tilde{S}_y = 0$. If one of these symmetries is violated by a proper choice of parameters it is possible to observe a static magnetization along x or y directions. We will be interested mostly in the case 3) when $\nu \neq 0$. One of the possibilities is to choose $\gamma \neq 0$ to break the symmetry. We will pay our attention especially to the $\tilde{S}_y$ component, since there is no magnetic field applied in the y direction.

In Fig. 1(a) $\tilde{S}_y(\omega)$ is shown. The maximum obtained value is $\tilde{S}_y \approx 0.015$, which is rather small compared to the spin value $\frac{1}{2}$.

4. Two interacting spins

Next we consider the system of two interacting spins $s = \frac{1}{2}$. and study the role of the interaction between spins, which provides with additional parameters by means of exchange coefficients J_x, J_y and J_z. It is well-known that the eigenstates of this system can be represented as a triplet (spin $s = 1$) and a singlet ($s = 0$). The singlet does not interact with the triplet states and thus does not contribute to nonzero dc magnetization. The triplet has three energy levels and the structure of these levels depends on the values of the exchange coefficients.

For isotropic exchange $J_x = J_y = J_z \equiv J$ we obtain a mere renormalization of the result for two noninteracting spins up to a maximum of 25% increase[3].

The case of anisotropic exchange is more interesting. It is possible to find such values of exchange coefficients when there is an avoided crossing between the triplet levels upon changing the value of a frozen external ac field (see Fig. 1(b)). We observed the enhancement of the maximum $\tilde{S}_y$ value exactly near the avoided-crossing (see Fig. 1(a)). It has a resonant behaviour[3] on ν.

In addition we would like to note, that the frequency of the maximum response corresponds to the distance between the levels E_1 and E_3. The observed enhancement in $\tilde{S}_y$ is due to narrow avoided-crossing between levels E_1 and E_2.

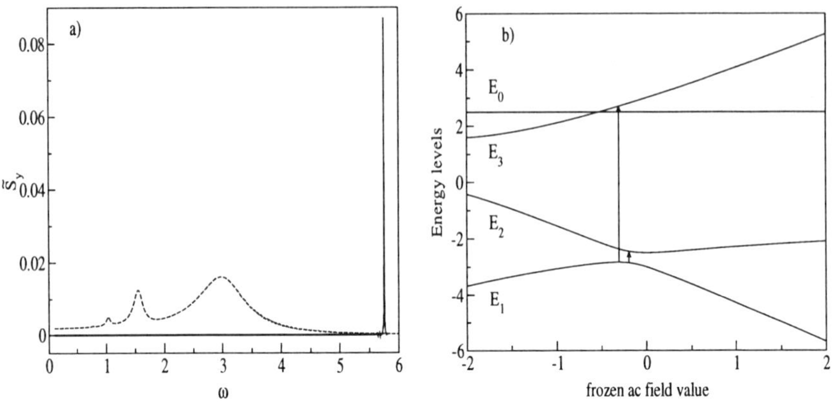

Figure 1. a) $\tilde{S}_y$ dependence versus ω for single spin (dashed line) and two spins (solid line) at the optimum value of $\nu = 0.1$ (single spin) and $\nu = 10^{-3}$ (two spins). Due to exchange symmetry the curve for $\tilde{S}_y$ in the case of two spins is shown per one spin. b) Energy level dependence on frozen ac field. The values of exchange coefficients are $J_x = J_y = 5.0$ and $J_z = 0$. Arrows indicate two possible resonant excitations. Others parameters are $h_0 = 3$, $\phi = \pi/4$, $\beta = 10$, $h(t) = \sqrt{2}\cos(\omega t)$.

References

1. N. Bloembergen, *Nonlinear optics* W. A. Benjamin Inc., (1965).
2. S. Flach and A. A. Ovchinnikov, *Physica* **A292**, 268 (2001).
3. S. Flach, A. E. Miroshnochenko, and A. A. Ovchinnikov, *Phys. Rev.* **B65**, 104438 (2002).
4. F. Bloch , *Phys. Rev.* **102**, 104 (1956).

THERMAL ACTIVATION OF BREATHERS IN 2D NON–LINEAR LATTICES

F. PIAZZA* AND R. LIVI

Dipartimento di Fisica and INFM UdR di Firenze
Via G. Sansone 1 I–50019 Sesto F.no (FI), ITALY

S. LEPRI

Dipartimento di Energetica "S. Stecco"
Via S. Marta 3 I–50139 Firenze and INFM UdR di Firenze, Italy

We study the energy relaxation process produced by damping 2D lattices of classical anharmonic oscillators at the edges. Spontaneous emergence of localised vibrations dramatically slows down dissipation and gives rise to quasi–stationary residual states where energy is trapped in the form of a gas of weakly interacting discrete breathers. We show that the existence of a gap in the breather spectrum in 2D causes the localisation process to become activated. We investigate such a mechanism by studying the localisation time and the average density of localised objects.

Nonlinearity has revealed one of the key ingredients for describing many relevant features of different states of matter. Recently, considerable efforts have been devoted to the study of periodic, localised, non–linear lattice excitations named "breathers" [1], which emerge in non–linear systems from the interplay of non–linearity and space discreteness [2]. The role of discrete breathers in non–equilibrium dynamics seems to be particularly fascinating. An example is the relaxation to energy equipartition of short–wavelength fluctuations [3]. Another interesting scenario where breathers are found to emerge spontaneously is observed upon cooling a thermalised lattice at its boundaries [4,5,6,7].

The mechanisms leading to spontaneous localisation are intimately related to how dissipation acts on linear modes of different wavelengths [6]. In particular, modulation instability of short lattice waves [8] is the mechanism underlying the birth of breathers from an interacting gas of solitons [3,9]. In

*e-mail: piazza@fi.infn.it

2D the result is a multi–breather quasi–stationary state, where breathers are static and close–packed in a "random" lattice, and whose decay is exponential with a huge time constant [7]. In addition, the existence of an energy activation threshold for breather solutions [10] leads to conjecture that a thermalised lattice may be cooled down to the residual state only above some initial energy, making spontaneous localisation of energy in 2D a fluctuations–activated process.

In this contribution, we concentrate on the characteristics of the relaxation dynamics that confirm the thermal activation hypothesis. We consider a $N \times N$ lattice with one degree of freedom per site and damping on all edges. The atoms are labelled by the indexes $i, j = 0, 1, \ldots, N-1$ and we denote with $u_{i,j}$ the displacement of the particle at site (i, j) from its equilibrium position. The lattice dynamics is given by the following equations of motion

$$\begin{aligned} \ddot{u}_{i,j} = &\; V'(u_{i+1,j} - u_{i,j}) - V'(u_{i,j} - u_{i-1,j}) + \\ &\; V'(u_{i,j+1} - u_{i,j}) - V'(u_{i,j} - u_{i,j-1}) - \sum_{p,q=0}^{N-1} \Gamma_{i,j}^{p,q} \dot{u}_{p,q} \quad . \end{aligned} \tag{1}$$

Here $V(x) = x^2/2 + x^4/4$ is the Fermi-Pasta-Ulam (FPU) potential and $\Gamma_{i,j}^{p,q} = \gamma[g_{i,p}\delta_{j,q} + \delta_{i,p}g_{j,q} - g_{i,p}g_{j,q}]$, with $g_{i,p} = \delta_{i,p}[\delta_{p,0} + \delta_{p,N-1}]$, is the damping matrix, γ being the damping rate. We take here free–ends boundary conditions (BC), since localisation is strongly inhibited by fixed–ends BC [6].

It is possible to confirm the thermal activation hypothesis by performing a statistical analysis of the residual state. In particular, we find that the average density of localised objects follows an Arrhenius law of the form [7]

$$\langle n_B \rangle \propto \exp(-\beta\Delta) \quad , \tag{2}$$

where $1/\beta$ is proportional to the initial energy density $e_0 = E(0)/N^2$ (Fig. 1 (a))

A further confirmation of the thermal activation hypothesis comes from a phenomenological analysis of the localisation pathway. The transient regime leading to the residual state can be studied by looking at the localisation parameter L, defined as

$$L(t) = N^2 \sum_{i,j} h_{i,j}^2(t) \Bigg/ \left[\sum_{i,j} h_{i,j}(t) \right]^2 \quad , \tag{3}$$

where $h_{i,j}$ are the symmetrised site energies. When the lattice energy is highly localised L is of order N^2, while L order 1 means that the energy is

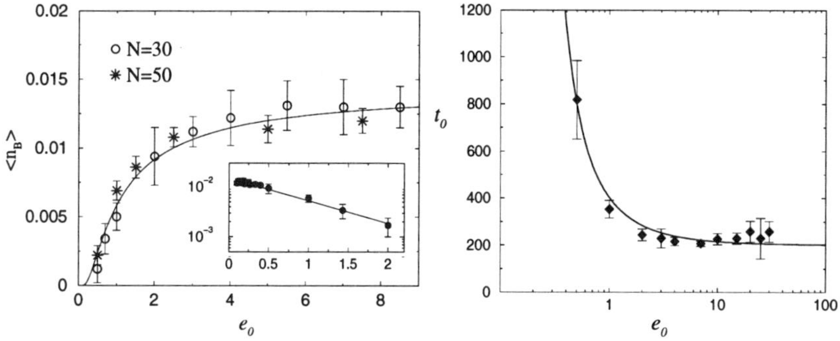

(a) (b)

Figure 1. 2D FPU lattice, $\gamma = 0.1$. (a) Average breather density $\langle n_B \rangle$ vs initial energy density for $N = 30$ and $N = 50$ and Arrhenius plot. The inset shows the average density measured in the $N = 50$ system vs $1/e_0$ in lin–log scale, and an exponential fit. (b) Localisation time t_0 vs initial energy density (symbols) and fit with Eq. (5)

evenly spread over the whole lattice.

We can introduce a localisation time t_0 by fitting the localisation parameter curve with the empirical function

$$L(t) = [L_0 + L_\infty (t/t_0)^\sigma]/[1 + (t/t_0)^\sigma] \quad , \tag{4}$$

where σ is a suitable exponent and $L_0 \approx 1$ and $L_\infty \approx N^2$ are the equilibrium and asymptotic values of $L(t)$, respectively (Fig. 2). From an energy–dependent analysis of the localisation curves we find that $t_0 \propto L_\infty$. On the other hand, it is not difficult to realise that in the residual state one must have

$$L_\infty \propto \langle \epsilon_B \rangle^2 \langle n_B \rangle / (\langle \epsilon_B \rangle \langle n_B \rangle)^2 \propto 1/\langle n_B \rangle \propto \exp(\beta \Delta) \quad , \tag{5}$$

where we have introduced the average breather energy $\langle \epsilon_B \rangle$. We show in Fig. 2 a fit with Eq. (5) to the experimental values of the localisation time for an FPU lattice with $N = 50$.

Conclusions

We study the phenomenon of spontaneous energy localisation upon cooling in 2D lattices. We perform a statistical analysis of the pseudo–stationary state, where spontaneously emerged breathers arrange on a static random

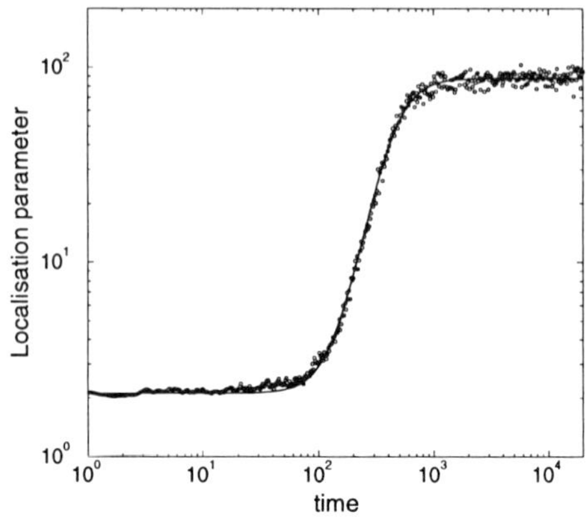

Figure 2. 2D FPU lattice with $N = 50$, $e_0 = 1$ and $\gamma = 0.1$. Localisation parameter vs time (circles) and fit with Eq. (4). Best–fit values of the parameters are: $L_0 = 2.1$, $L_\infty = 87.8$, $t_0 = 440.6$ and $\sigma = 3.1$.

lattice. We determine the average breather density as a function of the initial energy of the lattice and show that it is well described by an Arrhenius law. This conclusion confirms that spontaneous localisation of energy in 2D is a thermally–activated process, in accordance with the presence of an energy threshold for breather solutions in 2D. This conclusion is also confirmed by an analysis of the localisation time.

Acknowledgments

This work has been supported by the European Union under the RTN project LOCNET, Contract No. HPRN-CT-1999-00163.

References

1. S. Flach and C. R. Willis, *Phys. Rep.* **295**, 181 (1998).
2. S. Aubry and R. S. MacKay, *Nonlinearity* **7**, 1623 (1994).
3. T. Cretegny, T. Dauxois, S. Ruffo and A. Torcini, *Physica D* **121**, 109 (1998).
4. G. P. Tsironis and S. Aubry, *Phys. Rev. Lett.* **77** (26), 5225 (1996).
5. A. Bikaki, N. K. Voulgarakis, S. Aubry and G. P. Tsironis, *Phys. Rev. E* **59** (1), 1234 (1999).
6. F. Piazza, S. Lepri and R. Livi, *J. Phys. A* **34**, 9803 (2001).
7. F. Piazza, S. Lepri and R. Livi, to appear in *Chaos* — Focus Issue *Nonlinear localised modes; Fundamental Concepts and Applications*, **13** (2), (2003), cond–mat/0210027.
8. I. Daumont, T. Dauxois, M. Peyrard, *Nonlinearity* **10**, 617 (1997).
9. Yu. A. Kosevich and S. Lepri, *Phys. Rev. B* **61**, 6 (2000).
10. S. Flach, K. Kladko and R. S. MacKay, *Phys. Rev. Lett.* **78**, 1207 (1997).

QUASILOCAL MODES AND THE BREAKING OF INTEGRABILITY

NIURKA R. QUINTERO*

Departamento de Física Aplicada I, Escuela Universitaria Politécnica, Universidad de Sevilla, Virgen de África 7, 41011, Sevilla, Spain; and Instituto CARLOS I *de Física Teórica y Computacional. Universidad de Granada E-18071 Granada, Spain.*
E-mail: niurka@euler.us.es

P.G. KEVREKIDIS †

Department of Mathematics and Statistics, University of Massachusetts, Amherst MA 01003-4515, USA
E-mail: kevrekid@math.umass.edu

We present a numerical study of the spectral properties of the linearized problem around the kink solution in the parametrically modified sine-Gordon equation (pmsG). We show that for $r > 0$ the appearance of a quasi-local mode in the linearization spectrum around a kink suggests the non-integrability properties of this model.

The inelasticity of collisions between solitary waves has been studied extensively in the nonlinear Klein-Gordon models[1]. In particular, it has been found that, contrary to the integrable case (sine-Gordon equation) in which the velocity and the shape of the soliton are preserved, the final state[a] is very sensitively dependent on the initial velocity, v_i. For certain values of v_i the reflection or transmission can occur upon a second bounce (two-bounce windows or resonance windows) or even after three or more collisions, when the appropriate resonance condition is satisfied[1]. In this case, the (kinetic) energy of the translational mode (TM) is stored as internal mode (IM) energy upon the first interaction and during the second collision the energy is

*This research is supported by Ministerio de Ciencia y Tecnología of Spain under contract BFM2001-3878-C02-01 and by the Junta de Andalucía under the project FQM-0207.
†Work partially supported by the University of Massachusetts through a Faculty Research Grant, and by the NSF through DMS-0204585.
[a]The waves can be reflected/transmitted, or trapped depending on the nonlinearity and the initial velocity.

partially transferred back from IM to TM, providing enough kinetic energy to the waves to escape each other's attraction. Due to the nonlinearity, harmonics of the IM frequency are also excited[2]. As a consequence, the energy of the IM is also partially transferred to extended waves (phonons). In the non-integrable systems that do not possess an IM, however, the inelastic collision (and consequently the non-integrability) cannot be explained by the above mechanism. One such example is the pmsG equation[3]

$$\phi_{tt} - \phi_{xx} = -\frac{(1-r^2)^2 \sin(\phi)}{[1+r^2+2r\cos(\phi)]^2}, \quad |r|<1, \quad \phi_x(\pm\infty)=0, \qquad (1)$$

for $0 < r < 1$. In this case the crucial question that arises is: what is the mechanism which, in the absence of IM, generates phonon radiation and causes the inelasticity of collisions in a near-integrable system?

To examine the above question, we analyze the linearization spectrum of the Eq. (1) around a kink in the following way[4]: firstly, we insert the expansion $\phi_0(x) + \epsilon\psi(x,t)$ in Eq. (1) and obtain that the small perturbation $\psi(x,t) = \mathrm{e}^{i\omega t} f(x)$ satisfies, to $O(\epsilon)$, the linearized equation (a Sturm-Liouville problem) for the eigenfrequencies and the corresponding eigenfunctions $\{\omega, f(x)\}$. Then, we compute the numerical static kink solution of Eq. (1). Finally, we approximate the eigenvalue problem by a numerical discretization (for free boundary conditions).

In Figure 1 (left) we show the variation of the first few eigenfrequencies close to the phonon band edge as a function of r. We have plotted the differences between the computed frequencies ω and the one of the band edge of the continuous spectrum, $\omega_{ph} = (1-r)/(1+r)$. If $\omega < \omega_{ph}$, we identify an IM ($\omega = \Omega_I$). This occurs here for $r < 0$ (see the circles joined by solid line). The bifurcation of this IM from the continuous spectrum for $r < 0$ can be quantified[5],

$$\Omega_I = \sqrt{1 - 4r - \left(\frac{16}{3}\right)^2 r^2} \approx \sqrt{\omega_{ph}^2 - \left(\frac{16}{3}\right)^2 r^2}. \qquad (2)$$

For small negative r, good agreement between the numerical data of Ω_I and the analytical prediction (2) is observed[4]. Notice that variation of r does not affect all the phonon frequencies, but rather *only* the first few phonons related to the kink's width[6] (see the plus signs joined by dot-dashed lines) which consequently move "faster" than the rest (stars connected by dashed lines). This is due to the fact that changes in r affect *crucially* the width of the exact static kink. Hence, in the linear spectrum around the solitary wave, the main change that can appear for different values of r should be related either to the internal modes or to the phonon modes that are able to modify the initial width of the kink. Such changes are evident for the odd phonon modes in their respective spatial profiles.

In Figure 1, we also observe that as r increases, passing from negative to positive values, the internal mode Ω_I (that was outside the band for $r < 0$) disappears inside the band. Notice that, in this case, also a redistribution of the phonons occurs. Indeed, as the mode "enters" the band, becoming the lowest phonon mode, the rest of the phonons "feel its presence" and some of them (again the ones with the same parity as the –disappearing– IM) are repelled by it moving further away from the lower band edge.

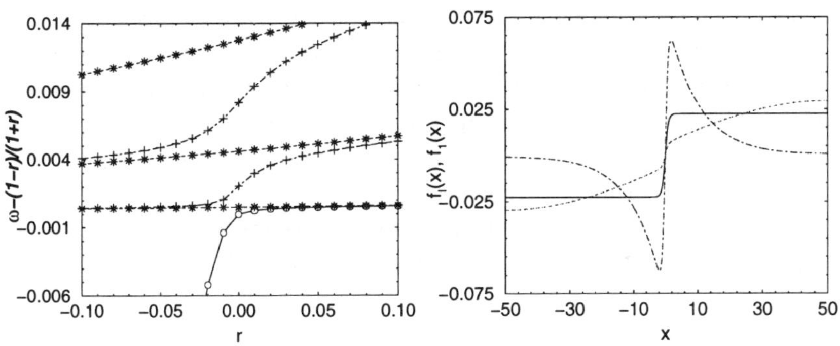

Figure 1. Left: We plot the difference between the computed eigenfrequencies and the band edge eigenfrequency $\omega_{ph} = (1-r)/(1+r)$ versus r. Right: The spatial profiles of the IM (for $r = -0.02$; dot-dashed line) and of the first phonon mode (for $r = 0$ by solid line and for $r = 0.02$ by dashed line) are shown.

Having shown that for $r > 0$, the internal mode "disappears" in the continuum spectrum, we now analyze how the spatial profiles of the eigenfunctions change as a function of r. Successive modes have alternating parity between them. The first eigenfunction (even mode) corresponds to the translational mode and remains (practically) unchanged for different values of r. The second eigenfunction (odd mode) is represented in Figure 1 (right) for 3 different values of r. Notice that, contrary to the case for the second e.f. for $r = 0$ (solid line), this mode is completely localized for $r = -0.02$ (dot-dashed line) and is a quasilocal mode[7] for $r = 0.02$ (dashed line). In the case of negative values of r it is the IM, whereas for positive values of r, it is the lowest phonon mode. We have also observed that the even phonon modes when $r \neq 0$ behave (exactly) as they do in the sG case[4].

We would like to stress that generally the linear spectrum around a solitary wave determines the response of the system to small perturbations. For example, due to the presence of the IM, phonons or of an impurity mode, similar mechanisms of energy transfer arise during the interaction of the kink with impurities[8] as well as when the kink is excited by an external periodic field[9]. We have argued through the resonance picture (when the model is driven by an external ac-drive), as well as the collision picture (of quasi-resonances present in a kink-antikink collision) that the "penetration" of the mode inside the band (e.g., for $r > 0$) affects the properties of nearby modes of the same parity (their frequency as well as the width of the corresponding eigenvectors).

We have shown by numerical analysis of the spectral properties of the linearized problem around the static kink solution in the pmsG that when r is positive there is no IM to support the non-integrability properties of the system. Instead, the localized spatial profile of the IM, which is present for $r < 0$, is converted to a quasilocal mode for $r > 0$. Then, in the interaction between kink and antikink, this quasilocal mode and the same parity modes in its vicinity are excited and the energy of these modes is dispersed, hence the system cannot be integrable.

Similar considerations are expected to generalize to other systems such as the discrete nonlinear Schrödinger equation (DNLS)[10].

References

1. David K. Campbell, Jonathan F. Schonfeld and Charles A. Wingate. *Physica* **9D**, 1, (1983); M. Peyrard and D. K. Campbell. *Physica* **9D**, 33, (1983). David K. Campbell and Michel Peyrard. *Physica* **18D**, 47, (1986); D.K. Campbell, M. Peyrard and P. Sodano, *Physica* **19D**, 165 (1986).
2. P.G. Kevrekidis, *Phys. Lett.* **A285**, 383 (2001).
3. M. Remoissenet and M. Peyrard. *J. Phys. C: Solid State Phys.* **14**, L481, (1981); *Phys. Rev.* **B26**, 2886 (1982).
4. Niurka R. Quintero and Panayotis G. Kevrekidis *Physica* **D170**, 31, (2002).
5. Yu.S. Kivshar, D. Pelinovsky, T. Cretegny and M. Peyrard, *Phys. Rev. Lett.* **80**, 5032 (1998).
6. N.R. Quintero and P.G. Kevrekidis, *Phys. Rev.* **E64**, 056608 (2001).
7. Similar mode have been observed in B. A. Ivanov, H. J. Schnitzer and F. G. Mertens, *Phys. Rev.* **B58**, 8464, (1998).
8. Yu. S. Kivshar, F. Zhang, and L. Vázquez, Phys. *Rev. Lett.* **67**, 1177 (1991).
9. Z. Fei, V. V. Konotop, M. Peyrard and L. Vázquez, *Phys. Rev.* **E48**, 548 (1993); N.R. Quintero, A. Sánchez and F.G. Mertens, *Phys. Rev. Lett.* **84**, 871 (2000); *Phys. Rev.* **E62** Rapid Comm., R60 (2000).
10. P.G. Kevrekidis, K.Ø. Rasmussen and A.R. Bishop, *Intn. J. Mod. Phys.* **B15**, 2833 (2001).

CONTROLLING THE ENERGY FLOW IN NONLINEAR LATTICES: A MODEL FOR A THERMAL RECTIFIER

M.TERRANEO

Laboratoire de Physique Quantique, Université P.Sabatier
F-31062 Toulouse Cedex 4, France
E-mail: terraneo@irsamc.ups-tlse.fr

M.PEYRARD

Laboratoire de Physique, Ecole Normale Supérieure de Lyon,
69364 Lyon Cedex 07, France

G.CASATI

International Center for the study of Dynamical Systems,
Universitá dell'Insubria,
Via Valleggio 11, 22100 Como, Italy

We discuss the possibility of controlling the heat flux in nonlinear one-dimensional systems. We show that, by modifying the on-site potential in a small region we can control the amount of heat flowing through the system. Moreover, we can build a system which displays an insulator-conductor transition by raising both the thermal bath temperatures of the same amount. Our analysis leads us to propose a system which acts as a a thermal rectifier.

In the last decades, a large effort has been made into the study of conductance properties in classical dynamical systems. The main aim of this works was to establish the validity conditions for the Fourier law on purely dynamical grounds[1,2]. In this paper we also consider transport properties but we address a different problem, namely the possibility to control the heat flux through the systems. A first report on this work has been published in Ref. 3.

We consider a nonlinear 1-D system coupled at both edges to heat reservoirs at different temperatures. Such a non-equilibrium configuration leads to heat transport along the chain. We show how the nonlinear potential can be chosen in order to lower or to increase the energy transfer. Finally, our analysis leads us to propose a system that acts as a thermal rectifier, i.e. a system which conduces heat mainly in one direction.

We chose the Hamiltonian

$$H = \sum_{n=1,N} \frac{p_n^2}{2m} + V_n(y_n) + \frac{1}{2}K(y_n - y_{n-1})^2 \tag{1}$$

which describes a 1-D system of N particles with nonlinear on-site potential V_n and harmonic nearest-neighbor coupling with constant K. The on-site potential is the Morse one, $V_n = D_n(e^{-\alpha_n y_n} - 1)^2$. Such a model is known as the Peyrard-Bishop model, originally introduced to describe the dynamics and the thermodynamics of the DNA molecule [4,5,6]. In this context m is the reduced mass of a base pair, y_n is its stretching from the equilibrium position and p_n its momentum.

We work in a non-equilibrium regime, where the first and last L sites of the chain are coupled to heat reservoirs at two different temperatures T_1, T_2. We model the heat reservoirs by using the Nosé-Hoover approach[7,8] and, in cases very far from equilibrium, a Langevin model for the thermostats. We perform the simulations by numerically integrating the equations of motion for the whole system (chain+reservoirs) by a 4th order Runge-Kutta method[9]. Then, we compute temperature profiles and local heat flux from their microscopic definition[2]: $T_i = m\langle \dot{y}_i^2 \rangle$ and $J_i = K\langle \dot{y}_i(y_i - y_{i-1})\rangle$, where the average $\langle\ \rangle$ is over the time. Simulations are performed until the non equilibrium steady state was reached, i.e. the local heat flux is uniform $J_i = J$ along the chain and the average local temperature T_i does not evolve anymore.

We divide our system into three parts, where parameters α_n and D_n can take different values. In the homogeneous case, $\alpha_n = \alpha$, $D_n = D$ for each n, the system behaves as a conductor: the temperature profile shows a gradient (Fig.1-squares) and the heat flux obeys the Fourier law $J \sim 1/N$. By changing the parameters in the central region i.e. by considering non-homogeneous systems, the transport properties change dramatically.

If we take $D_n = D_1 \neq D$ in a small central slice of the chain ($M = 8$ sites, $\sim 10\%$), the system displays a transition from a conducting (squares) to an insulating (diamonds) behavior. In the insulating case the heat flux is reduced to $\sim 1\%$ of the conducting case (Fig.1, inset).

Such results can be explained by considering an harmonic system obtained by linearizing the Morse Potential

$$H = \sum_{n=1,N} \frac{p_n^2}{2m} + D_n\alpha_n^2 y_n^2 + \frac{1}{2}K(y_n - y_{n-1})^2 \tag{2}$$

Simulations performed with the Hamiltonian (2) show the same transition (Fig.1, inset), and it is even sharper in the linearized system. This sug-

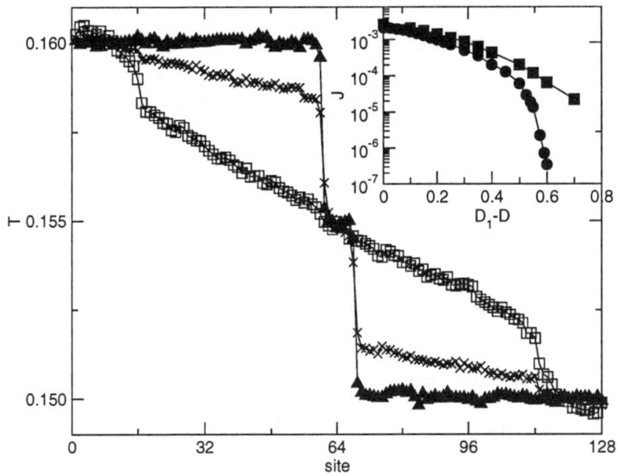

Figure 1. Temperature profile of the chain with parameters $D = 0.5$, $N = 128$, $\alpha = 1$, $M = 8$,$L = 16$, $K = 0.3$, $T_1 = 0.16$, $T_2 = 0.15$. D_1 takes values $D_1 = D = 0.5$ (squares - homogeneous system), $D_1 = 0.8$ (+) and $D_1 = 1.2$ diamonds. Inset: heat flux as a function of the parameter $D_1 - D$; the squares refer to the Morse potential, while the circles refer to the corresponding harmonic limit.

gests that an explanation based on the propagation of plane waves may be sufficient to understand the numerical findings of Fig.1.

In a homogeneous harmonic system, a plane wave with frequency ω and wavenumber κ, $y_n = Ae^{i(\kappa n - \omega t)}$ can propagate only if the dispersion relation $\omega^2 = 2K + 2D\alpha^2 - 2K \cos \kappa$ is satisfied. So the frequencies must belong to a finite range, the so called *phonon band.* If we split our system into different parts, each region has its own band. In order to freely propagate, the frequencies of the excitation must belong to each phonon band. But, if the bands cover different frequencies in different regions, as in our simulations, only a few frequencies can propagate. The fraction of such frequencies decays as $|D_1 - D|$ increases. The ω values which do not belong to all the bands give rise to exponentially damped states, which contribute only slightly to the heat transfer. If $|D_1 - D| > 2K$ (band separation larger than bandwidth) all the frequencies give rise to exponentially damped states. So the total heat flux is expected to be exponentially small.

Such an argument, which explains the findings for the linear case, can be applied to the nonlinear systems as well. Here, since plane waves are not exact solution for the equations of motion, the argument is only approximate, but it can explain the results of Fig.1. Notice that the insulator behavior (diamonds) is obtained as soon as $D_1 - D \geq 2K$.

A further step in order to take into account the nonlinear interactions

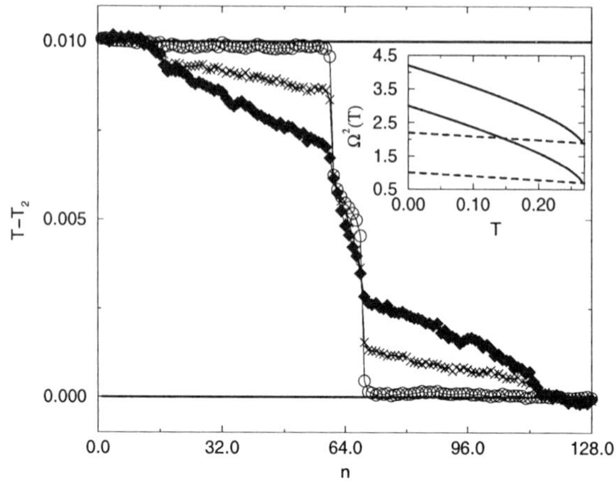

Figure 2. Temperature profiles for the system (1) at three different values of T_2 for the same temperature gradient $T_2 - T_1$. Parameters are $N = 128$, $M = 8$, $\alpha_1 = 2$, $\alpha = 1$, $D_1 = 0.375$, $D = 0.5$; $T_2 = 0.09$ (circles), $T_2 = 0.19$ (x) and $T_2 = 0.29$ (diamonds). Inset: Phonon band profiles vs. temperature; the continuous lines represent the phonon band for $\alpha = 2$, $D = 0.375$; the dashed lines $D = 0.5$, $\alpha = 1$. The full phonon band is shown with bandwidth $4K = 1.2$

is the so called *effective phonons method* [5]. It consists in considering an harmonic Hamiltonian, which parameters are temperature-dependent. By a self-consistent procedure, one can determine the temperature dependency.

Thus, at fixed T, we can approximate our nonlinear Hamiltonian with a linear one which contains the typical frequencies of the system. Since our system is not homogeneous, we have different effective phonon bands, with a different temperature dependence. Therefore, by estimating the overlap of the effective bands, we can devise a system which conduces heat, if the bands overlap, or which does not, if the bands are separated.

For instance, we considered a system where the (effective) bands overlap at high T and are separated at low T (Fig.2). So the system should conduct heat at high T and behave as an insulator at low T. Such a behavior is shown in Fig.2: raising the temperature T_1, T_2 and keeping $T_1 - T_2$ fixed, the system displays a transition from an insulating to a conducting behavior. The heat flux increases, although we do not get a clean gradient for the conducting case.

This proves that the effective bands overlap is a rather good criterion to predict the conducting properties of a system. On such grounds, we can depict a system which can control the heat flow, by allowing only one-way energy transport.

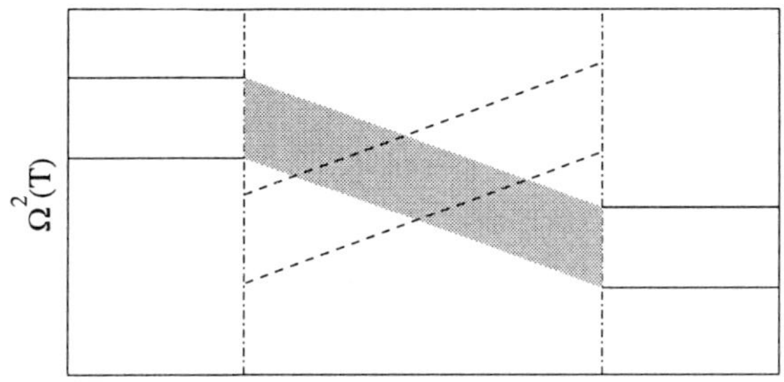

Figure 3. Schematic picture of the phonon bands in a "thermal rectifier" for two directions of the temperature gradient. The bands in the left and right weakly anharmonic regions do not change significantly with the orientation of the gradient. In the central part, when the high temperature side is on the right, the band evolves in space as shown by the shaded region, while, if the high temperature side is on the left, the band evolves according to the dashed lines.

We chose the asymmetric architecture shown in Fig.3: the phonon bands in the left and right regions do not match. Moreover, we chose the potential in order to have bands which depends only slightly on temperature in the left and right regions. This can be achieved by choosing an almost harmonic on-site potential in these two regions.

On the contrary, the bands in the central region are chosen to be strongly T-dependent. Since through the central region the local temperature T_i decreases from T_1 to T_2 from one side to another, the local effective frequencies change with space as well. If the central region is large enough it ensures a smooth variation from one value to another along this region, which can be used to control the matching with the bands in the left and right regions.

Fig.3 shows the principle of the proposed device. If the higher temperature is on the right, the frequencies in the central region evolve according to the gray band. This results in a good coupling at both edges, so plane waves injected from the sides can move along the system. This is the conducting direction. The behavior is quite different if the temperature gradient is reversed (dashed band, Fig.3). In the latter case, there is a band mismatch, and this shows up in a difficult heat transmission. Thus the system should conduct heat only in one direction, when $T_2 > T_1$. We numerically simulated the behavior of such a system. Indeed, the heat flux computed after reversing the gradient was about 50% of the original one. So the simulations confirmed that our system can work as a thermal rectifier, although the efficiency reached is quite low.

There are several reasons for such an imperfect functioning. First of all, we based our analysis on the self consistent phonon method, which is not expected to hold perfectly in our case. In fact, the approximation had been proved to work in equilibrium systems, while here we work in strongly non equilibrium ones. Moreover, the frequencies should evolve in space along the chain, so we expect that our rectifier would work better if the local temperature T_i only change very slowly with space. Finally, we considered only the on-site Morse potential and a different on-site potential could perhaps yield better results. We stress that we chose the PB model since several results were available, especially those concerning the self-consistent phonon method but a systematic search of the optimal system to make a rectifier has not been done. What is interesting is that the arguments hold for a generic nonlinear potential whose typical frequencies decrease as the energy increases (weakening of the phonon band).

In spite of the low efficiency of the present rectifier, the numerical data clearly show that the idea works. Building a thermal rectifier is in principle possible, even if improving the efficiency is necessary before thinking about any possible application. Moreover, the simple mechanism we depicted can be a working hypothesis to understand better the phenomena that occurs in some biological systems. In fact, biomolecules need to control the heat transfer: energy needs to be stored and then used in proper places and at the right time. For instance, in ref.10 it was shown that during muscles contraction the energy of ATP hydrolysis is locally released and used in several steps. So the myosin molecule itself controls the heat transfer.

Such experimental findings haven't found a clear theoretical explanation yet. On the other side, our system displays a clear and simple mechanism to control the energy transfer. This may suggest that we can try to apply the ideas above discussed to the study of heat transport in biological systems. Even if our model is clearly too rough to explain such findings, we showed that the presence of nonlinearities can achieve a control on the energy flow. Biological molecules are highly nonlinear systems because they undergo large conformational changes, but the relevance of any mechanism such as the ones we described above in the energy control by biomolecules still remains an open interesting problem.

Acknowledgments

Support from EU contract No. HPRN-CT-1999-00163 (LOCNET network) is acknowledged.

References

1. F.Bonetto, J.L.Lebowitz and L. Ray Bellet, in A. Fokas, A. Grigoryan, T. Kibble and B. Zegarlins Eds., *Mathematical Physics* 2000, Imperial College,

London 2000, and *math-phys/0002052*
2. S. Lepri, R. Livi and A. Politi, *cond-mat/0112193*
3. M.Terraneo, M.Peyrard and G.Casati, *Phys. Rev. Lett.* **88**, 094302 (2002)
4. M.Peyrard and A.R. Bishop, *Phys. Rev. Lett.* **62**, 2755 (1989)
5. T. Dauxois, M.Peyrard and A.R. Bishop, *Phys. Rev.E* **47**, 684 (1993)
6. N. Theodorakopoulos, T.Dauxois and M.Peyrard, *Phys.Rev.Lett.* **85**, 6 (2000)
7. S. Nosé, *J.Chem. Phys.* **81**, 511 (1984), W.G.Hoover, *Phys.Rev.A* **31**, 1695 (1985)
8. G.Martyna, M.L.Klein and M.Tuckerman, *J. Chem. Phys.* **97**, 2635 (1992)
9. W.H. Press, S.A. Teukolsky, W.T. Vetterling and B.P. Flannery, *Numerical Recipes*, Cambridge University Press (1992)
10. K. Kitamura, M. Tokunaga, A.H. Iwane and T. Yanagida, *Nature* **397** 129 (1999)

FRACTIONAL DERIVATIVE: A NEW FORMULATION FOR DAMPED SYSTEMS.

D. USERO
Dpto. de Matemática Aplicada, Facultad de Informática, Universidad Complutense de Madrid. 28040 Madrid (SPAIN)

L. VÁZQUEZ
Dpto. de Matemática Aplicada, Facultad de Informática, Universidad Complutense de Madrid. 28040 Madrid (SPAIN)

Centro de Astrobiología (CSIC/INTA) Carretera de Ajalvir km. 4, 28850 Torrejón de Ardoz, Madrid, SPAIN

Traditionally the damping in physical models has been studied as a velocity dependant term introduced in the equations of the system under study. In some cases the introduction of this term is fully justified, but in some others this is done in a heuristic way, in order to model some behaviour, or to avoid explosions of some quantity. Here I propose a different way to introduce damping in the system in terms of a nonlocal generalization of the usual derivative called Fractional Derivative. This type of derivatives has been studied long time ago, but recently they have been applied in many fields in science giving interesting results.

1. Basis of Fractional Calculus

Similar to standard Calculus, Fractional Calculus is a part of Mathematical Analysis which study the properties of integrals and derivatives of fractional arbitrary index. The name *fractional* is related with the non integer order, but does not refer necessarily to fractions but to real numbers of any kind.

This field of Analysis is as old as normal Calculus since it start with speculations of mathematicians like G. W. Leibniz (1695,1697) and L. Euler (1730) and it has developed up to nowadays. See [1] and [2] for a detailed historical review.

There are different ways to define a fractional derivative, but two of them are the most widely used. Both of them are based on the Cauchy Integral formula to calculate the n-th primitive of a function. According to Riemann-Liouville it is also possible to define a fractional primitive of order α ($0 < \alpha \leq 1$) of a function as follows:

$$J_\alpha f(t) = \frac{1}{\Gamma(\alpha)} \int_0^t (t-\tau)^{\alpha-1} f(\tau)\, d\tau. \tag{1}$$

Being $J_0 = \mathbf{I}$. It is easy to probe that these operators J_α verify the semigroup property $J_\alpha J_\beta = J_{\alpha+\beta}$.

There are two possibilities to define a fractional derivative of order α:

$$D_\alpha = D\, J_{1-\alpha} \qquad D^*_\alpha = J_{1-\alpha}\, D, \tag{2}$$

where operator D refers to first derivative.

The first definition is called *Abel-Riemann* fractional derivative and it can be also defined as the left inverse of operator J_α since $D_\alpha\, J_\alpha = \mathbf{I}$.

Second definition correspond to the *Caputo* fractional derivative, which is the right inverse of J_α. Both derivatives are not equivalent and can be probed that:

$$(D_\alpha - D^*_\alpha) f(t) = \frac{t^{-\alpha}}{\Gamma(1-\alpha)} f(0). \tag{3}$$

It is clear from eqs. (3) and (2) that fractional derivative does not fulfill two important rules of derivative: **Chain rule** and **Product rule**. Fractional derivative problems should be treated with special care in order to avoid mistakes.

In this work I have used the Caputo fractional derivative. This choice is important from a mathematical point of view since initial conditions to be used are not the same. Caputo's definition admit the usual initial conditions but Abel-Riemann's needs initial conditions over the fractional derivatives of the function.

Another important difference between them is about analytical properties of the functions. Caputo's definition implies derivability of the functions just as normal derivative does. This is because derivative operator acts first. In the definition of Abel-Riemann the first operator is the fractional integral and then we just require this integral of the function to be derivable, not the function itself.

2. Fractional Harmonic Oscillator

In a previous work[4] we have studied the properties of Hamiltonian systems subject to fractional time derivatives. Following the standard Hamiltonian formalism, we consider a one-dimensional oscillator, with mass equal to one, defined by $H = \frac{1}{2}(p^2 + \omega_0^2 x^2)$. It is possible to obtain fractional equations by generalizing the equations of motion for this system.

$$\begin{cases} D_\alpha^* x = \dfrac{\partial H}{\partial p} = p \\ \\ D_\alpha^* p = -\dfrac{\partial H}{\partial x} = -\omega_0^2 x \end{cases} \tag{4}$$

Denoting with $\mathcal{L}$ the Laplace transform it is easy to show:

$$\mathcal{L}(J_\alpha f) = \frac{\mathcal{L}(f)}{s^\alpha}, \quad \mathcal{L}(D_\alpha f) = s^\alpha \mathcal{L}(f), \quad \mathcal{L}(D_\alpha^* f) = s^\alpha \mathcal{L}(f) - s^{\alpha-1} f(0). \tag{5}$$

And introducing the complex variable $\Omega = \sqrt{\omega_0}\, x - i\, p/\sqrt{\omega_0}$ we can write the equations of motion (4) as $D_\alpha^* \Omega = i\omega_0 \Omega$. Taking Laplace transform:

$$\mathcal{L}(\Omega) = \frac{s^{\alpha-1}}{s^\alpha - i\,\omega_0}\Omega(0). \tag{6}$$

Solution for equation (6) for $\alpha = 1$ is the usual for harmonic oscillator $\Omega(t) = \Omega(0)\, e^{i\omega_0 t}$, and for any other value of α the solution can be given in terms of an extension of the exponential function called **Mittag-Leffler function** (see [1], [2] and [3]). This function can be represented in series as:

$$E_\alpha(z) = \sum_{n=0}^{\infty} \frac{z^n}{\Gamma(n\alpha + 1)}, \tag{7}$$

and again for $\alpha = 1$ we recover the exponential. Other generalized function of Mittag-Leffler type is

$$E_{\alpha,\beta}(z) = \sum_{n=0}^{\infty} \frac{z^n}{\Gamma(n\alpha + \beta)}, \tag{8}$$

where it is clear that $E_\alpha = E_{\alpha,1}$.

In terms of these functions, solution for equation (6) reads

$$\Omega(t) = \Omega(0)\, E_\alpha(i\omega_0 t^\alpha) \tag{9}$$

We show in Fig.1 the evolution of $x(t)$ compared with the harmonic oscillator ($\alpha = 1$). This evolution is shown for three different values of α ($\alpha = 0.99$, 0.9 and 0.5).

From Fig.1 it is that the fractional oscillator seems to be a damped oscillator.

This possibility is investigated in a previous work[4] and finally is obtained that for values of α close to one, equation (4) is similar to a damped oscillator with damping constant:

$$k = \omega_0^{1/\alpha} \cos\left(\frac{\pi}{2\alpha}\right). \tag{10}$$

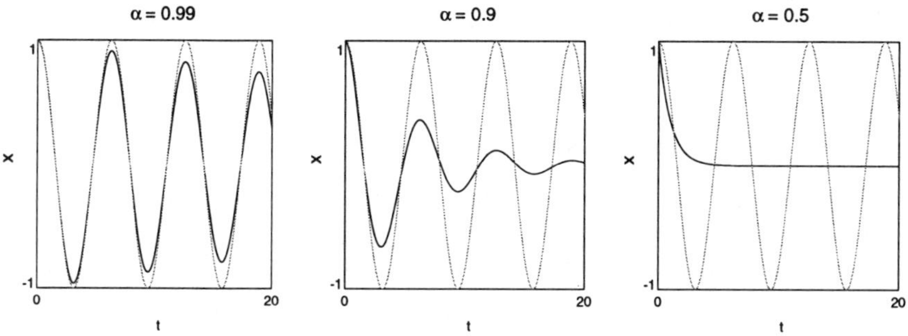

Figure 1. Time evolution of the Fractional Oscillator $x(t)$ for $\alpha = 0.99$, 0.9 and 0.5. Grey line shows the behaviour of the Harmonic Oscillator.

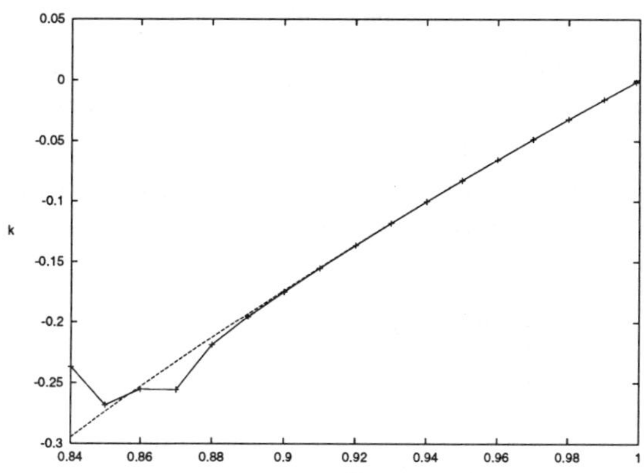

Figure 2. Comparison of the numerical damping constant k and the theoretical given by eq. (10).

In Fig.2 it is shown the numerical values of the damping constant compared with the theoretical approximation. Both coincides satisfactorily up to $\alpha \approx 0.9$.

An important feature of fractional systems that make them different from classical damped system is that the exponential decay shown in (10) is valid only for $t \leq (1-\alpha)^{-1}$ while for $t \geq (1-\alpha)^{-2}$ there is a **power law decay**. Following the same arguments given in these work[4], it is possible to write

$$E_\alpha(i\omega_0 t^\alpha) = \frac{1}{\alpha}\exp\left(\omega_0^{1/\alpha} t e^{\frac{i\pi}{2\alpha}}\right) + \frac{i}{\omega_0}\frac{1-\alpha}{\Gamma(2-\alpha)}\frac{1}{t^\alpha} + O(t^{-2\alpha}) \qquad (11)$$

The power law decay is shown in certain systems and traditionally was explained by complicated arguments. In the present case this behaviour is obtained in a natural way, there is no need to use any artificial term in the hamiltonian.

3. Forced Fractional Oscillator

Another interesting problem to consider is the response of the system under the action of an external periodic forcing term.

Damped Harmonic oscillator shows very well known behaviour and the competence between damping and forcing results in a periodic solution of frequency equal to the forcing and an amplitude (A_{res}) given by (see [5] for instance)

$$A_{res}(\omega) = \frac{A_0}{\sqrt{(\omega^2 - \omega_0^2)^2 + (k\omega)^2}}, \tag{12}$$

where A_0 is the amplitude of the forcing, ω is the forcing frequency, ω_0 is the oscillator frequency and k is the damping constant. This amplitude does not depend on the initial condition of the system.

It is interesting to study the resonant behaviour of fractional systems in the presence of an external forcing of this kind. Under such forcing the equations of motion are:

$$\begin{cases} D_\alpha^* x = p \\ \\ D_\alpha^* p = -\omega_0^2 x + A_0 \cos(\omega t) \end{cases} \tag{13}$$

In terms of the complex variable defined in the previous section the equation of motion is $D_\alpha^* \Omega = i\omega_0 \Omega - iA_0 \cos(\omega t)$. Again using the Laplace transform it is possible to write:

$$\mathcal{L}(\Omega) = \frac{s^{\alpha-1}}{s^\alpha - i\,\omega_0} \Omega(0) + iA_0 \frac{s}{(\omega^2 + s^2)(s^\alpha - i\omega_0)}, \tag{14}$$

according to [6] and using the properties of the transform, the second term of (14) can be written as a convolution between the forcing and the derivative of the Mittag-Leffler function.

Equating the resulting series it is possible to calculate the inverse Laplace Transform of expression (14) in terms of generalized Mittag-Leffler functions (8) obtaining

$$\Omega(t) = \Omega(0) E_\alpha(i\omega_0 t^\alpha) - \frac{iA_0}{\sqrt{\omega_0}} \sum_{k=0}^{\infty} (i\omega_0)^k t^{\alpha(k+1)} E_{2,\alpha(k+1)+1}(-\omega^2 t^2). \tag{15}$$

This complicated expression can be understood easily with the help of Fig.3. There, it is shown the solution of the forced problem compared with

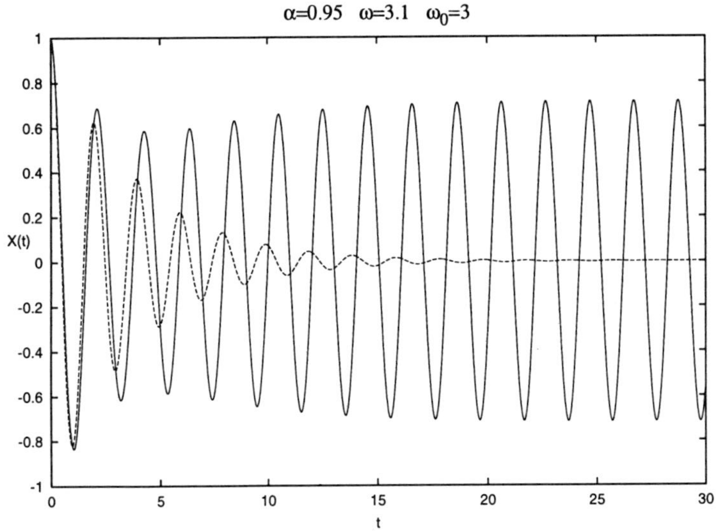

Figure 3. Time evolution of the Forced Fractional Oscillator $x(t)$ for $\alpha = 0.95$, $\omega_0 = 3$ and $\omega = 3.1$.

the decaying solution without forcing. Initially both coincide since the last term of eq. (15) vanishes, so the solution is the usual of the harmonic oscillator. But after a short time, oscillations of frequency ω_0 decays to zero and the last term of equation (15) becomes dominant forcing the system to oscillate with frequency ω.

The limit cycle,*i. e.* the final oscillating state of the system is given by

$$\Omega(t) \rightarrow \frac{-iA_0}{2\sqrt{\omega_0}} \frac{e^{(i\omega t+\delta)}}{(i\omega)^\alpha - i\omega}, \tag{16}$$

which is obtained in [6]. Then, any initial condition will evolve to an oscillation of frequency ω and amplitude

$$A_{res}(\omega) = \frac{A_0}{2\omega_0} \left| \frac{-i}{(i\omega)^\alpha - i\omega} \right|, \tag{17}$$

and only the phase δ of eq. (16) depends on the initial condition.

In Fig.4 we can see two plots with the amplitude of the stationary oscillating solution for two values of α. Points have been calculated numerically while solid lines are given by $A_{res}(\omega)$ form eq.(17).

4. Conclusions

As it has been shown in the previous sections Fractional Derivative systems seem to behave as damped systems and numerical simulations confirm this

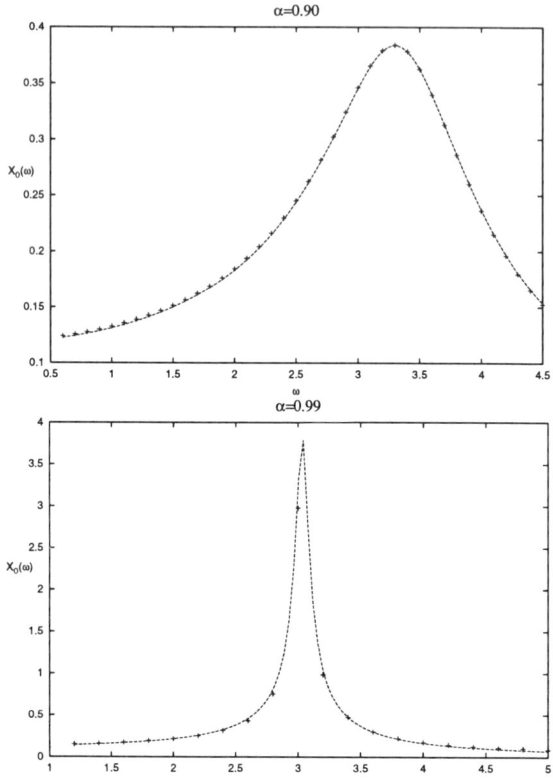

Figure 4. Amplitude of oscillatory solution of the forced fractional oscillator calculated numerically with $\omega_0 = 3$, $A_0 = 1$ and different values of the forcing frequency ω. Solid line correspond to theoretical estimations.

supposition. But the real fact is that the fractional derivative is not only a damping, and there are several differences between them.

The first difference is the way of introducing this damping. In this case the hamiltonian form is unchanged and the equations of motion of the system can be obtained generalizing the Hamilton-Jacobi formalism. There is no need to introduce extra terms.

Other significant difference appears in the asymptotic behaviour of the solutions. By using a damping term only exponential decay can be obtained, but with the fractional derivative is also possible to describe algebraic decays.

On the other hand with the fractional derivative it is possible to reproduce similar results of damped systems, and solutions can be written, at least in the easiest cases, in terms of generalized exponentials called

Mittag-Leffler functions.

Main difficulty of working with fractional derivative operators is that they do not conmute and usual manipulation becomes of great difficulty.

Future lines of investigation should include the search for real systems apart from viscoelasticity and anomalous diffusion problems where the fractional derivative can explain the observed behaviour.

It is also interesting to observe coupled systems of oscillators, nonlinearity, systems with temperature, etc.

Ackowledgements

This work has been partially supported by the Comisión Interministerial de Ciencia y Tecnología of Spain under the grant PB98-0850.

References

1. S. G. Samko, A. A. Kilbas, O. I. Marichev, *Fractional Integrals and Derivatives. Theory and Applications.* Gordon and Breach Science Publisher (1993).
2. I. Podlubny, *Fractional Differential Equations Mathematics in Science and Tecnology* **v. 198** Academic Press (1999).
3. R. Gorenflo, F. Mainardi, *Fractional Calculus: Integral and diffeential equations of fractional order* in *Fractals and Fractional Calculus in Continuum Mechanics*, A. Carpinteri and F. Mainardi eds. Springer-Verlag, Wien, New York, pp 223-276.
4. G. Turchetti, D. Usero, L. Vázquez, *Hamiltonial Systems with Fractional Time Derivative*, Tam. Oxf. J. Math. Phys. **18(1)**(2002) 31-44.
5. A. P. French, *Vibraciones y Ondas* Ed. Reverté (2000). Spanish version of *Vibrations and Waves* MIT publications.
6. D. Usero, L. Vázquez, *Forced Hamiltonian Systems with Fractional Time Derivative* (to appear).
7. L. Vázquez, *Fractional Diffusion Equations with Internal Degrees of Freedom* J. Comp. Math. (2002, in press).
8. L. Vázquez, R. Vilela-Mendes, *Fractionally Coupled Solutions of the Diffusion Equation*, App. Math. Comp. (2002, in press).

SPONTANEOUS PATTERN RETRIEVAL IN A NEURAL NETWORK.

M.-P. ZORZANO
Centro de Astrobiología
(CSIC/INTA Associated to NASA Astrobiology Institute)
Carretera de Ajalvir km. 4, 28850 Torrejón de Ardoz, Madrid, SPAIN
E-mail: zorzanomm@inta.es

A fully connected network of FitzHugh-Nagumo oscillators is used as a toy model for memory networks. This system which is able to distinguish and store different patterns, can recover them spontaneously when excited by noise. Unpaired tuning of the level of neurotransmitters elicited by an excited neuron (the coupling) may result in too much or too little excitability and therefore in miss-functioning of the network.

1. Introduction.

We want to study the spontaneous retrieval of memories and explore in which way deviations of the inner structure of the neural tissue produce abnormal results in the biological functions of the neural network.

We will consider a standard connectionist model of memory where information is represented through a connected network of units (the units are each individually meaningless and information is represented only in a distributed fashion, as a function of the simultaneous activation of multiple units). In particular we will use a popular attractor network, the Hopfield net [1,2], in which each node is connected to every other node and where stimuli are presented as patterns of activation. This network reproduces the main features of neural networks: memorization and pattern recognition or completion, and it is often used to investigate psychopathologies involving persistent repetition of certain states, such as repeated recall of particular memories [3].

The impulse transmission in a single neuron can be essentially described with the Hodgking-Huxley (HH) model [4] or its simplified version, the FitzHugh-Nagumo (FHN) oscillator [5,6]. It has been shown that a set of fully connected FHN oscillators oscillates synchronously when the system is excited by noise within a certain range of noise intensities [7]. Applying this idea we will let a local random discharge recall a pattern that has been stored in a set of fully connected FHN oscillators.

2. The model.

In the FHN model all the dynamical variables of the neuron are reduced to two quantities with two dynamical time scales: the voltage of the membrane u (fast variable) and the recovery variable v which corresponds to the refractory properties of the membrane (slow variable).

$$\dot{u_i} = c\left(-v_i + u_i - \frac{u_i^3}{3} + I_i(t) + \xi_i(t)\right) \qquad (1)$$

$$+\sum_{j=1}^{N} \frac{w}{N} J_{ij}\left[u_j(t-\Delta) - u_{eq}\right]$$

$$\dot{v_i} = u_i - bv_i + a \qquad (2)$$

$$< \xi_i(t)\xi_j(t') > = 2D\delta(t-t')\delta_{ij} \qquad (3)$$

Here the subindex i denotes any of the neurons $i = 1...N$ in the neural network. The term J_{ij} represents the weight of the synapse connection from neuron j to neuron i and determines the network connectivity. Neuron i may be excited by an external coherent input I_i or by a fluctuating force which we describe as white Gaussian noise $\xi_i(t)$, with noise strength D. The neuron is said to "fire" or emit an "action potential" when $u > 0$. The synaptic communication depends on the propagation of these action potentials between neurons, often with appreciable distances between them. This, in turn, induces transmission delays Δ. As in any system of coupled oscillators, the time delay must be properly tuned for the network to synchronize optimally. Once the impulse gets to the end of the axon it elicits the emission of neurotransmitters which excite the next neuron. The coupling strength w/N, corresponds to the number of neurotransmitters emitted by every neuron of this network when it elicits an action potential.

For all the cases described below we will study this system for $c = 10, a = 0.7, b = 0.8$ and $N = 200$. The FHN model with the above parameters shows a typical characteristic of a neuron, namely it has a stable rest and with an appropiate amount of disturbance it henerates a pulse with a characteristic height and width. This model is a reduced version of the HH model of the dynamics of the ionic concentration gradient accross the neuron membrane. Finally (u_{eq}, v_{eq}) are the values of the variables u and v when the system is at rest, which for the chosen parameters are $(u_{eq}, v_{eq}) \approx (-1.199, -0.624)$.

2.1. *Learning phase*

The synapse connection from neuron j to neuron i, J_{ij}, is modified in the learning phase to simulate the Hebbian learning rule (or un-supervised learning) which essentially says that "neurons that fire together wire together". We simulate this in a network where all the initial connections are set to $J^0_{ij} = 1$ and let the network "learn" the pattern with a symmetric negative feedback rule.

The initial state is a set of $N = 200$ fully connected neurons ($J_{ij} = 1, J_{ii} = 0$), at rest $(u, v) = (u_{eq}, v_{eq})$. As an input $I_i(t)$ is fed into the net we update the connections between neurons every time step ($dt = 0.01$) of the learning phase $t < 30$. Let T_i^{last} be the last firing time of neuron i, i.e. the last time that neuron i changed its state from non-excited ($u < 0$) to excited ($u > 0$). The learning rule acts at the connection from j to i (and from i to j) from time step $(n-1)$ to (n) as:

$$J_{ij}(n) = J_{ji}(n) = J_{ij}(n-1)\exp\left(-|T_i^{last} - T_j^{last}|/10\right). \tag{4}$$

We test this for an arbitrary input, such as:

$$I_i(t < 30) = \theta(t-30)[\xi_i(1) \times 2 \times I + \xi_i(2) \times 1 \times I + \xi_i(3) \times 1.5 \times I]. \tag{5}$$

where $\theta(t-30)$ is the heaviside step function and

$$\begin{aligned} \xi_i(1) &= 1,\ 1 \le i \le 30,\ 100 \le i \le 130 && (6)\\ &= 0, \text{otherwise} \\ \xi_i(2) &= 1,\ 50 \le i \le 80,\ 150 \le i \le 180, && (7)\\ &= 0, \text{otherwise} \\ \xi_i(3) &= 1,\ 30 \le i \le 50,\ 80 \le i \le 100,\ 130 \le i \le 150, && (8)\\ &= 0, \text{otherwise} \end{aligned}$$

$$(9)$$

and $I = 0.2$. Because of the FHN nonlinear dynamics, neurons excited with different intensities respond with different frequency and in turn will be excited at different phases. Due to the negative feedback learning rule neurons in the same phase will strengthen their synapses with respect to the asynchronous ones. This is used to teach the network to recognize the spatial distribution of the input intensity.

We show in figure 1 the behavior of the system over a time window of a few seconds during the learning phase. In this simulation the delay has been set to $\Delta = 3.0$ and $w = 3.2$. Neurons excited with the same intensity I, fire in phase, with the same frequency. We plot a dot every time that neuron i changes from $u < 0$ to $u > 0$ (it fires an action potential).

This explains the lines that can be seen in the plot, corresponding to the synchronous excitation of the neurons involved in a pattern. Therefore all the units within the same pattern synchronize strongly and their synapsis connections J_{ij} will almost be equal to 1 by the end of the learning phase (after $t = 30$). Whereas neurons belonging to different patterns (which are excited with different intensities and therefore emit action potentials with different frequency) will be connected by much weaker synapsis J_{ij}.

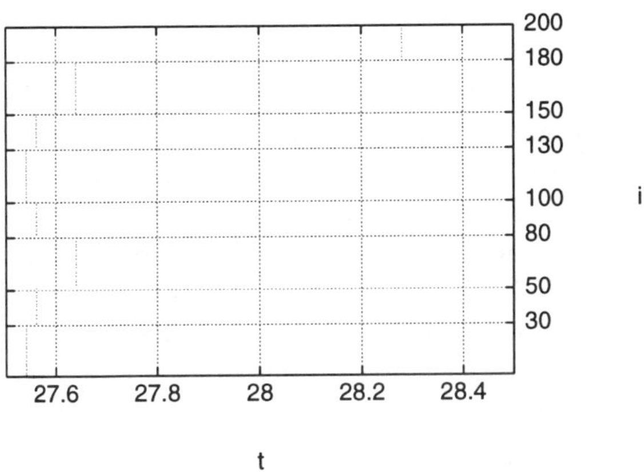

Figure 1. Learning phase: time response of a network of FHN oscillators, with $i = 1..N$, excited with different intensities in the patterns $\xi(1) = (0-30, 100-130), \xi(2) = (50-80, 150-180)$ and $\xi(3) = (30-50, 80-100, 130-150)$. We plot a dot every time that neuron i changes from $u < 0$ to $u > 0$ (it fires an action potential). The network distinguishes the three different patterns and stores them by modifying the connections.

Following this procedure the information of the patterns has been stored in the connectivity matrix J_{ij}. After $t = 30$ the Hebbian learning rule is turned off and the connectivity matrix is kept constant. The state of the neurons is set back to the resting state (u_{eq}, v_{eq}) and the network is ready to retrieve the stored patterns.

3. Selective recall of patterns induced by local noise.

If we excite now this network by a fraction of any of the three stored memories the network will retrieve the rest. But we are interested in the spontaneous retrieval of patterns that takes place in processes such as dreaming or creativity. In this case the pattern retrieval is not triggered by any external coherent input. A feasible internal mechanism to recover a pattern would be a random discharge of a set of neurons acting on the area where the memory was stored. We simulate this by applying white Gaussian noise of intensity D locally, in the area where the pattern has been stored. Depending on the global parameter w the recovery induced by noise can recall or not the desired pattern.

In figures 2 and 3, we show a typical noise realization of this system for $w = 0.8, 1.6$ and 3.2. The first 30 ms correspond to the learning phase explained above. In the retrieval phase we excite some neurons ($i = 0..20$) (which were involved in pattern ξ_1) with a random input of strength $D = 0.002$. If the level of neurotransmitters is too low, as for $w = 0.8$ in Fig. 2-(a), no pattern is recovered. If the level of neurotransmitters is properly tuned, as for the case $w = 1.6$ shown in Fig. 2-(b), pattern ξ_1 is recalled completely. Finally, if the neurotransmitters level is too high, the recovery can even go beyond the desired pattern, see case $w = 3.2$ in Fig. 3-(a) or an enlarged view in Fig. 3-(b). The three patterns can be clearly distinguished as a set of neurons oscillating in different phase and following the spatial profile of the stored memories. In this case the local pattern is first excited stochastically but, due to the high level of neurotransmitters defined by w, even the neurons connected by weaker synapses J_{ij} receive inputs above their excitability threshold and finally all the stored patterns are retrieved.

A similar scenario may appear in schizophrenic cases: although the precise cause of schizophrenia remains unknown, it has been shown that in a brain with schizophrenia far more neurotransmitters (higher w) are released between neurons than in a normal brain.

4. Conclusions.

We have used an attractor network with the Hebbian rule to investigate the spontanous activation of patterns. We have verified that noise alone, with a sufficiently strong intensity and in the optimized conditions of coupling, can produce the spontanous recall of a pre-recorded pattern in a robust, selective and precise way.

In the retrieval of associated patterns it is determinant the strength of the synapsis transmission. Unpaired tuning of this parameter may result

(a) (b)

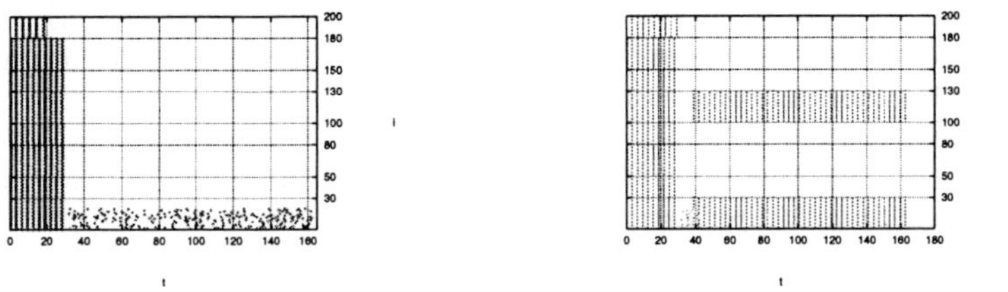

Figure 2. Retrieval phase:after $t = 30$ the external input is turned off and the network is excited locally by noise. (a) If the neurotransmitter level is too low ($w = 0.8$) no pattern is retrieved and only the noisy excitation can be seen.(b) For $w = 1.6$ only the desired pattern $\xi_1 = (0 - 30, 100 - 130)$ which is stored in this area is recalled.

(a) (b)

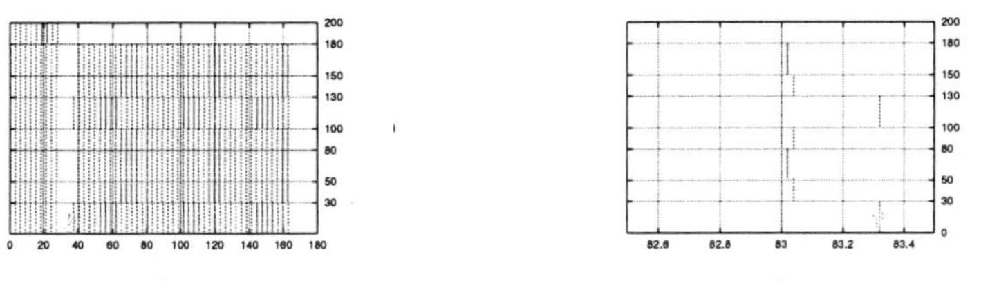

Figure 3. (a) For $w = 3.2$ all the associated patterns are recalled. (b)Enlarged view of the retrieval phase for $w = 3.2$. The level of neurotransmitters is too high and in addition to ξ_1, the undesired patterns ξ_2 and ξ_3 are also retrieved.

in too much or too little excitability and therefore in missfunctioning of the network (which might be associated to pathological performances of the net, such as delusion and hallucination characteristic to schizophreny).

5. Acknowledgement

The author whishes to thank L. Vázquez for scientific discussions and comments on the manuscript.

References

1. Hopfield, J. (1982). Neural networks and physical systems with emergent collective computational properties. Proceedings of the National Academy of Sci-

ences of the USA, 79:2554 - 2588.
2. Hopfield, J. J. (1984). Neurons with graded response have collective computational properties like those of two-state neurons. In Proceedings of the National Academy of Sciences, pp. 81:3088-3092. National Academy of Sciences.
3. Neural Networks and Psychopathology Connectionist Models in Practice and Research. Edited by Dan J. Stein, Jacques Ludik New York: Cambridge University Press 1998.
4. A.L. Hodkin and A.F.Huxley, *A Quantitative Description of Membrane Current and its Application to Conduction and Excitation in Nerve.* J. Physiol. (london) 117, 500 (1952).
5. R. FitzHugh *Impulse and physiological states in models of nerve membrane* 1961. Biophysics J., Vol. 1: 445-466.
6. J.S. Nagumo, S. Arimoto, S. Yoshizawa (1962). *An active pulse transmission line stimulating nerve axon.* 1962. Proc. IRE: vol.50:
7. M.-P. Zorzano, L. Vazquez. Emergence of Synchronous Oscillations in Neural Networks Excited by Noise. To be published in Physica D, 2003.

RESONANT BREATHER IN COUPLED JOSEPHSON JUNCTIONS SYSTEM

A. BENABDALLAH
Institut für Physik, TU-Chemnitz, D-09107 Chemnitz, Germany.
E-mail: benabdallah@physik.tu-chemnitz.de

Since the dynamics of the breathers in a Josephson ladder is complex [3,4,5] we carry out in Ref. [1,2] a consistent study of a *single plaquette model* containing three junctions. It consists as shown in the left panel of figure 1 of two *vertical* Josephson junctions parallel to the bias direction and a *horizontal* one that is transversal to the bias. Such a single plaquette model is the simplest system which supports permutationally broken states in the presence of a homogeneous dc bias γ and magnetic field H.
The dynamics of a single plaquette of three Josephson junctions is determined by the time dependent Josephson phases of vertical junctions $\phi_{1,2}^v$, and the horizontal junction ϕ^h [1]

$$N(\phi_1^v) = \gamma + I_m \;, \quad N(\phi_2^v) = \gamma - I_m + I_{local} \;, \quad N(\phi^h) = I_m/\eta \quad (1)$$
$$\phi_1^v - \phi_2^v + \phi^h + 2\pi f = -\beta_L I_m \;,$$

where the nonlinear operator $N(\phi) = \ddot{\phi} + \alpha\dot{\phi} + \sin(\phi)$, α is the damping. In the Eq. (1) γ is the bias current, β_L is the self-inductance of the cell, η is the anisotropy and f is the frustration due to the external magnetic field. In the absence of this later on can expect in a time-averaged only four different states to be realized: superconducting state (SS) .i.e. all phase are static, homogeneous whirling state (HWS) i.e. ϕ_1^v, ϕ_2^v are in the resistive state ϕ^h is static and finally two breather states (BS) .i.e one of the vertical phases (ϕ_1^v or ϕ_2^v) oscillates, the other one and ϕ^h are in the resistive state. Using a *dc analysis* [1] one can obtain a condition of switching from (BS) to (HWS) $\gamma_{b,HWS} > \frac{1+\eta}{1+2\eta}$ and from (BS) to (SS) $\gamma_{b,r} < \frac{4}{\pi}\alpha(1+\eta)$ (See the right panel of the figure 1).
The linearization of the Eq. (1) around the breather state yield to electromagnetic oscillations frequency $\omega_{\pm}$. The necessary condition of the appearance of the resonances in the *I*-*V* curves is the matching of the breather frequency Ω or its higher harmonics with the frequencies of EO's, $\omega_{\pm}$ (see Fig. 1). In Ref. [2] we show that the applied magnetic field allows to control the strength of the resonance. For example the primary resonance ($\omega_{\pm} = m\Omega$) can be enhanced or enlarges by application of the external

magnetic field (see the resonance labeld B in Fig.2). The magnitude of the resonant step is proportional to

$$\Delta\gamma = \left|\sin\left[\frac{1}{2}\text{asin}\left(\frac{1+2\eta}{1+\eta}\gamma\right) + \frac{\beta_L\gamma\eta}{2(1+\eta)} + \pi f\right]\right| .$$

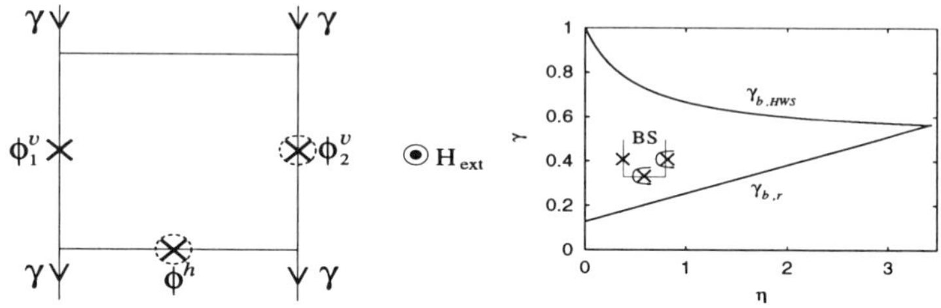

Figure 1. Left panel: Sketch of the plaquette with three Josephson junctions (marked by crosses) in the presence of uniform dc bias γ and an externally applied magnetic field H_{ext}. Dashed circles denote junctions in the resistive (whirling) state. Right panel: Analytic boundaries of the breathers existence in (γ, η) plane for $\beta_L = 0$, $\alpha = 0.1$. The dc analysis was used.

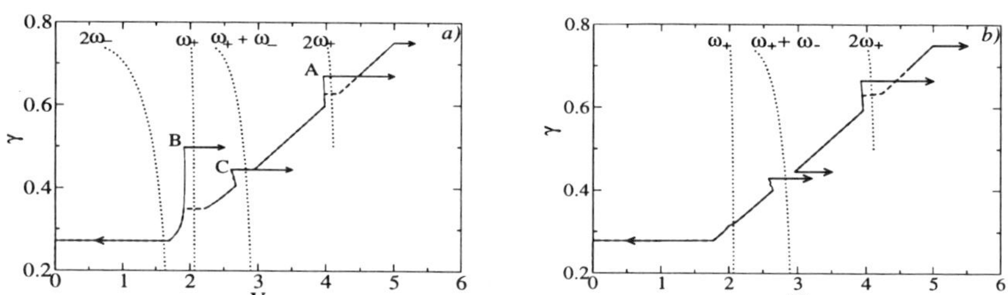

Figure 2. *I-V* characteristics of a breather state for $\beta_L = 1$, $\alpha = 0.1$ and $\eta = 0.5$. The values of the magnetic field are (a) $f = 0$; (b) $f = -0.1$. The dotted lines show the dependence of the characteristic frequency combinations of EOs on the dc bias γ . Arrows indicate the various switching processes.

References

1. A. Benabdallah, M. V. Fistul and S. Flach, *Physica D* **159** 202 (2001).
2. M. V. Fistul, S. Flach and A. Benabdallah, *Phys. Rev. E* **65** 046616 (2002).
3. P. Binder et al. *Phys. Rev. Lett.* **84** (2000) 745.
4. E. Trías et al. *Physica D* **156** (2001) 98.
5. A. E. Miroshnichenko et al. *Phys. Rev. E* **64** (2001) 066601.
6. F. Pignatelli and A. V. Ustinov, submitted to Phys. Rev. E (2002).

APPROXIMATION OF BREATHERS IN 1-DIMENSIONAL LATTICES USING HOMOCLINIC ORBITS

J.M. BERGAMIN

Department of Mathematics,
University of Patras,
26500 Patras, Greece
E-mail: jeroen_bergamin@hotmail.com

Homoclinic orbits of maps and breathers are intimately related. To obtain low-dimensional maps whose homoclinic orbits approximate breather profiles a more accurate alternative to the rotating wave approximation is proposed, building on the work by G.P. Tsironis.

1. General idea

The equations of motion considered are

$$\ddot{u}_n + V'(u_n) = W'(u_{n+1} - u_n) - W'(u_n - u_{n-1})$$

where the u_n denote the amplitudes of oscillators on a 1-dimensional lattice with on-site potential $V(x)$ and coupling potential $W(x)$.

Making the change of variables $u_n = a_n T_n(t)$ (with $T_n(0) = 1$ and $\dot{T}_n(0) = 0$) a discrete breather is defined as a solution for which $a_n \to 0$ as $|n| \to \infty$ (i.e. spatially localized, definition of a homoclinic orbit) and $T_n(t + \tau) = T_n(t)$ (i.e. time-periodic with period τ).

The amplitudes a_n are related through an infinite dimensional map[1]. Approximations lead to low-dimensional maps. One can for instance look at only a single Fourier mode (the rotating wave approximation, see e.g. Bountis et al.[2]). Here, I will show how an extension of a method by Tsironis[3] leads to more accurate approximations to breather profiles.

To calculate u_n we assume that variables can be separated as follows[3]: 1) $u_{n+1} - u_n \approx (a_{n+1} - a_n)T_n$ and 2) $u_n - u_{n-1} \approx (a_n - a_{n-1})T_n$. A logical thought might be to write this as $T_{n+1} \approx T_{n-1} \approx T_n$ but this is not fulfilled by general breather solutions. In a forthcoming article an alternative justification for this approximation will be presented[5].

Using this approximation the differential equations are reduced to

$$a_n\ddot{T}_n + V'(a_nT_n) = W'((a_{n+1} - a_n)T_n) - W'((a_n - a_{n-1})T_n)$$

which is an ordinary differential equation for T_n.

We propose the following strategy. Since we choose the period τ, knowledge of a_n and a_{n-1} (or a_{n+1}) is enough to find the value of a_{n+1} (or a_{n-1}) for which the time-function T_n has period τ. Thus, we have an invertible mapping $a_{n+1} = f(a_n, a_{n-1}; \tau)$ and we are able to obtain homoclinic orbits using the methods presented in a recent article[4].

2. An application: Klein-Gordon lattice with ϕ^4 potential

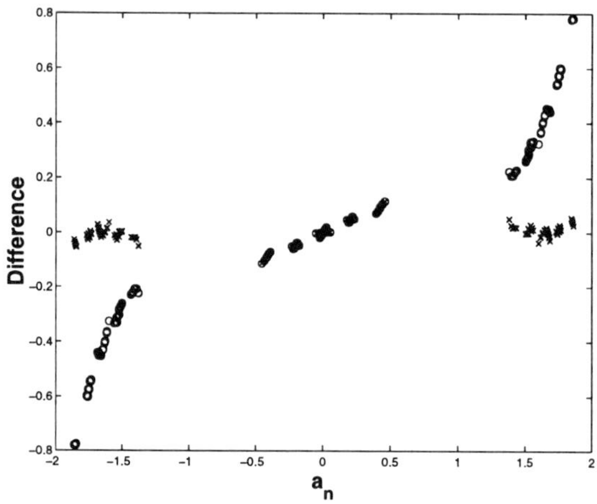

Figure 1. Homoclinic orbits were calculated for the potentials $V(x) = \frac{3}{20}x^2 + \frac{1}{4}x^4$, $W(x) = \frac{1}{8}x^2$ and period $\tau = \frac{2\pi}{\sqrt{2.8}}$. We show the difference between the function values $f(u_n(0))$ predicted by our method (crosses) and the rotating wave approximation (circles) compared to the actual breather solutions obtained by using a Newton method with the homoclinic orbits as initial approximations. The values predicted with the new method are within 5% of the data for large amplitudes, whereas the maximum error made in the rotating wave approximation is about 40%.

References

1. S. Flach, *Phys. Rev. E* **51** 3579 (1995).
2. T. Bountis, H.W. Capel, M. Kollmann, J.C. Ross, J.M. Bergamin and J.P. van der Weele, *Phys. Lett. A* **268** 50 (2000).
3. G.P. Tsironis, *J. Phys. A* **35** 951 (2002).
4. J.M. Bergamin, T. Bountis and M.N. Vrahatis, *Nonlinearity* **15** (2002) 1603.
5. J.M. Bergamin, "*Numerical approximation of breathers in lattices with nearest-neighbor interactions*", to appear in *Phys. Rev. E* (2003).

QUANTUM STATISTICAL MECHANICS OF FRENKEL–KONTOROVA MODELS

NUNO R CATARINO & ROBERT S MACKAY

Mathematics Institute, University of Warwick,
Gibbet Hill Road, Coventry CV4 7AL, UK
E-mail: catarino@maths.warwick.ac.uk, E-mail: mackay@maths.warwick.ac.uk

We generalise the classical Transition by Breaking of Analyticity in the Frenkel–Kontorova model to non-zero Planck's constant and temperature. This analysis is based on the study of a renormalization operator for the case of irrational mean spacing using Feynman's functional integral approach.

1. Introduction

The *Frenkel–Kontorova model* (FK) is a one-dimensional lattice model exhibiting incommensurate structures. It is a system of elastically coupled particles in an external periodic potential (a discrete version of sine–Gordon) with Lagrangian

$$L[x] = \sum_{n \in \mathsf{Z}} \left\{ \frac{m\dot{x}_n^2}{2} - \mathsf{v}[x_n, x_{n+1}] \right\}, \tag{1}$$

where

$$\mathsf{v}[x, x'] = \frac{w}{2}(x'-x)^2 + p(x'-x) + \frac{u}{4\pi^2}\cos[2\pi x]. \tag{2}$$

Here, the period of the onsite potential has been scaled to one. The mass m and coupling constant w can also be scaled to one.

For $u = 0$ its *minimum energy configurations* (i.e. $x \in \mathsf{R}^{\mathsf{Z}}$ such that $\forall M < m < N, V_{M,N}(x) \equiv \sum_{n=M}^{N-1} \mathsf{v}(x_n, x_{n+1})$ is minimum for all variations of x_m) are arrays of equally spaced particles and the *mean spacing* ρ is simply $-p/w$. In the opposite limit, $w = 0$, all particles are in the minima of the onsite potential and the mean spacing is the closest integer to $-p/w$ (or any value in between the two if non-unique).

An interesting set of codimension-2 critical points occur between these two regimes (often called *Transition by Breaking of Analyticity*, TBA): in the space of parameters (u, p), for each irrational ρ there there is a curve $p = p_c(u)$ of mean spacing ρ containing one critical value u_c (for other potentials there may be more than one u_c). The regime for u less than u_c is called *subcritical* (or *sliding phase*) and above u_c *supercritical* (or *pinned phase*) [1].

We call successive pairs (x_n, x_{n+1}) *bonds* and $\mathsf{v}(x_n, x_{n+1})$ the *energy of the bond.* Considering in particular the case of mean spacing $\rho = \gamma^{-1}$ where $\gamma = \frac{1+\sqrt{5}}{2}$ is the golden mean, relabeling the bonds v appropriately [2] as τ and υ results in a Fibonacci sequence of τs and υs. The TBA point can be viewed as a fixed point of the renormalization operator that minimises the energy of the sum of successive υ and τ-bonds with suitably chosen space and energy scalings, $\alpha, J \in \mathsf{R}$ respectively (depending on (υ, τ)) (actually it is also necessary to subtract a constant and a quadratic coboundary but we will suppress reference to these inessential terms).

The renormalization operator has a nontrivial fixed point (υ^*, τ^*) with $\alpha \simeq -1.414\,836\,0$ and $J \simeq 4.399\,143\,9$. This fixed point corresponds to the critical u_c along the curve in (u, p) for $\rho = \gamma^{-1}$, the transition point between the subcritical and supercritical regimes. It has two unstable directions: one along the curve of constant mean spacing, with eigenvalue $\delta \simeq 1.627\,950\,0$ and the other transverse to this curve in the (u, p) plane (which we'll call the p direction), with eigenvalue $\eta = -J/\gamma \simeq -2.681\,738\,4$ [2].

2. Phonon Spectrum

The *phonon spectrum* of a configuration is the set of frequencies of the linearised vibrations about it. For the FK model with irrational ρ the phonon spectrum includes 0 in the subcritical regime, but the minimum phonon frequency becomes positive above the critical point [1] u_c.

The classical action for time-periodic solutions of the FK model (of period T) is

$$S(x;T) = \sum_{n \in \mathsf{Z}} s(x_n, x_{n+1}; T), \qquad s(x, x'; T) = T \int_0^1 \left\{ \frac{\dot{x}^2}{2T^2} - \mathsf{v}[x, x'] \right\} dt, \quad (3)$$

where time, t, has been rescaled to make the period of the loops be 1. The equations of motion are given by stationarising the action with respect to changes in $x(t)$, keeping boundaries fixed, and phonons are (time) periodic solutions of the linearisation of these equations. Consider $S\left(x; \frac{2\pi}{\Omega}\right)$, where Ω is a potential phonon frequency and x now represents a period-1 loop in R^{Z}. To extend the renormalization method to study the phonon spectrum, $\mathcal{R}$ will now have to act on pairs of functionals $s\left(x, x'; \frac{2\pi}{\Omega}\right)$, so call an s-bond a τ or an υ according to the minimum energy case. The renormalization operator $\mathcal{R}$ can be extended (including one further scale, the phonon frequency scale) to

$$\mathcal{R}\left[(\tau, \upsilon)(x, x'; \Omega)\right] = J\varepsilon\left(\upsilon \oplus \tau, \tau\right)\left(x/\alpha, x'/\alpha; \Omega/\varepsilon\right). \tag{4}$$

where the operator $\oplus$ is

$$(\upsilon \oplus \tau)(x, x'; \Omega) = \min_{z}\left[\upsilon(x, z; \Omega) + \tau(z, x'; \Omega)\right], \tag{5}$$

and z now runs over period-1 functions $z(t)$.

Apart from the phonon spectrum scaling, ε, in the limit of $\Omega \to 0$ the operator is MacKay's minimum energy renormalization operator (applied pointwise to loops). It has therefore a fixed point in the $\Omega = 0$ subspace consisting of $\tau(x, x') = \int_0^1 dt\, \tau^*(x(t), x'(t))$ and $\upsilon(x, x') = \int_0^1 dt\, z\upsilon^*(x(t), x'(t))$. It was found numerically to have an additional unstable direction with eigenvalue [2] $\varepsilon = 1.649\,415$ (affecting the linearised dynamics), which we will call the Ω direction, whose origins we don't yet fully understand.

Thus the phonon spectrum (as a function of u, p) is asymptotically self-similar under scaling by ε, δ and η in the three directions (Ω, u, p). In particular the *phonon gap* (i.e. the minimum of the phonon spectrum) reduces by a factor ε when u approaches u_c by a factor δ from above along the curve $\rho =$ constant.

3. Quantum Effects

3.1. *Zero temperature*

The trace of the propagator for the lowest frequency phonons in the FK model is

$$\int \mathcal{D}[x]\, e^{\frac{i}{\hbar}\sum_n s\left(x_n, x_{n+1}; \frac{2\pi}{\Omega}\right)}, \tag{6}$$

where the Feynman functional integration is over closed loops and includes the endpoints. In stationary phase approximation (small $\hbar$) the integral is dominated by contributions of the stationary points of S; these are those functions which satisfy the classical equations of motion and are periodic with period T.

Consider pairs of functionals υ and τ as before, extend the operator $\oplus$ to non-zero $\hbar$ by

$$(\upsilon \oplus \tau)(x, x'; \Omega) = -i\hbar \log \int \mathcal{D}[z]\, e^{\frac{i}{\hbar}\left[\upsilon(x, z; \Omega) + \tau\left(z, x'; \Omega\right)\right]}, \tag{7}$$

and extend also $\mathcal{R}$ to non-zero $\hbar$ with this new $\oplus$. It can be easily seen (keeping only the first terms in stationary phase) that in the limit $\hbar \to 0$

the operator $\oplus$ becomes the previous one for the phonon spectrum. If τ, υ and $\hbar$ are multiplied by a constant λ, then $\upsilon \oplus \tau$ is simply multiplied by the same constant λ, so the fixed point of the renormalization operator for the phonon spectrum extends to one at $\hbar = 0$ for this new $\mathcal{R}$ with a new unstable eigenvalue $\kappa = J\varepsilon \simeq 7.157\,051$ in the quantum direction. So the effect of including the quantum direction, close to the fixed point, is that $\mathcal{R}$ scales asymptotically like

$$\mathcal{R}(\tau, \upsilon)\left(\Delta u, \Delta p, \Omega^{-1}, \hbar\right) = (\tau, \upsilon)\left(\delta\Delta u, \eta\Delta p, \varepsilon\Omega^{-1}, \kappa\hbar\right), \tag{8}$$

for $\Delta u \equiv u - u_c$ and $\Delta p \equiv p - p_c(u)$ and $\Omega, \hbar$ small.

3.2. *Statistical mechanics*

The quantum partition function can be easily obtained from the trace of the quantum propagator by Wick rotation [3] (i.e. putting $T = -i\beta\hbar$ in equation (6)). The quantum partition function is then

$$Z(\beta, \hbar) = \int \mathcal{D}[x]\, e^{-\beta S_E(x;\beta\hbar)}, \tag{9}$$

where the Euclidean action S_E for the FK model is now the sum (instead of the difference) of the 'kinetic' and potential contributions. This gives another possible direction to which renormalization can be extended, the temperature direction, $\Theta = 1/\beta$. Redefine then the operator $\oplus$ by 'Wick rotating' the operator $\oplus$ for the trace of the propagator:

$$(\upsilon \oplus \tau)(x, x'; \beta, \hbar) = -\beta \log \int \mathcal{D}[z]\, e^{-\beta^{-1}\left[\upsilon(x,z;\beta\hbar)+\tau\left(z,x';\beta\hbar\right)\right]}, \tag{10}$$

Therefore, close to the fixed point of the associated renormalization operator, we have

$$\mathcal{R}(\tau, \upsilon)(\Delta u, \Delta p, \beta\hbar, \hbar) \sim (\tau, \upsilon)(\delta\Delta u, \eta\Delta p, \varepsilon\beta\hbar, \kappa\hbar), \tag{11}$$

so under the renormalization $\mathcal{R}$ there is a fixed point at $\hbar = 0$, $\beta\hbar = \infty$, with an unstable eigenvalue of $\kappa/\varepsilon = J \simeq 4.339\,143\,9$ in the temperature direction. This agrees with the result found by MacKay [4] for the classical statistical mechanics.

Acknowledgments

NRC was funded by the Portuguese institution 'Fundação para a Ciência e Tecnologia'.

References

1. S.Aubry & P.Y.Le Daeron, *Physica D* **8**, 381 (1983)
2. R.S.MacKay, *Physica D* **50**, 71 (1991)
3. J.W.Negele & H.Orland *Quantum Many-Particle Systems* (Perseus, 1998)
4. R.S.MacKay, *J. Stat. Phys.* **80**, 45 (1995)

DISCRETE BREATHERS IN 2D JOSEPHSON ARRAYS

J. J. MAZO

Departamento de Física de la Materia Condensada and ICMA
CSIC- Universidad de Zaragoza
E-50009 Zaragoza, Spain
E-mail: juanjo@posta.unizar.es

We have predicted theoretically and confirmed numerically the existence of discrete breathers (intrinsic localised modes) in the dynamics of a two-dimensional Josephson-junction array biased by radio-frequency fields. The solutions are linearly stable in the framework of the Floquet theory and robust in the presence of thermal fluctuations. We have also discussed the conditions for realizing an experimental detection of these modes.

Two-dimensional Josephson-junction arrays (2DJJA) are paradigmatic experimental systems for the study of many physical phenomena [1]. As realizations of the XY model, they have been designed for the study of phase transitions in unfrustrated and frustrated two-dimensional systems. Since they are tailored arrays, they have helped us to understand the role of geometry and disorder in granular superconductors. Modeled by coupled pendula, they are important to understand problems of synchronization of oscillators in complex lattices. Since they present vortices and anti-vortices, from these arrays we have learnt on the behavior of those types of nonlinear coherent excitations in the presence of ac and/or dc perturbations, and their role in equilibrium and non-equilibrium phase transitions.

A different type of coherent localized excitations in nonlinear lattices are the so called discrete breathers (DB's) or intrinsic localized modes [2]. Physically they are dynamical solutions for which energy remains sharply localized in a few sites of the array. Then, there is not significant radiation of this energy to the rest of the lattice.

DB's have been mainly studied in lattices in one dimension and only experimentally found in some quasi-one dimensional systems. One of such system is an underdamped JJ ladder array biased by dc external currents. In these superconducting networks, the localized states are localized voltage solutions: not all of the junctions have the same voltage although they are all coupled, and driven by the same current. Following theoretical predictions [3], DB's in the ladder were shown to exist for a wide range of parameter values, excited at will and detected by local voltage measurements and a low temperature laser scanning microscopy [4].

An interesting issue to address is the role and the possible detection of such excitations in the dynamics of 2D lattices. We have studied this problem and showed numerical evidence for the existence of DB's in the ac dynamics of a 2DJJA [5] (see figures). We proposed this system as an adequate device to carry out the experimental detection of DB's in a 2D system. About the possible experimental detection of such modes, arrays with the desired parameters can be easily made and the external driving is similar to that used in standard voltage designs. We have also proposed, and checked numerically, a method to excite such solutions in the array. The experimental detection of the mode may be done by the use of local voltage probes.

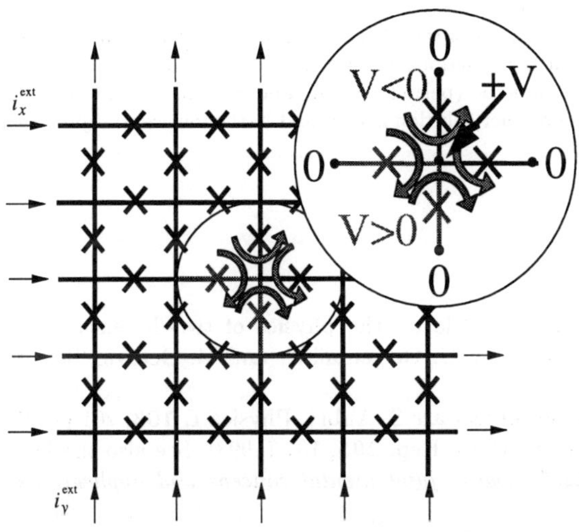

Figure 1. Sketch of the 2DJJA with breather. Crosses represent junctions in the array. The breather is a localized voltage solution: four junctions sited in the bulk of the array are in the resistive state, two with voltages $+V$ and two with $-V$, while the rest follow the ac field in a oscillating state of mean voltage $V = 0$.

Acknowledgments

Financial support is acknowledged to DGES PB98-1592, MCyT and FEDER BFM 2002-00113 and European Network LOCNET HPRN-CT-

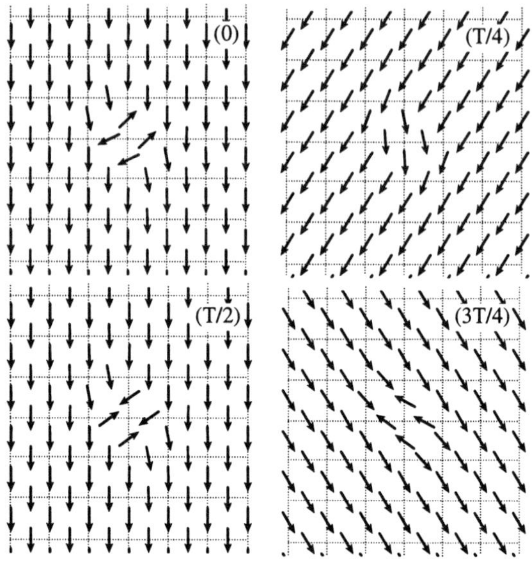

Figure 2. Four snapshots of the phase evolution for a DB solution in a 2DJJA. Arrows represent phases of the junctions and we show the central part of a 11 × 11 array. The solutions are time periodic (period T) and time increases form 0 (arbitrary) to $3T/4$ as labeled. The rotobreather solution is localized in the four central junctions which rotate.

1999-00163.

References

1. A comprehensive article on the physics of two-dimensional arrays is R. S. Newrock, C. J. Lobb, U. Geigenmüller and M. Octavio, Sol. State Phys. **54**, 263 (2000).
2. Reviews on the subject are S. Aubry, Physica D **103**, 201 (1997) and S. Flach and C. R. Willis, Phys. Rep. **295**, 181 (1998). See also the Focus Issue, *Nonlinear localized modes: fundamental conceps and applications*, Chaos **13**(2) (2003).
3. L. M. Floría, J. L. Marín, P. J. Martínez, F. Falo and S. Aubry, Europhys. Lett. **36**, 539 (1996). S. Flach and M. Spicci, J. Phys.: Condens. Matt. **11**, 321 (1999). J. J. Mazo, E. Trías and T. P. Orlando, Phys. Rev. B **59** 13604 (1999).
4. E. Trías, J. J. Mazo and T. P. Orlando, Phys. Rev. Lett. **84**, 741 (2000). P. Binder, D. Abraimov, A. V. Ustinov, S. Flach, and Y. Zolotaryuk, Phys. Rev. Lett. **84**, 745 (2000). P. Binder, D. Abraimov and A. V. Ustinov, Phys. Rev. E **62**, 2858 (2000). E. Trías, J. J. Mazo, A. Brinkman and T. P. Orlando, Physica D **156**, 98 (2001).
5. J. J. Mazo. Phys. Rev. Lett. **89**, 234101 (2002).

FLUXON RATCHET POTENTIALS

J. J. MAZO AND F. FALO
Departamento de Física de la Materia Condensada and ICMA
CSIC- Universidad de Zaragoza
E-50009 Zaragoza, Spain
E-mail: juanjo@posta.unizar.es

T. P. ORLANDO
Department of Electrical Engineering and Computer Science,
Massachusetts Institute of Technology,
Cambridge, Massachusetts 02139

The study of fluxons in Josephson-junction parallel arrays provides an excellent opportunity to check the predictions of the theory about directional motion of particles in ratchet potentials. We have shown that quite simple designs on such a JJ arrays produce controllable potential profiles. The experiments nicely fit the main predictions of the theory.

The possibility of producing a directional motion of particles without unbalanced fields, is an active field of research which is rapidly moving from basic to applied physics [1]. Although first attempts have been done to apply the underlying ideas to biomolecular motors, the most promising applications belong to the world of nano- and micro- technologies. At molecular level some interesting designs have been proposed but it seems that an easier experimental realization could be achieved at the scale of micrometers. Towards this aim, devices based on the Josephson effect are being proposed and studied [2,3]. In most cases the effect is realized by a "ratchet" potential, that is, an spatially asymmetric potential.

Josephson arrays are excellent systems for the predictions of non-linear dynamics theory. For example, parallel arrays are experimental realizations of the Frenkel-Kontorova or discrete sine-Gordon model. Then, a fluxon trapped in the array behaves like a kink in this system and experiences a spatially periodic pinning potential. The feasibility to fabricate micro-electronic circuits with any desired geometry and a broad range of physical parameters, allows one to obtain different potential profiles for the fluxon, for instance ratchet ones.

Consider a fluxon trapped in a one-dimensional parallel array of Josephson junctions with the geometry of a ring. Different configurations of critical currents and cell areas result in different profiles for the fluxon potential. We have studied the minimal conditions to achieve an effective potential for the fluxon in which mirror symmetry is absent, *a fluxon ratchet potential*, designed such arrays and experimentally confirm the theoretical predictions [3]. It is important to highlight that all interactions in the system are symmetric in the field variables (superconducting phases). Thus, the inversion symmetry breaking has to be geometrical, i.e. using the possibility of spatial variations of the array parameters. We have available two set of parameters to play with: plaquette self-inductances and junction critical currents. One possibility for designing arrays which display fluxon ratchets potentials corresponds to the choice of junctions with two different critical currents and cells with two different areas. We calculated the dynamical properties of a fluxon in such device, designed and fabricated the array and the ratchet character of the potential was experimentally shown [3].

Acknowledgments

Financial support is acknowledged to DGES PB98-1592, MCyT and FEDER BFM 2002-00111, European Network LOCNET HPRN-CT-1999-00163, the Commission for Cultural, Educational and Scientific Exchange between the United States of America and Spain and NSF Grant DMR-9988832.

References

1. A recent popular article about ratchets is: R. D. Astumian, Scientific American, p.57, July 2001. For a review see P. Reimann, Phys. Rep. **361**, 57 (2002). For minireviews and an impression of the state-of-the-art see the special issue "Ratchets and Brownian motors: Basics, experiments and applications", Appl. Phys. A **75**, 167-352 (2002).
2. I. Zapata, R. Bartussek, F. Sols and P. Hänggi, Phys. Rev. Lett. **77**, 2292 (1996). G. Carapella and G. Costabile, Phys. Rev. Lett. **87**, 077002 (2001). S. Weiss, D. Koelle, J. Müller, R. Gross, and K. Barthel, Europhys. Lett. **51**, 499 (2000). C.S. Lee, B. Janko, I. Derényi and A.L. Barabasi, Nature **400**, 337 (1999). J. F. Waunbaugh, C. Reichardt, C. J. Olson, F. Marchesoni and F. Nori. Phys. Rev. Lett. **83**, 5106 (1999).
3. F. Falo, P.J. Martínez, J.J. Mazo, and S. Cilla. Europhys. Lett. **45**, 700 (1999). E. Trías, J.J. Mazo, F. Falo y T.P. Orlando. Phys. Rev. E **61**, 2257 (2000). F. Falo, P.J. Martínez, J. J. Mazo, T. P. Orlando, K. Segall and E. Trías . Appl. Phys. A **75**, 263 (2002).

REGIONS OF STABILITY FOR AN EXTENDED DNLS EQUATION

MICHAEL ÖSTER* AND MAGNUS JOHANSSON†

Department of Physics and Measurement Technology (IFM), Linköping University, S-581 83 Linköping, Sweden

A realistic model equation governing the amplitude of the electric field in a one-dimensional array of nonlinear optical waveguides with nearest-neigbour coupling is derived. The equation is an extension of the discrete nonlinear Schrödinger equation, which previously has been the main model for such systems. Attention is turned towards localised solutions and investigations are made from the viewpoint of the theory of discrete breathers. Calculations for one-site and two-site stationary discrete breathers are made for the model equation and the linear stability is investigated resulting in maps of different regions of stability. Boundaries of stability inversion are discovered, suggesting the existence of high-mobility narrow solutions with potential application to switching.

1. Introduction

We consider a one-dimensional array of optical waveguides embedded in a nonlinear Kerr material, i.e. a material where the refractive index depends on the intensity of the field. Such systems have previously mainly been studied using the discrete nonlinear Schrödinger (DNLS) equation, or variants thereof, as a model equation (see the paper by Eilbeck and Johansson in these proceedings for a survey on the DNLS equation). Both self-trapped solutions and travelling localised solutions, which for example can be used for optical switching, have been found [1,2,3,4,5,6]. Going beyond the approximations leading to the DNLS equation and including inter-site nonlinearities, we derive a realistic model equation governing the amplitude of the electric field in the waveguides.

*e-mail: micos@ifm.liu.se

†e-mail: mjn@ifm.liu.se

2. Model

Consider an array of waveguides and the modes guiding the amplitude of the fields in each waveguide. If the fields have a preferred direction of polarisation and are decaying sufficiently fast outside the waveguides we can assume that the mode of each waveguide is real and make a nearest-neighbour approximation. Using coupled-mode theory [5,7] to relate the fields and the polarisation in the nonlinear material the resulting equation is [8]

$$\begin{aligned} i\frac{\mathrm{d}A_n}{\mathrm{d}z} &= Q_1 A_n + Q_2\left(A_{n-1}+A_{n+1}\right) + 2Q_3 A_n|A_n|^2 \\ &\quad +2Q_4\left(2A_n\left(|A_{n-1}|^2+|A_{n+1}|^2\right) + A_n^*\left(A_{n-1}^2+A_{n+1}^2\right)\right) \\ &\quad +2Q_5\left(2|A_n|^2\left(A_{n-1}+A_{n+1}\right)+A_n^2\left(A_{n-1}^*+A_{n+1}^*\right)\right. \\ &\qquad \left.+A_{n-1}|A_{n-1}|^2+A_{n+1}|A_{n+1}|^2\right). \end{aligned} \tag{1}$$

Here A_n is the complex amplitude of the field in waveguide n, z is a variable measuring distance along the waveguides and $\{Q_i\}$ are constants depending on the geometry of the waveguides and the overlap of the modes. For $Q_4 = Q_5 = 0$ the equation reduces to the DNLS equation. These constants can however not be neglected when the penetration length of the electric field in the nonlinear material is large or the waveguides are closely spaced.

An important feature of Eq. (1) is that it possesses conserved quantities. One is the Hamiltonian [8,9]

$$\mathcal{H} = \mathcal{H}_1 + \mathcal{H}_2 + \mathcal{H}_2^* = \mathcal{H}_1 + 2\Re\left(\mathcal{H}_2\right), \tag{2}$$

$$\mathcal{H}_1 = \sum_n \left[Q_1|A_n|^2 + Q_3|A_n|^4\right], \tag{3}$$

$$\begin{aligned} \mathcal{H}_2 = \sum_n &\left[Q_2 A_n A_{n+1}^* + Q_4\left(2|A_n|^2|A_{n+1}|^2 + A_n^2 A_{n+1}^{*\,2}\right)\right. \\ &\left.+2Q_5 A_n A_{n+1}\left(A_n^{*\,2}+A_{n+1}^{*\,2}\right)\right], \end{aligned} \tag{4}$$

from which Eq. (1) can be derived by introducing the complex canonical variables (A_n, iA_n^*) and the Hamilton equations of motion,

$$i\frac{\mathrm{d}A_n}{\mathrm{d}z} = \frac{\partial\mathcal{H}}{\partial A_n^*}, \qquad -i\frac{\mathrm{d}A_n^*}{\mathrm{d}z} = \frac{\partial\mathcal{H}}{\partial A_n}. \tag{5}$$

A second conserved quantity is the norm

$$\mathcal{N} = \sum_n |A_n|^2, \tag{6}$$

which corresponds to conservation of (Poynting) power along the waveguides. The conservation of norm is connected to the phase invariance of

Eq. (1), i.e. the fact that if $\{A_n\}$ is a solution so is $\{A_n e^{i\phi}\}$ for $\forall\phi \in \mathbb{R}$. As a consequence the constant Q_1, which connects the norm and the Hamiltonian, can be made to vanish by the simple substitution $A_n \mapsto A_n e^{-iQ_1 z}$. Because of the phase invariance Eq. (1) also supports an important class of solutions with harmonically oscillating amplitude. For a solution with fixed frequency ω we make a transformation to a rotating frame of reference. The transformation $A_n(z) = \sqrt{|w|/2|Q_3|}B_n e^{i(\omega - Q_1)z}$ will do this, as well as reduce the number of independent parameters. The resulting stationary equation is

$$\begin{aligned} 0 = {} & B_n + K_2\left(B_{n-1} + B_{n+1}\right) - B_n|B_n|^2 \\ & +2K_4\left(2B_n\left(|B_{n-1}|^2 + |B_{n+1}|^2\right) + B_n^*\left(B_{n-1}^2 + B_{n+1}^2\right)\right) \\ & +2K_5\left(2|B_n|^2\left(B_{n-1} + B_{n+1}\right) + B_n^2\left(B_{n-1}^* + B_{n+1}^*\right)\right. \\ & \qquad \left. +B_{n-1}|B_{n-1}|^2 + B_{n+1}|B_{n+1}|^2\right) , \end{aligned} \tag{7}$$

where it has been assumed that ω and Q_3 have different signs. The constants are given by

$$K_2 = \frac{Q_2}{\omega}, \qquad K_4 = \frac{Q_4}{2|Q_3|}\operatorname{sgn}(\omega), \qquad K_5 = \frac{Q_5}{2|Q_3|}\operatorname{sgn}(\omega)\,. \tag{8}$$

Noteworthy is also that a solution $\{B_n\}$ for (K_2, K_4, K_5) can be transformed to a solution for $(-K_2, K_4, -K_5)$ by the substitution $B_n \mapsto (-1)^n B_n$. This reduces the part of parameter space that must be investigated to get a complete picture of the equation.

3. Results

Here we will restrict our attention to localised solutions and especially discrete breathers (DB), which are time-periodic spatially localised solutions. The time-dependence is here replaced with the distance z along the waveguides. DB exist in Hamiltonian systems provided an anharmonicity condition and a non-resonance condition with linear phonons is fulfilled [10], which for Eq. (7) corresponds to the inequality $|K_2| < 1/2$. Concerning the property of the solutions a lower bound on the amplitude of non-zero stationary solutions of Eq. (7) can be derived using a contraction mapping theorem [9],

$$\sup_{n\in\mathbb{Z}} |B_n| \geq \sqrt{\frac{1 - 2|K_2|}{3 + 36|K_4| + 48|K_5|}}\,. \tag{9}$$

Note that the bound becomes zero at the edge of the linear spectrum, suggesting that localised solutions can approach zero in this limit. Further it is possible to obtain exact *compact* stationary solutions, i.e. solutions which are strictly zero outside an interval. For such solutions, where $B_{m-1} = 0$ and $B_m \neq 0$, we find from Eq. (7) at site $n = m - 1$,

$$K_{\text{eff}} \stackrel{def}{=} K_2 + 2K_5|B_m|^2 = 0\,. \tag{10}$$

The quantity K_{eff}, when zero, describes a decoupling of sites $n < m$ from sites $n \geq m$. By direct substitution it is clear that a unit amplitude one-site-only solution satisfies Eq. (7) when $K_2 + 2K_5 = 0$. On either side of this plane in parameter space there will in general be solutions corresponding to different effective coupling. It is noted that the sign of the effective coupling can easily be determined by the criterion $\text{sgn}(K_{\text{eff}}) = -\text{sgn}(B_{m-1}/B_m)$, whenever $|B_{m-1}|^2 \ll |B_m|^2$. Also two-site-only (and more) solutions can be obtained by direct substitution in Eq. (7). Compact solutions have previously been observed in DNLS-type models, but only in models with apparently little physical relevance [11].

For numerical calculation of DB we use the method of continuation from the anti-continuous limit [12], which is $K_j = 0, j = 2,4,5$ for Eq. (7). Extensive calculations, covering a large portion of parameter space, have been made for 1-site and symmetric/anti-symmetric 2-site solutions, i.e. solutions that in the anti-continuous limit have a single site excited or two sites excited with phases aligned or anti-aligned respectively. Primarily the stability properties have been investigated. The problem of deciding linear stability of a stationary solution is by standard methods easily reduced to a matrix eigenvalue problem [4,9]. Since the system is Hamiltonian all eigenvalues must lie on the imaginary axis for linear stability.

In Fig. 1(a) the stability regions (shaded) for the 1-site solutions are shown for $K_2 = 0.2$. All instabilities in the figure are due to eigenvalues leaving the imaginary axis at the origin along the real axis, but following the compact solution to smaller K_4 a small region of complex (Krein) instabilities is encountered. The dotted line indicates where the compact 1-site solution exists, i.e. where $K_{\text{eff}} = 0$. In region I only solutions with $B_{m\pm1}/B_m > 0$ exist and in region II only solutions with $B_{m\pm1}/B_m < 0$. In region III there exist three different solutions all corresponding to the same solution in the anti-continuous limit (see Fig. 1(b)). The three solutions originate from either region I or II or from the compact solution at the dotted line. The latter solution coincides with the former at the dash-dotted lines, which indicates where continuation is stopped numerically. At the left boundary solutions with $B_{m\pm1}/B_m < 0$ are lost coming from the right and at the right boundary solutions with $B_{m\pm1}/B_m > 0$ are lost coming from the left.

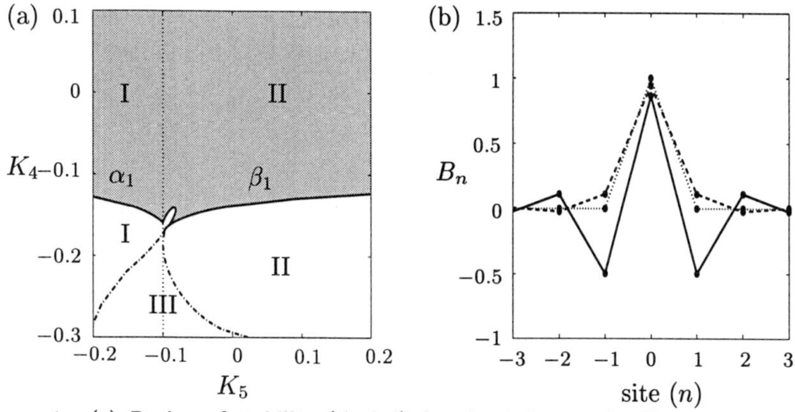

Figure 1. (a) Region of stability (shaded) for the 1-site DB for $K_2 = 0.2$. The solid line represents the onset of real instabilities and the dotted line marks the existence of compact 1-site solutions. The dash-dotted lines are boundaries where continuation is stopped. (b) Three solutions for the same parameter values inside region III, $K_2 = 0.2$, $K_4 = -0.2$ and $K_5 = -0.1$. The sign of the effective coupling, $\text{sgn}(K_{\text{eff}}) = -\text{sgn}(B_{\pm 1}/B_0)$, is consistent with the region from which the solution is continued.

In Fig. 2 the regions of stability (shaded) for the symmetric and anti-symmetric 2-site solutions are depicted for $K_2 = 0.2$. Both real (solid lines) and complex (dashed lines) instabilities are present as well as boundaries where the continuation is stopped (dash-dotted lines). Note that there is a region around $K_4 = 0.1$ and $K_5 = 0$ where both solutions are stable, as well as the 1-site solution. Such simultaneous stability is not observed for the DNLS equation.

An important feature of the stability maps is that some stability boundaries nearly coincide, with a resulting stability inversion. Such inversions occur both between 1-site and symmetric 2-site DB, at the boundaries labelled α_1 and α_2, and between 1-site and anti-symmetric 2-site DB, at the boundaries labelled β_1 and β_2. The boundaries do not exactly coincide, but there are points where they cross. Note that the line of zero effective coupling for 1-site solutions (dotted line in Fig. 1(a)) marks the separation between the two inversion boundaries. Note also that the equivalent line for the symmetric 2-site solution (dotted line in Fig. 2(a)) will intersect with the boundary of stability inversion when continued outside the figure at $K_4 \approx -0.11$ and $K_5 \approx -0.3$. Hence the solutions involved in the stability inversion at this point will be particularly narrow.

To get a picture of the behaviour of the respective solutions in the vicinity of the inversion boundaries we look at the growing modes, i.e. the eigenvectors corresponding to eigenvalues with positive real part, when the

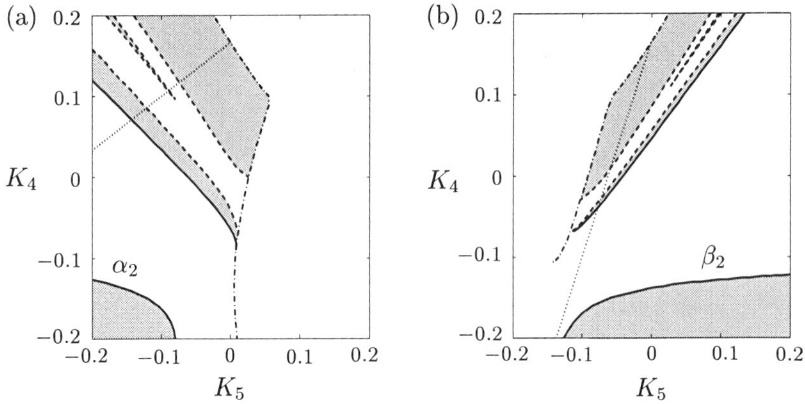

Figure 2. Regions of stability (shaded) for (a) the symmetric and (b) the anti-symmetric 2-site DB for $K_2 = 0.2$. The solid lines represent the onset of real instabilities and the dashed lines the onset of complex (Krein) instabilities. The dash-dotted lines are boundaries where continuation is stopped. It is possible to reach the region to the left of this boundary in (b) by following paths around the edge of the boundary (has not been done). The dotted lines indicate where exact compact solutions of either type exist.

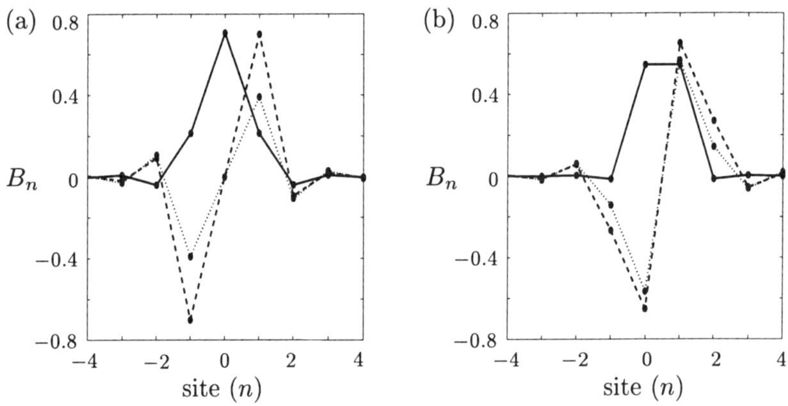

Figure 3. The solution (solid) and the growing mode with the perturbation to the real part (dashed) and imaginary part (dotted) of the amplitude, for (a) the 1-site DB and (b) the symmetric 2-site DB. The imaginary part has been scaled a factor 50. The parameter values are $K_2 = 0.2$, $K_4 = -0.1104$ and $K_5 = -0.3$.

solutions are unstable. In Fig. 3 the growing modes of a 1-site DB and a symmetric 2-site DB are shown at a point along the boundary α_1/α_2 where both solutions are unstable. The modes are such that the 1-site DB will change into something similar to the 2-site DB and vice versa. It is likely that narrow moving DB exist in the vicinity of these boundaries, corresponding to a repeated transformation between stationary 1-site and 2-site DB along the transverse direction of the array. In Klein-Gordon models it has been demonstrated that a symmetry-breaking perturbation to a stationary solution near a stability boundary can induce a moving localised solution [13,14]. This is to our knowledge the first report of stability inversion in a DNLS-type model.

4. Discussion

The derived model equation for arrays of nonlinear coupled waveguides, going beyond the approximations leading to the DNLS equation, shows a number of new phenomena of potential interest for optical multiport switching, the most prominent being stability inversion between both symmetric/antisymmetric 2-site DB and 1-site DB, respectively. The exact behaviour at the inversion as well as the dynamics of any mobile solutions will be analysed elsewhere [15]. The prospect for switching is particularly good for mobile solutions in connection with compact symmetric 2-site DB, since these will be extremely narrow. This is a sought-after property, allowing better selection of output channel.

Acknowledgements

MJ acknowledges support from the Swedish Research Council.

References

1. O. Bang and P. D. Miller, *Physica Scripta* **T67**, 26 (1996).
2. L. J. Bernstein, *Opt. Comm.* **94**, 406 (1992).
3. D. N. Christodoulides and R. I. Joseph, *Opt. Lett.* **13**, 794 (1988).
4. J.C. Eilbeck, P.S. Lomdahl and A.C. Scott, *Physica D* **16**, 318 (1985).
5. S.M. Jensen, *IEEE J. Quantum Eletron.* **18**, 1580 (1982).
6. M.I. Molina and G.P. Tsironis, *Physica D* **65**, 267 (1993).
7. A. Yariv, *Optical Electronics in Modern Communications 5th ed.* (Oxford University Press, New York 1997).
8. A. Eriksson, Master's Thesis LiU-IFM-Fysik-Ex-468, Linköping University (1995).
9. M. Öster, Master's Thesis LiTH-IFM-Ex-1064, Linköping University (2002).
10. R. S. MacKay and S. Aubry, *Nonlinearity* **7**, 1623 (1994).
11. P. G. Kevrekidis and V. V. Konotop, *Phys. Rev. E* **65**, 066614 (2002).
12. J.C. Eilbeck, P.S. Lomdahl and A.C. Scott, *Phys. Rev. B* **30**, 4703 (1984); J.L. Marín and S. Aubry, *Nonlinearity* **9**, 1501 (1996).

13. S. Aubry and T. Cretegny, *Physica D* **119**, 34 (1998).
14. D. Chen, S. Aubry and G. P. Tsironis, *Phys. Rev. Lett.* **77**, 4776 (1996).
15. M. Öster, M. Johansson and A. Eriksson, *Enhanced mobility of strongly localized modes in waveguide arrays by inversion of stability*, (submitted).

BREATHERS IN A MODEL OF A POLYMER WITH SECONDARY STRUCTURE

MICHAEL KASTNER AND J. TOMÁS LÁZARO*

I.N.F.M., UdR di Firenze,
Via G. Sansone 1,
50019 Sesto Fiorentino, Italy
E-mail: kastner@fi.infn.it, jose.tomas.lazaro@upc.es

A simple model of a polymer is considered: a chain of (different) point masses, connected by harmonic springs, embedded in two dimensional space. In order to determine conditions for existence and stability of breather excitations, the method of numerical continuation of a breather solution from the anticontinuous limit is employed. Approaching the limit of equal masses, stable breather solutions are found only within an extremely narrow band of frequencies.

1. Introduction

The search for discrete breather solutions in simple models of polymers with a secondary structure might be regarded as a step towards an understanding of the role that local excitations might play for the functionality of proteins. A minimal model should include the following two features:

(1) a one-dimensional system (chain) is embedded in d-dimensional space ($d > 1$),

(2) inter-particle interactions involve at least d neighbours in order to obtain a secondary structure.

2. The model

We use a slightly modified version of a model proposed by Zolotaryuk *et al.*:[1] a chain of N classical particles (point masses) m_i, $i = 1, ..., N$, which interact by means of linear forces between nearest and next-nearest neighbours. The Hamiltonian function of the system is

$$\mathcal{H} : \mathbb{R}^{4N} \to \mathbb{R}$$

*permanent address: departament de matemática aplicada i, universitat politécnica de catalunya, diagonal 647, e-08028 barcelona, spain.

$$(\mathbf{q_1},\ldots,\mathbf{q_N},\mathbf{p_1},\ldots,\mathbf{p_N}) \mapsto \sum_{i=1}^{N} \frac{\mathbf{p_i}^2}{2m_i} + \\ + \sum_{i=1}^{N-1} U(|\mathbf{q_i} - \mathbf{q_{i+1}}|) + \\ + \sum_{i=1}^{N-2} V(|\mathbf{q_i} - \mathbf{q_{i+2}}|).$$

The particles are allowed to move in the Euklidian plane, i.e., $\mathbf{q_i}, \mathbf{p_i} \in \mathbb{R}^2$. For closest similarity to real polymers, free boundary conditions are employed. In contrast to the original model of Zolotaryuk *et al.*, we allow different values for the masses m_i. This facilitates the construction of an 'anticontinuous limit' as explained (and required) in Sec. 3.1.

For simplicity, harmonic interaction potentials $U(x) = V(x) = \frac{1}{2}x^2$ are chosen between nearest as well as next-nearest neighbouring particles. Despite the linearity (in the inter-particle distance) of the forces, the geometry of the chain gives rise to effective nonlinearities in the equations of motion, and therefore existence of breather solutions is not ruled out from the outset.

The system displays a secondary structure, for example a zig-zag chain (or 2-helix) as illustrated below. (Note, however, that the 'equilibrium configuration' of the system is not unique!)

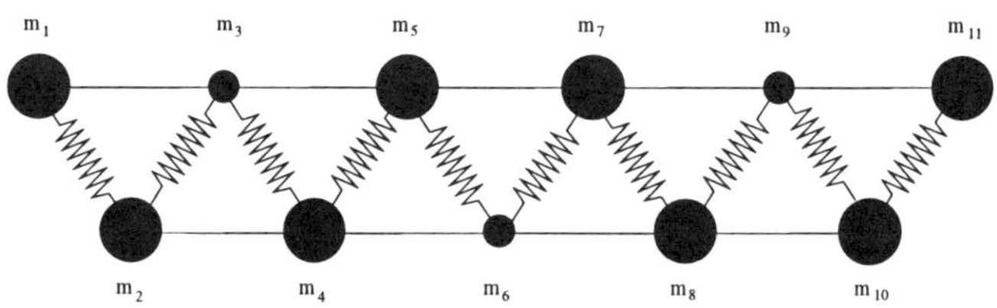

Figure 1. Zig-zag chain.

3. Existence of breathers and stability conditions

We are interested in the following question: *In which way do the parameters of the system (like oscillation frequencies or particle masses) have to be tuned in order to find breather solutions?*

3.1. *Numerical calculation from the anticontinuous limit*

A numerical method due to Marín and Aubry[2] is used in order to determine the existence region of breather solutions: A (known) breather solution of the system for certain parameter values is continued to other (not too different) parameter values by means of Newton's method. Such a continuation is possible, in principle also analytically, for breather oscillation frequencies which are non-resonant with the spectrum of the linearized equations of motion of the system. This allows us to obtain breather solutions for, e.g., arbitrary mass ratios and frequencies, provided that

(1) such solutions exist, and
(2) a breather solution is known from which continuation up to the desired parameter values is feasible.

As a starting point for such a continuation, a certain limiting value (the so-called anticontinuous limit) of a parameter is taken, for which existence of breather solutions is known. Here, a diatomic chain is considered, where two particles of mass $M \geq 1$ are followed by one particle of mass $m = 1$:

$$MMmMMmMMmMMm...$$

Existence of localized oscillations (breathers) is obvious in the limit of zero mass ratio, $\frac{m}{M} \to 0$, for example: all particles are at rest but a single one of mass m, which is coupled only to particles of mass $M \to \infty$. This solution can be used as a starting point for continuation.

3.2. *An "existence and stability diagram"*

The number of parameters present in the system is still quite large. As an example, existence and stability regions of discrete breathers are investigated for a particular choice of parameters:

- A chain of 23 particles.
- The set of initial conditions in the anticontinuous limit is restricted: all but the centre particle are at rest. The centre particle moves on the reflection symmetry line of the zig-zag chain. (Consider a vertical oscillation of particle 6 in Figure 1, while all the other particles are at rest.)
- For an oscillating particle in the anticontinuous limit, the frequency is a non-monotonic function of the energy (see Fig. 2). We restrict the initial conditions to energies corresponding to the decreasing (left hand) part of $f(energy)$.

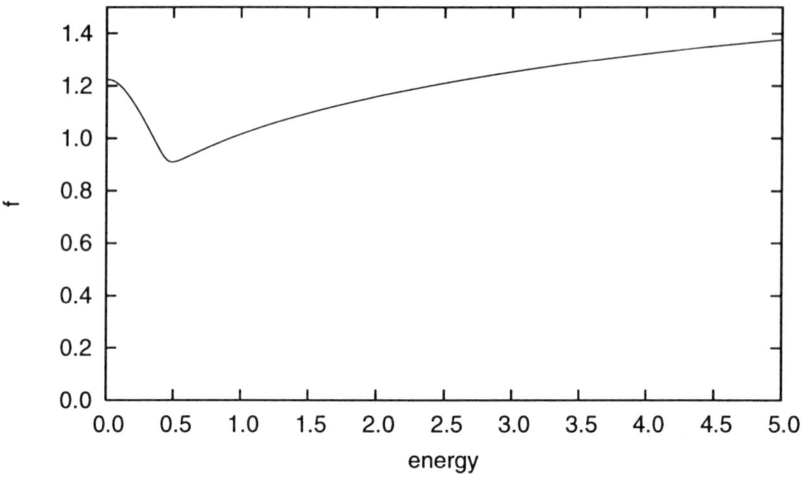

Figure 2. Frequency vs. energy of a single particle of mass $m = 1$ in the anticontinuous limit.

In order to visualize such a breather solution, the orbits of the particles of a breather solution are plotted in Fig. 3 for mass ratio $\frac{m}{M} = \frac{2}{7}$ and frequency $f = 0.212$.

Exemplarily for the above choices of particle number and initial conditions, Figure 4 shows the regions in the parameter space of frequencies and mass ratios, for which symmetric breather solutions can be found.

3.3. *Stability of breathers*

In the anticontinuous limit, all breather solutions are stable. For non-zero mass ratios $\frac{m}{M}$, there are regions of stable and unstable breathers. Note

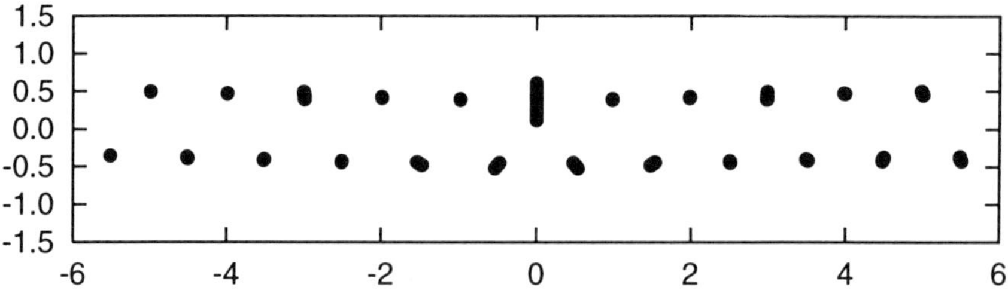

Figure 3. Orbits of the particles of a symmetric breather solution located at the central position for mass ratio $\frac{m}{M} = \frac{2}{7}$ and frequency $f = 0.212$.

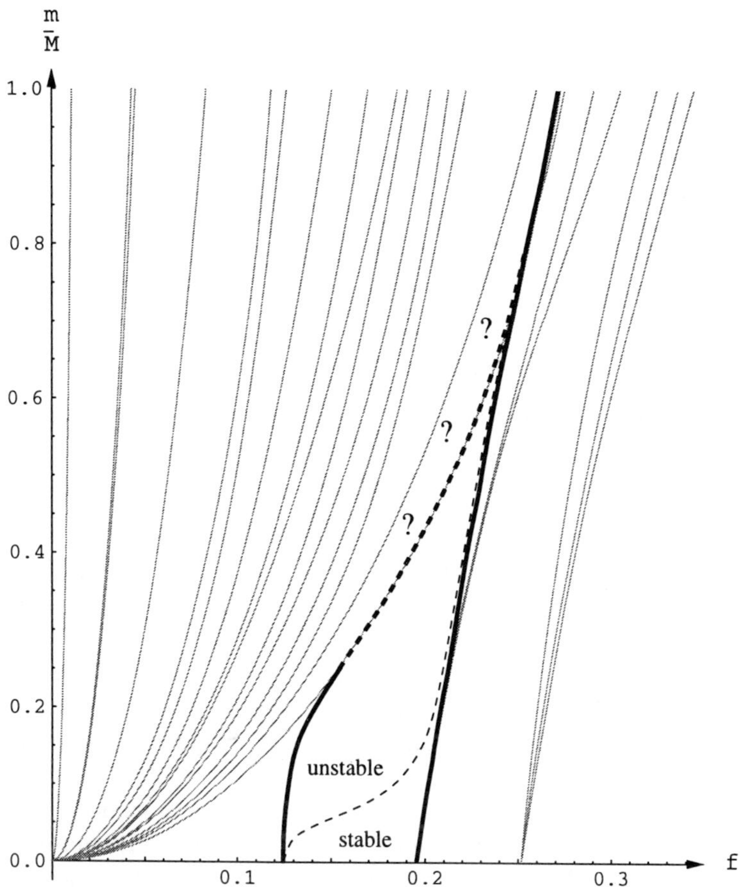

Figure 4. Existence and stability regions of symmetric breather solutions in the parameter space of breather frequencies f and mass ratios $\frac{m}{M}$. The bottom line corresponds to the anticontinuous limit. The thin gray lines are the eigenfrequencies of the linearized equations of motion. Under the conditions specified above, breather solutions are found for parameter values lying between the solid black lines.
The solid black line on the right hand side coincides with an eigenfrequency of the linearized equations of motion. Approaching this boundary, the breather amplitude goes to zero. On the left hand side, the region of breather existence is more difficult to determine, mainly because of the following reason: the concept of localization is somewhat ill-defined in finite systems. Numerically, breather solutions can be continued across the bold dashed line coinciding, again, with an eigenfrequency of the linearized equations of motion. The localization of the solutions found in this region, however, is so weak, that a distinction between extended and localized modes does not seem feasible.
In between these two boundaries breather solutions exist, although, for larger mass ratios $\frac{m}{M}$, stable breathers appear only within an extremely narrow frequency range.

that both, stable and unstable breathers, are obtained by continuation from the same solution in the anticontinuous limit, and continuation across the stability boundary does not cause any numerical difficulties. The regions of (linearly) stable and unstable breathers are indicated in Figure 4.

In contrast to the original work [2], the numerical continuation was performed with respect to *two parameters*, namely the mass ratio and the frequency. From Figure 4 it becomes clear, why this process is essential: for masses $M \lesssim 10$, stable breather solutions cannot be found by continuing a solution from the anticontinuous limit while keeping the frequency fixed.

3.4. *Other parameter values*

Varying some parameters of the model under investigation, quantitative changes occur, while no qualitative modifications of the phase diagram have been observed:

- Due to the localized character of the oscillation, increasing the number of particles does not lead to significant changes (maybe apart from the case of mass ratio $\frac{m}{M}$ close to unity, where localization is less distinct).
- Enlarging the coupling constant of the nearest neighbour interaction (while keeping the next-nearest neighbour coupling constant to unity), the lower bound on the accessible frequencies can be shifted towards zero.
- Breather existence and stability can be strongly influenced by the shape of the interaction potentials U and V in the Hamiltonian. In particular the softening (as in our case) or hardening property of the potentials often gives rise to entirely different behaviour of the respective systems.

Acknowledgments

We would like to thank Roberto Livi for arousing our interest in the topic of discrete breathers and for stimulating discussions. The work was supported by EU contract HPRN-CT-1999-00163 (LOCNET network).

References

1. A. V. Zolotaryuk, P. L. Christiansen, and A. V. Savin, *Two-dimensional dynamics of a free molecular chain with a secondary structure*, Phys. Rev. E **54**:3881–3894 (1996).
2. J. L. Marín and S. Aubry, *Breathers in nonlinear lattices: numerical calculation from the anticontinuous limit*, Nonlinearity **9**:1501–1528 (1996).

OBSERVATION OF BREATHER-LIKE STATES AND RESONANCES IN A SINGLE JOSEPHSON CELL

F. PIGNATELLI AND A. V. USTINOV

Physikalisches Institut III, Universität Erlangen-Nürnberg,
Erwin-Rommel-str. 1, D-91058 Erlangen, Germany.
E-mail: pignatelli@physik.uni-erlangen.de

We present an experimental study of broken-symmetry states (breather-like) in a single Josephson cell. The dependence of the region of existence of the breather-like states on the damping is measured. On the current-voltage characteristic of the broken-symmetry states we observed resonant steps due to interaction between the breather and the electromagnetic oscillations in the cell.

The study of breather states in smaller structures as single Josephson cells has been proposed to allow an insight in localized states of more complex systems as ladders and two dimensional Josephson junction arrays [1,2]. Particularly the system we studied is a superconducting loop (hole area $4 \times 4\ \mu m^2$) with three small underdamped Nb/Al-AlO_x/Nb Josephson junctions (JJs). Two JJs of same area ($7 \times 7\ \mu m^2$), called vertical, are placed along the same direction of the bias current, I, and one, horizontal, of different area ($3 \times 4\ \mu m^2$), is placed on the transversal branch of the cell, see inset (a) of Fig. 1. The coupling between vertical JJ is described by the ratio between the area of the horizontal JJ and the vertical one, that is the anisotropy $\eta = A_h/A_v$ (in our system $\eta = 0.24$). The inductive coupling, that depends on the geometry of the cell, is given by the self-inductance $\beta_L = 2\pi L I_{cv}/\Phi_0$ (in our case $\beta_L = 1.47$ at 4.2 K). Finally the damping of the system, $\alpha = \sqrt{\Phi_0/(2\pi I_{cv} R^2 C)}$, is evaluated from a linear fit of the sub-gap branch of the current-voltage characteristic ($\alpha \simeq 0.023$ at 4.2 K). Broken-symmetry states in single Josephson cells were previously theoretically studied and proposed to have characteristics similar to discrete breathers in Josephson ladders [1,2,3]. Because of the reduced dimension of the system, we call these localized states as breather-like states. Particularly we are interested in the two equivalent states with the horizontal JJ and one of the vertical JJs in the resistive state while the other vertical JJ remains in the superconducting state, see inset (b) of Fig. 1.

The existence of these states is a consequence of the hysteretic behavior of small underdamped JJs, that in a range of current below I_c allows for two possible states, the superconducting state (voltage on the JJ, $V = 0$) and the resistive state ($V \neq 0$). Breather-like states can be experimentally generated in the single cell by biasing the system by a local current, I_{loc}, through the transverse branch, 4-2 in inset (a) Fig. 1, and rising this current until the system switches to a broken-symmetry state. Afterwards, in order to keep the system in the generated state in presence of uniform bias current, we increase the bias current I and simultaneously decrease I_{loc} to zero. The measurements were performed in a ^{4}He cryostat in a temperature range between 4 K and 8 K, so that the parameters of the system α and β_L could be varied. We generated and observed stable breather-like states in the range of temperatures from 4 K to 7.75 K [4]. No broken-symmetry state was found stable for temperatures above 7.75 K ($\alpha > 0.35$). The measured dependence of the region of existence and stability of the breather-like states on the damping is found to be in good agreement with the theoretical expectations [2,3], see Fig. 1. The damping here is evaluated from the retrapping current of the homogeneous whirling state. The experimental current-voltage characteristics of the broken-symmetry state between 6 K and 7 K, showed resonances due to interaction between the breather and the electromagnetic oscillations of the cell [2,3], see Fig. 2. The experimental dependence of the magnitude of these resonances on the external magnetic field was also studied in detail [4] and found coherent with theory [3].

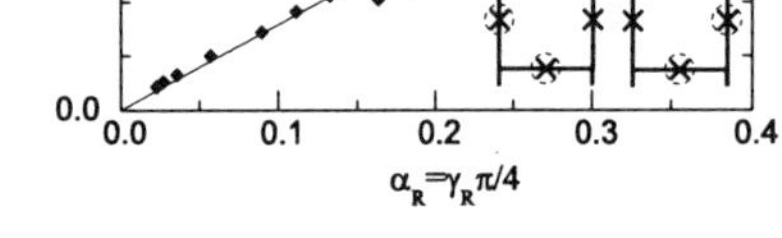

Figure 1. Maximum (solid squares) and minimum (solid diamonds) currents for the breather-like state. The solid lines show the theoretical expectations. In inset (a) the single cell system (V_1 and V_2 are the vertical JJs and H the horizontal) and (b) the two breather-like states.

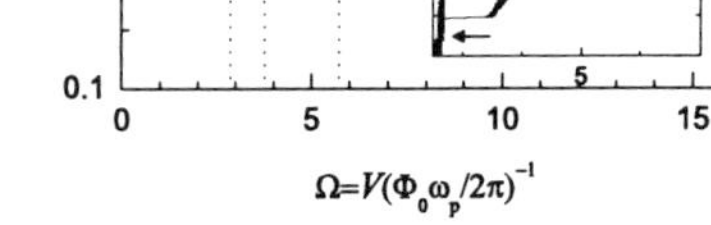

Figure 2. Measured current-voltage characteristic of the breather-like state at temperature $T = 6.65$ K. The dotted lines show the theoretical expectations for the resonances on the current voltage characteristic.

References

1. J. J. Mazo, E. Trías, and T. P. Orlando, *Phys. Rev. B* **59**, 13604, (1999).
2. A. Benabdallah, M. V. Fistul and S. Flach, *Physica D* **159**, 202 (2001).
3. M. V. Fistul, S. Flach and A. Benabdallah, *Phys. Rev. E* **65**, 046616, (2002).
4. F. Pignatelli and A. V. Ustinov, cond-mat/0208605, (2002).

BREATHERS IN FPU SYSTEMS, NEAR AND FAR FROM THE PHONON BAND

B. SÁNCHEZ-REY†, JFR. ARCHILLA†, G JAMES‡, AND J. CUEVAS†
†*Nonlinear Physics Group, University of Sevilla, Spain*
‡*Département de Génie Mathématique, INSA de Toulouse, France*
Email:bernardo@us.es

Introduction. This work is motivated by a recent breathers existence proof in the one dimensional FPU system, given by the equations:

$$\ddot{x}_n = V'(x_{n+1} - x_n) - V'(x_n - x_{n-1}) \,, \quad n \in \mathbb{Z} \,, \tag{1}$$

where V is a smooth interaction potential satisfying $V'(0) = 0$ and $V''(0) > 0$. Using a center manifold technique[2], one can prove the existence of small amplitude breathers (SAB) with frequencies ω_b slightly above the phonon band if $B = \frac{1}{2}V''(0)V^{(4)}(0) - (V^{(3)}(0))^2 > 0$, and their non-existence for $B < 0$. Our aim is to test numerically the range of validity of this theoretical result and to explore new phenomena. For this purpose we shall fix $V(u) = u^2/2 + a\,u^3 + \frac{1}{4}\,u^4$, which yields $B = 3(1 - 12a^2)$.

We work with the difference variables $u_n = x_n - x_{n-1}$ more suitable for the use of our numerical method. We also use periodic boundary conditions $u_{n+2p}(t) = u_n(t)$ so that the maximum frequency of the linear phonons is exactly 2 as in the infinite lattice. Our computations are performed using a numerical scheme based on the anti-continuous limit and Newton method[3].

Test and range of validity. First, we have computed numerically SAB (i.e. breathers whose amplitudes go to zero when $w_b \to 2^+$) in the case $B > 0$. We have obtained breathers with symmetries $u_n(t) = u_{-n}(t)$ (Page mode) and $u_n(t) = u_{-n-1}(t + T_b/2)$ (Sievers-Takeno mode), where $T_b = 2\pi/\omega_b$ is the breather period. The force $y_n = V'(u_n)$ is the variable used in reference [2]. In Fig.1 (left) it is shown that the maxima of the force are of order $\mu^{1/2}$ when $\mu = w_b - 2 \to 0^+$, as predicted by the theory, up to relatively large values. Thus if $B > 0$ breathers exist for any small value of energy in our FPU system (1).

Another property of these SAB is that their width diverges when $w_b \to 2^+$. More precisely the theory predicts that their spatial extend is of order $\mu^{-1/2}$, which is in accordance with our numerical observations.

Other numerical observations. For $B > 0$, we have numerically continued the SAB as ω_b goes away from the phonon band. We have found that the maximum amplitudes of the oscillations are also approximately linear functions of $\mu^{1/2}$. This is expected for small μ, since $u_n = y_n + O(y_n^2)$, but it occurs surprisingly far from the phonon band, at least until values of $\mu \approx 1$ (see fig.1, left). We have also checked that the Page mode fits very well to the NLS soliton $u_n(t) = \alpha\sqrt{\mu}\,(-1)^n \cos(\omega_b t)\,[\cosh(\beta\sqrt{\mu}\,n)]^{-1}$, even far from the top of the phonon band.

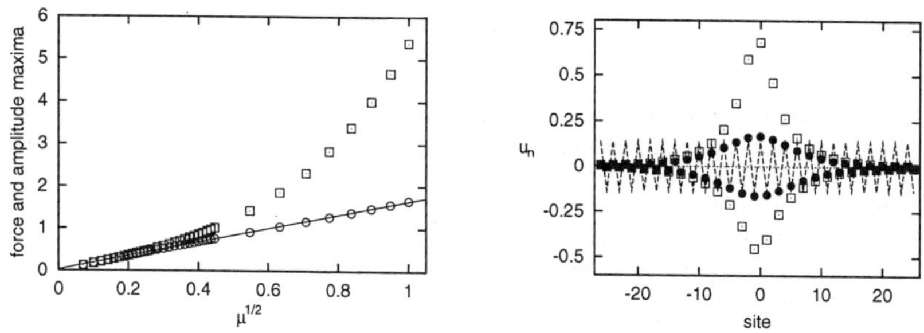

Figure 1. Left: Force (squares) and amplitudes (circles) maxima versus $\mu^{1/2}$. The cubic coefficient in V is $a = -0.1$ ($B = 2.64$). Right: Comparison between a SAB (full circles) for $a = -0.1$ and a LAB (blank squares) for $a = -1/3$ ($B = -1$) having the same frequency $w_b = 2.01$. The dashed line represents the linear phonon with frequency 2.

For $B < 0$ and V strictly convex ($\frac{1}{\sqrt{12}} < |a| < \frac{1}{\sqrt{3}}$), breathers exist near the top of the phonon band but they are large amplitude breathers[1] (LAB), i.e. their amplitudes do not go to zero when $w_b \to 2^+$. As a consequence there is an energy gap for breathers creation in these FPU systems. In figure 1 (right) we compare a SAB and a LAB having the same frequency $w_b = 2.01$. We have found LAB with the same symmetries as SAB (Page and Sievers-Takeno modes). The Page mode fits very well to an exponential profile having the form $u_n(t) = \alpha(\omega_b)\,(-1)^n \cos(\omega_b t)\,|\sigma(\omega_b)|^{|n|}$ where $\sigma(\omega_b) = 1 - (\omega_b^2)/2 + (\omega_b/2)(\omega_b^2 - 4)^{1/2} \in (-1, 0)$. As (1) is formulated as a mapping in a loop space[2] and $\omega_b > 2$, the linearized operator has a purely hyperbolic spectrum and the constant $\sigma(\omega_b)$ is the closest eigenvalue to -1 (with $\sigma(2) = -1$). Consequently, for $\omega_b \approx 2$ one can ask if the iterated map admits a global center manifold containing these LAB.

References

1. S Aubry, G Kopidakis, and V Kadelburg. *Discrete and Continuous Dynamical Systems* , serie B (DCDS-B) **1** (2001) 271-298.
2. G James. *C. R. Acad. Sci. Paris*, 332(1):581–586, 2001.
3. JL Marin and S Aubry. *Nonlinearity* 9 (1996), p. 1501-1528.

EXPERIMENTS ON RESONANT ROTOBREATHERS IN JOSEPHSON LADDERS

M. SCHUSTER, D. ABRAIMOV, A. P. ZHURAVEL AND A. V. USTINOV

Physikalisches Institut III, Universität Erlangen-Nürnberg
Erwin-Rommel-Str. 1, 91058 Erlangen, Germany
E-mail: schuster@physik.uni-erlangen.de

We study the properties of roto-breathers in superconducting Josephson ladders, which are solid state implementations of classical rotor lattices. Discrete breathers form therein as spatially confined regions in which the superconducting phase is rotating, while the phase in the remaining system just oscillates. Breathers can be created and sustained against dissipation by externally applied currents, and their internal rotation frequency can be measured as a dc voltage. By tuning the external currents we create breathers with a rotation frequency *inside* the frequency band of linear phase oscillations. We visualize the excited linear modes in experiment using Low Temperature Scanning Laser Microscopy (LTSLM).

After the theoretical predictions of the existence of discrete breathers[1], several groups were able to detect discrete breathers in various experimental systems[2,3]. Josephson ladders play a key role in the experimental study of discrete roto-breathers. They are formed by a ladder network of underdamped Josephson junctions and can be mathematically described similar to the one-dimensional discrete sine-Gordon chain, without conservation of topological charge. In the model, the superconducting phase differences across the Josephson junctions play the role of rotor angles, electric currents correspond to external forces (or torques), and voltages correspond to the average time derivative of the angles (hence, the rotation velocities).

In a lattice of rotors, a discrete roto-breather is formed if several lattice sites are in the rotational state, while the remaining lattice is silent. Since there is damping, a breather state would just decay; it is however possible to sustain the rotation of a rotor by applying a torque. Since the rotors have inertia, the minimum torque needed to *sustain* rotation is smaller than the torque needed to *start* rotation. Consequently, there is a range of torque values where both the rotation and the static state are possible. This fact is used to maintain a discrete roto-breather state in a Josephson ladder. We apply a uniform torque (i.e., current), small enough not to force the whole array into rotation, but large enough to keep a spatially confined region in a state of non-zero rotation. A roto-breather state can hence be detected in experiment by measuring dc voltages (which are proportional to the average rotation frequency) at several points of the array in the presence of an external bias current.

By plotting the voltage across a resistive junction versus the externally supplied current we can trace the breather rotation frequency versus the applied torque. If the breather frequency is far away from lattice eigenfrequencies, this dependency is linear. In recent experiments[4] we have shown the possibility to tune the breather frequency into the band of the linear lattice modes. When the breather frequency matches the frequency of an eigenmode, energy supplied by the external current will be delivered into this mode by the roto-breather. This energy loss by the breather decreases the ratio of voltage (rotation frequency) and current (torque), which is the measured differential resistance. Consequently, we detect the roto-breather in resonance with the linear lattice oscillations by observing steps in current vs. voltage characteristics.

In our current work, we pursue a different measurement scheme by using the method of Low Temperature Scanning Laser Microscopy (LTSLM). The experimental sample, consisting of Nb-AlO_x-Nb layered structures on a silicon wafer, is cooled down to liquid helium temperature (4.2 K) in an optical cryostat. A laser beam can be focused onto the lattice of Josephson junctions through a window in this cryostat. The dissipation at the illuminated lattice site is increased due to heating by the laser. Depending on the dynamic state of this site, the voltage measured across the array changes. Rastering the laser across the whole array and gray-scale-encoding the magnitude of the voltage change as a function of the beam coordinate reveals a two-dimensional picture, which maps the dynamic state of the whole array at a certain value of the external current and breather frequency. For more details on the LTSLM technique, see Ref. [3].

In Fig. 1, we present some of our first LTSLM data obtained from a *resonant* breather-like state in a Josephson ladder. The images were taken in the presence of an extended local current to support the finite voltage state (which we hence call "pseudo-breather"), but should be very similar to uniformly supported resonant breather states. On the upper left, a non-resonant state is visualized, superimposed by a sketch of the system. The bright areas are those of non-zero voltage, where the locally modified dissipation due to the laser irradiation leads to a voltage signal. The remaining area is dark since the junctions underneath are in the static (superconducting) state. If the current is increased, steps appear on the current voltage characteristic, as the breather frequency matches the different lattice eigen-frequencies. The LTSLM response is found now even for junctions *outside* the breather region, since their large amplitude oscillations are strongly influenced by the laser beam. Consequently, we are able to visualize an oscillation pattern in a lattice which is excited by a localized mode. In the future we will extend this technique to study the interaction between discrete breathers and linear waves.

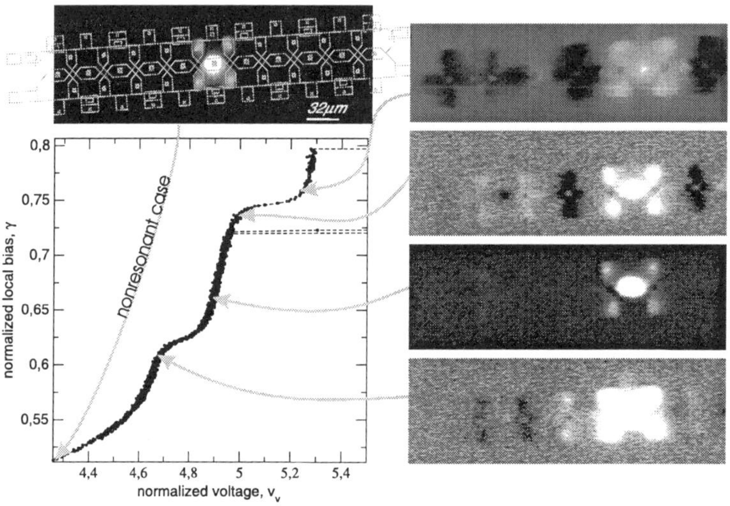

Figure 1. Current-voltage characteristic of a pseudo-breather state in a Josephson ladder. Resonances with the lattice eigen-modes lead to the steps on the curves. The gray-scale plots show the LTSLM response at different values of the current.

References

1. A. J. Sievers and S. Takeno, *Phys. Rev. Lett.* **61** 970 (1988).
2. B. I. Swanson *et al.*, *Phys. Rev. Lett.* **82** 3288 (1999); U. T. Schwarz *et al.*, *ibid.* **83** 223 (1999); E. Trías *et al.*, *ibid.* **84** 741 (2000).
3. P. Binder *et al.*, *Phys. Rev. Lett.* **84** 745 (2000).
4. M. Schuster, P. Binder, and A. V. Ustinov, *Phys. Rev. E.* **65** 016606 (2002).

LIST OF PARTICIPANTS

	NAME	INSTITUTION
1	Carlos Aguirre	Instituto de Ingeneriería del Conocimiento, Universidad Autónoma de Madrid, 28049 Madrid Spain Carlos.aguirre@iic.uam.es
2	Serge Aubry	Laboratoire Leon Brillouin CEA SACLAY 91191 Gif-sur-Yvette FRANCE aubry@llb.saclay.cea.fr
3	Azucena Alvarez Chillida	Departamento de Física Atómica, Molecular y Nuclear, Facultad de Física, Universidad de Sevilla, Avda. Reina Mercedes s/n 41012 Sevilla SPAIN azucena@us.es
4	Juan FR. Archilla	Departamento de Física Aplicada I, ETS Ingenieria Informática Avda. Reina Mercedes s/n 51012-Sevilla Spain archilla@us.es
5	Claude Baesens	Mathematics Institute University of Warwick Coventry CV4 7AL United Kingdom claude@maths.warwick.ac.uk
6	Mariette Barthes	GDPC CC26 Universite Montpellier 2- 34095 Montpellier CEDEX 5. FRANCE mariette@gdpc.univ-montp2.fr
7	Jeroen Bergamin	Department of Mathematics, University of Patras, 28500 Patras, Greece jeroen_bergamin@hotmail.com
8	Abdelhadi Benabdallah	Max Planck Institute for the Physics of Complex Systems Noethnitzer Str. 38 D-01187 Dresden Germany abenab@mpipks-dresden.mpg.de
9	Ioana Bena	Universitat de Barcelona, Facultat de Física, Diagonal 647 08028 Barcelona ioana@ecm.ub.es
10	V.A. Brazhnyi	Centro de Fïsica de Materia Condensada, Universidad de Lisboa, Complexo Interdisciplinar Av. Prof. Gama Pinto 2, 1649-003 Lisboa, PORTUGAL brazhnyi@cii.fc.ul.pt
11	Gianluca Castellani	15 Moultrie Street #2, 02124 Dorchester (MA) USA gcastell@lynx.dac.neu.edu
12	Nuno Catarino	Mathematics Institute University of Warwick Coventry CV4 7AL United Kingdom catarino@maths.warwick.ac.uk

13	Peter L. Christiansen	Informatics and Mathematical Modellin,g Building 321 Richard Petersens Plads. The Technical University of Denmark DK-2800 Kgs. Lyngby Denmark plc@imm.dtu.dk
14	Joseph Choi	Max-Planck-Institut für Physik komplexer Systeme Nöthnitzer Str. 38 01187 Dresden, Germany
15	Jean Christophe Comte	Physics Department University and Foundation for Researcher and Technology-Hellas P.O. Box 2208, 71002 Heraklion, Crete, GREECE comte@physics.uoc.gr
16	Leonor Cruzeiro-Hansson	FCT University of Algarve, Campus de Gambelas, 8000 Faro, Portugal lhansson@ualg.pt
17	Jesus Cuevas Maraver	Dpto. Física Aplicada I, Universidad de Sevilla. ETS Ingeniería Informática. Avda. Reina Mercedes s/n 41012 Sevilla SPAIN jcuevas@us.es
18	Luke Donev	Anlagenstasse 5, Bis Frau Leppart 91054 Erlangen, Deutschland donev@physik.uni-erlangen.de
19	Jerome Dorignac	Max Planck Institute for the Physics of Complex Systems Noethnitzer Str. 38 D-01187 Dresden Germany dorignac@mpipks-dresden.mpg.de
20	Chris Eilbeck	Department of Mathematics, Heriot-Watt university, Edinburgh EH14 5AS UK chris@ma.hw.ac.uk
21	Francois Fillaux	Ladir CNRS 2 rue H. Dunant, 94320 Thiais France fillaux@glvt-cnrs.fr
22	Sergei Flach	Max Planck Institute for the Physics of Complex Systems Noethnitzer Str. 38 D-01187 Dresden Germany flach@mpipks-dresden.mpg.de
24	Daniel Graham	205 Eddy St. #1 Ithaca, NY 14850 USA djg45@cornell.edu
25	Guillaume James	Institut National des Sciences Appliquees, departement GMM, 135 avenue de Rangueil, 31077 Toulouse Cedex 4, France james@gmm.insa-tlse.fr
26	Salvador Jimenez	Universidad Alfonso X el Sabio, Madrid, SPAIN sjimenez@uax.es
27	Magnus Johansson	Department of Physics and Measurement Technology, Linkoping University, S-581 Linkoping, Sweden majoh@ifm.liu.se

28	Michael Kastner	Diparimento di Fisica, Universitá di Firenze. Via Sansone 1, 50019 Sesto Fiorentino ITALY kastner@fi.infn.it
29	Vladimir Konotop	CFMC Universidade de Lisboa, Complex Interdisciplinar, Av. Prof. Gama Pinto 2, 1649-003 Lisboa Portugal konotop@cii.fc.ul.pt
30	Vladimir Kuzmiak	Dept. Of Physics and Astronomy, University of California, IRVINE, CA 92697 USA Vkuzmiak@duey.ps.uci.edu
31	J. Tomás Lazaro	Diparimento di Fisica, Universitá di Firenze. Via Sansone 1, 50019 Sesto Fiorentino ITALY lazaro@fi.infn.it
32	Stefano Lepri	Diparimento di Fisica, Universitá di Firenze. Via Sansone 1, 50019 Sesto Fiorentino ITALY stefano.lepri@unifi.it
33	Anna Litvak-Hinenzon	Mathematics Institute University of Warwick Coventry Hill Road, CV4 7AL United Kingdom litvak@maths.warwick.ac.uk
34	Roberto Livi	Diparimento di Fisica, Universitá di Firenze. Via Sansone 1, 50019 Sesto Fiorentino ITALY livi@fi.infn.it
35	Luis LLuna Reig	Dpto. Análisis Matemático, Facultad de Matemáticas, Universidad de La Laguna, 38271 LA LAGUNA (Tenerife) SPAIN llluna@ull.es
36	Robert S. MacKay	Mathematics Institute University of Warwick Coventry CV4 7AL United Kingdom mackay@mail.maths.warwick.ac.uk
37	Panagiotis Maniadis	Laboratoire Leon Brillouin CEA SACLAY 91191 Gif-sur-Yvette FRANCE maniadis@llb.saclay.cea.fr
38	Juan Jose Mazo	Dpto. Física Materia Condensada, Facultad de Ciencias, Universidad de Zaragoza,/Pedro Cerbuna 12, 50009 Zaragoza SPAIN juanjo@posta.unizar.es
39	Matthias Meister	Dpto. Física Materia Condensada, Facultad de Ciencias, Universidad de Zaragoza,/Pedro Cerbuna 12, 50009 Zaragoza SPAIN matthias@posta.unizar.es
40	Andrey Miroshnichenko	Max Planck Institute for the Physics of Complex Systems Noethnitzer Str. 38 D-01187 Dresden Germany andreym@mpipks-dresden.mpg.de

41	Pascal Noble	Laboratoire MIP, Université Paul Sabatier Toulouse 3, 118 route de Narbonne, 31062 Toulouse Cedex 4, France noble@mip.ups-tlse.fr
42	Michael Oester	Department of Physics and Measurement Technology, Linkoping University, S-581 Linkoping, Sweden micos@ifm.liu.se
43	Pedro Pascual	Instituto de Ingeneriería del Conocimiento, Universidad Autónoma de Madrid, 28049 Madrid Spain pedro.pascual@iic.uam.es
44	Victor Perez Garcia	Departamento de Matemáticas, ETSI Industriales Universidad de Castilla-La Mancha 13071 Ciudad Real SPAIN Victor.PerezGarcia@uclm.es
45	Francesco Piazza	Diparimento di Fisica, Universitá di Firenze. Via Sansone 1, 50019 Sesto Fiorentino ITALY piazza@fi.infn.it
46	Francesca Pignatelli	Physicalishes Institute III des Universitaet Erlangen-Nuernberg Erwin-Rommel-str. 1 91058 Erlangen, Germany pignatelli@physik.uni-erlangen.de
47	Tatiana Popova	Department of Analytical and Medical Biochemistry and Microbiology, Voronezh State University, Universitetskaya sq. 1, 394006 Voronezh RUSSIA tatiana@po.vsu.ru
48	Niurka Rodriguez Quintero	EUP. Virgen de Africa 7. 41011. Departamento de Fisica Aplicada I. Escuela Universitaria Politécnica. Universidad de Sevilla. Sevilla SPAIN niurka@us.es
49	Bernardo Sanchez Rey	EUP. Virgen de Africa 7. 41011. Departamento de Fisica Aplicada I. Escuela Universitaria Politécnica. Universidad de Sevilla. Sevilla SPAIN bernardo@us.es
50	Jose Maria Sancho	Universitat de Barcelona, Facultat de Física, Diagonal 647 08028 Barcelona SPAIN jmsancho@ecm.ub.es
51	Marcus Schuster	Physicalishes Institute III des Universitaet Erlangen-Nuernberg Erwin-Rommel-str. 1 91058 Erlangen, Germany schuster@physik.uni-erlanger.de
52	Jacques-Alexandre Sepulchre	Institute non lineaire de Nice, 1361 route des lucioles, 06560 Valbonne-Sophia Antipolis FRANCE Jacques-Alexandre.Sepulchre@inln.cnrs.fr

53	Marcello Terraneo	118 Route de Narbonne, Labo Physique Quantique, Université Paul Sabatier, 31062 Toulouse CEDEX FRANCE terraneo@irsamc.ups-tlse.fr
54	Nikos Theodorakopoulos	Theoretical and physical chemistry institute. National Hellenic Research Foundation. Vas. Constantinou 48, Athens 11635 Greece nth@eie.gr
55	G.P. Tsironis	Department of Physics, University of Crete and Forth, PO Box 2208 71003 Heraklion, Crete, Greece gts@physics.uoc.gr
56	David Usero	Departamento de Matemática Aplicada, Facultad de Informática, Universidad Complutense de Madrid, 28040 Madrid SPAIN umdavid@fis.ucm.es
57	Luis Vázquez	Departamento de Matemática Aplicada, Facultad de Informática, Universidad Complutense de Madrid, 28040 Madrid SPAIN lvazquez@fdi.ucm.es
58	Maria-Paz Zorzano	Centro de Astrobiologia (CSIC/INTA), Associated to NASA Astrobiology Institute.Carretera de Ajalvir km 4, 28850 Torrejón de Ardoz, Madrid, SPAIN zorzanomm@inta.es